INTEGRATED DEVICES IN DIGITAL CIRCUIT DESIGN

Integrated Devices in Digital Circuit Design

G. S. HOPE

Department of Electrical Engineering
University of Calgary

A WILEY-INTERSCIENCE PUBLICATION

JOHN WILEY & SONS

New York • Chichester • Brisbane • Toronto

Library of Congress Cataloging in Publication Data:

Hope, Gordon Sidney, 1931–
Integrated devices in digital circuit design.

"A Wiley-Interscience publication."
Includes bibliographical references and indexes.
1. Logic circuits. 2. Logic design. 3. Integrated circuits–Medium scale integration. 4. Microcomputers–Design and construction. 5. Digital electronics. I. Title.

TK7874.H66 621.381'73 80-17152
ISBN 0-471-07920-0

Printed in the United States of America

10 9 8 7 6 5 4 3 2 1

To Adery, Erik and Willow

Preface

The purpose of this book is to outline and introduce the subjects of logic design, computer architecture, microprogramming, and software-driven logic design. However, the major goal of the book is to present a method of designing, constructing, and building logic hardware circuits.

The general area of logic and digital computer design is rapidly expanding. There are a great number of books available that deal with these subjects, and many of them are excellent logic design textbooks. The question that may well be asked is, why does the author feel that yet another such textbook is needed? The first reason is that none of the books, as good as they are, has the same direction and objectives intended by the author. The second reason is that those books that have similar directions concentrate on classical logic design techniques which are complex and ill-suited to the construction of large hardware circuits. And the third reason is that although some books present modern methods of using MSI chips to design pulse-driven sequential circuits, there is nowhere presently available a book that explains MSI techniques to design event-driven circuits.

The methods of designing event-driven circuits presented in this book are claimed to be new in that the author has never seen them presented in any other book or paper. However, pulsed sequential circuits using MSI technology are presented by some authors. The use of MSI chips in logic hardware design simplifies the design procedures immensely. The number of chips required for a comprehensive design is reduced greatly, and as a result the effort required to make layouts, including artwork and soldering, is reduced.

In this book logic elements are considered to be ideal. Circuits designed using the new techniques work well under this assumption because real circuits always have a gate in the feedback path. This gate suppresses glitches and ringing.

The author assumes that the average reader is a second- or third-year student in engineering or science whose main interest is in real-time control of processes. Further it is assumed that such students have had basic courses in a man-oriented language like Fortran or PL1 and in electrical networks. However, this is not a strict requirement. The book is meant as preparatory material for a further course in microprocessor and assembler applications to process control. Thus the introduction of computer architecture, microprogramming, and software-driven logic circuits is included.

Large interactive minicomputer or microprocessor systems under the control of higher order languages and operating systems are seen as the ultimate technique for real-time applications. Therefore, one of the purposes of the developments in this book is to give the perspective and background needed to make more advanced courses meaningful.

It may help the reader to understand the intent and viewpoint of this book if he knows a little of the author's design interests. His introduction to the field of logic was through the door of power systems when working on the design of nuclear power stations.

Early application of logic was in the design of control for human access to high radiation areas and of control circuits for large electrical machines. The machines are in high radiation areas so that access is restricted. Logic circuits are required to guarantee the good health of these machines, since regular maintenance is out of the question.

More recent applications have been in the area of control of electrical generating stations and electrical power system protection. Expertise gained in this area has led to the design of circuitry for computers.

The point of all this is that the author views logic circuits strictly from a user viewpoint. The saving of components in the overall design is not considered to be an important factor. However, ease of design, ease of construction, and ability to debug the circuit are of primary concern. An intelligent circuit designed to take advantage of subtle points and thus reduce the simplicity of the circuit, and increase the complexity of debugging, is considered undesirable.

The author places a great deal of emphasis on the skills required in building circuits. Thus, laboratory work is considered to go hand in hand with theoretical development. Basic principles are first introduced. Conventional or classic methods of achieving this design are next described, and this is followed rapidly by an explanation of techniques using MSI or LSI systems.

Obviously there is material in this book that is of fundamental importance—and material that is included for completeness. The author has his suspicions as to which way the general field of logic and digital design will develop. But in reality there is no way of knowing the exact course. There also arises the problem or the consideration of specific and general usefulness of the book. The major intention is to develop knowledge, skills, and techniques directly applicable to the majority of students. However, there may be the rare individual who is faced with a different set of specifications in the design of his circuit. I have in mind here the designer of logic for a pumping station involving the control of valves, pumps, fans, compressors, and so on, where much of the logic should of necessity occur at reasonably high power levels. This designer would not use the shortcuts advocated in the following sections but would instead rely heavily on the classical techniques. Therefore, completeness is important. More specifically you may wish to read but not study Sections 3.1 and 3.3 to 3.12. However, the main results of this chapter, Sections 3.2, 3.13, and 3.14, are very important.

A course in logic should be a unique experience for you, particularly if you are in an engineering program. Most engineering courses have a common basis. The general problem-solving technique is to transfer a problem to a flow chart for solution. This transfer is usually accompanied by certain assumptions that make this practical. The solution then involves the application of standard engineering prin-

ciples, generally in the form of theory adopted from physics and the application of mathematics from the differential calculus. The synthesis technique allows for variations in approach and, in general, for a continued development of the engineering concepts and their application. Logic is different in that it applies the theories developed in Boolean algebra and the calculus of propositions. I suspect that you view logic as a subject in which the techniques that you learned are of little use, and therefore the subject presents the appearance of a blank wall that cannot be penetrated. The route through the wall, however, is really quite easy. One must first learn and become familiar with the new concepts. An illustration of how these concepts are used and some practice in their application follows, and the student rather suddenly sees the light. He is thus faced with a situation where almost momentarily he moves from a position where he is in a confusing maze to a position where everything is light and clarity. Once you have penetrated this position using the techniques advocated in this book, you will be able to tackle complex problems with ease. You will wonder what is difficult about logic and will rapidly forget the difficulties you faced in mastering this subject. You will therefore be left with a rather different experience. You will have totally mastered a subject to the degree that you now feel it is almost trivial. This should leave you with an extremely elated feeling.

G. S. HOPE

Calgary, Canada
September 1980

Acknowledgments

The author would especially like to thank his former students for encouragement that initiated work on this book. Appreciation is also expressed to members of the Department of Electrical Engineering for their support. Many others contributed support in a variety of ways. The individuals are too numerous to mention, but their contributions are appreciated.

The author would like to thank Linda Sherwood, Doreen Nordquist, Bess Nelson, and Florence Knight not only for their assistance in preparing this manuscript but also for their tolerance.

The help from the following friends is specially acknowledged. Demetrius Zissos has improved the author's approach to practical logic and has always provided support and buoyancy. Om Malik has made the task of writing easier by continuous effort over many years. George Csanyi-Fritz has worked for many hours and often, because it was the only time the computer could be shut down for tests, until early morning hours. He accepted and improved most of the novel sequential logic techniques proposed in this book. Garwin Hancock has worked conscientiously to build and improve the microcomputer system on which this book is based.

Contents

CHAPTER 1 COMBINATIONAL DESIGN 1

1.1 Logical Operations, 1
1.2 Truth Tables, 2
1.3 Karnaugh Maps as Input Output Expressions, 3
1.4 Minimum Forms, 4
1.5 Maximum Forms, 4
1.6 Minterm Forms, 7
1.7 Symbols, 8
1.8 Fundamental Relationships, 9
1.9 Karnaugh Maps as Design Tools, 13
1.10 Don't Care Conditions, 18
1.11 Implementation of Logic Functions, 18
1.12 Logic *nand* Implementation, 19
1.13 Logic *nor* Implementation, 19
1.14 Implementation Examples, 20
1.15 The *exclusive or* Function, 21
1.16 Symmetrical Forms, 22
1.17 Reduction, 23
1.18 Practical Circuits, 25
1.19 Problems, 30

CHAPTER 2 MULTIPLEXERS AND DEMULTIPLEXERS IN COMBINATIONAL CIRCUITS 34

2.1 Multiplexer Configuration, 34
2.2 The *mux* Design Notation, 35

2.3 Applications to Combinational Logic, 36

2.4 Direct Method, 38

2.5 Practical Circuits, 43

2.6 The Use of Large Scale Integration (LSI) Memories as Combinational Elements, 44

2.7 Demultiplexers, 44

2.8 Problems, 46

CHAPTER 3 FUNDAMENTAL MODE CIRCUITS 49

3.1 Fundamental Mode Design Conventions, 49

3.2 Concept of Stability, 50

3.3 Basic Operation of a Sequential Circuit, 51

3.4 Mode Diagram Design, 51

3.5 Flow Table, 51

3.6 Elimination of Redundant Stable States, 53

3.7 Merging Rows, 54

3.8 Secondary State Assignment, 55

3.9 Excitation Map (Y Map), 56

3.10 Design of Output Circuit, 57

3.11 Circuit Realization, 57

3.12 Synthesis of Fundamental Mode Sequential Circuits, 58

3.13 Design Features, 59

3.14 Buffer Circuit Design, 60

3.15 Problems, 63

CHAPTER 4 EVENT-DRIVEN SEQUENTIAL CIRCUITS 67

4.1 State Diagrams for Event-Driven Sequential Circuits, 67

4.2 RSFF, 68

4.3 Design Flexibility of RSFFs, 70

4.4 Steps in the Design of an Event-Driven Circuit Using RSFFs, 71

4.5 Operation of an Automatic Start-Up Shut-Down Circuit, 72

4.6 State Variable Equations for the Start-Up Shut-Down Circuit, 74

4.7 The Output Function, 79

4.8 Problems, 80

CHAPTER 5 EVENT-DRIVEN CIRCUIT IMPLEMENTATION USING MULTIPLEXERS 84

5.1 Multiplexers as a Memory Device, 84

5.2 Sequential Circuit Multiplexer Notation and Conventions, 85

5.3 State Diagrams in Multiplexer Notation, 86

5.4 Realization of an Annunciation Circuit Using Multiplexers, 89

5.5 Input Matrix for the Automatic Start-Up Shut-Down Circuit, 91

5.6 Application Guidelines, 97

5.7 Problems, 97

CHAPTER 6 CLOCK-DRIVEN SEQUENTIAL CIRCUITS 104

6.1 Activity in Synchronous Sequential Circuits, 104

6.2 Two-Phase Synchronous Logic, 106

6.3 Operation of the JKFF, 106

6.4 Design Flexibility of JKFFs, 108

6.5 Steps in the Design of Clock-Driven Sequential Circuits Using JKFFs, 108

6.6 Example of a Counter Design Using JKFFs, 109

6.7 Ripple Counters in Realization, 111

6.8 BCD Synchronous-Counter Design, 111

6.9 Multievent Pulsed Sequential Circuits, 114

6.10 Synthesis of Multievent Clock-Driven Sequential Circuits, 116

6.11 Programmable Circuits, 119

6.12 Nested Circuits, 121

6.13 Problems, 121

CHAPTER 7 COUNTERS, MULTIPLEXERS AND DEMULTIPLEXERS IN CLOCK-DRIVEN SEQUENTIAL CIRCUITS 128

7.1 Counters as Memory Devices, 128

7.2 Standard MSI Circuit for Clock-Driven Sequential Circuit Design, 129

7.3 State Diagrams in Counter Sequence Notation, 130

7.4 Single-Output Circuits, 135

7.5 Design for Flexibility in Clock-Driven Sequential Circuits, 137

7.6 Problems, 138

CHAPTER 8 STATE DIAGRAM CONSTRUCTION 142

8.1 Format, 142

8.2 A Cell-to-State Duality, 143

8.3 Construction of the Preliminary State Diagram, 143

8.4 Completion of the State Diagram for Event-Driven Sequential Circuits, 144

8.5 Completion of the State Diagram for Clock-Driven Circuits Using JKFFs, 145

8.6 Completion of the State Diagram for Pulse Sequential Circuits Using MSI Components, 145

8.7 Example of an Event-Driven Sequential Circuit State Diagram Construction, 146

8.8 Example of a Pulse Sequential Circuit State Diagram Construction, 152

8.9 Construction of State Diagrams for Pulse-Driven MSI Circuits, 154

8.10 A Generalized Approach to Logic Circuit Design, 154

8.11 The Design and Control of Completely Generalized State Diagrams, 155

8.12 Problems, 155

CHAPTER 9 REGISTERS IN LOGIC DESIGN 158

9.1 Data Storage Using Registers, 158

9.2 Shift Registers, 159

9.3 Generation of Codes Using Shift Registers, 161

9.4 Generation of Sequences Using Shift Registers, 166

9.5 Multifunction Registers, 168

9.6 Binary Counting Registers, 170

9.7 Complementing Register, 172

9.8 Superposition of Registers, 173

9.9 Problems, 174

CHAPTER 10 A DIGITAL SYSTEM 176

10.1 ROM and RAM, 177

10.2 The CPU, 178

10.2.1 ALU Module, 178
10.2.2 Instruction Decoder, 180
10.2.3 Working Registers, 180
10.2.4 Address Bus, 181
10.2.5 Data Bus, 181
10.2.6 Timing Circuit and Control, 181

10.3 State Diagram Illustration of Machine Cycles, 183

10.4 Input and Output (I/O), 188

10.5 Microcomputer System Studies, 190

10.5.1 The MC6800 CPU, 191
10.5.2 The Intel 8085A CPU, 193

CHAPTER 11 PROGRAMMING AND PROGRAMMING AIDS 196

11.1 Mnemonic Programming, 197

11.1.1 Assembler Programs, 198
11.1.2 Assembler Listing, 199

11.2 Assembler Formats, 199

11.2.1 The Mnemonic Field, 200
11.2.2 The Operand Field, 200
11.2.3 The Label Field, 200
11.2.4 The Comment Field, 200

11.3 Notations and Rules, 201

11.4 Memory Addressing, 204

11.5 Programming Models, 206

11.5.1 Motorola's MC6800 Programming Registers, 206
11.5.2 Intel 8085 Programming Model, 207

11.6 **The Instruction Set, 208**

11.6.1 Data Transfer Instructions, 210
11.6.2 Arithmetic Instruction, 215
11.6.3 Logic Instructions, 217
11.6.4 Index, Indirect and Stack Instructions, 223
11.6.5 Branch and Jump Instructions, 233
11.6.6 Status, I/O, and Special Instructions, 236

11.7 **Assembler Directives, 239**

11.7.1 Directive Notation and Rules, 542
11.7.2 Motorola Assembler Directives, 243
11.7.3 Intel Assembler Directives, 247

11.8 **Systems and System Executive Statements, 250**

11.8.1 Motorola Fundamental Module Configuration, 252
11.8.2 Intel Fundamental Module Configuration, 254

11.9 **Problems, 257**

CHAPTER 12 INPUT, OUTPUT TECHNIQUES 260

12.1 **General-Purpose Parallel Adapter, 261**

12.1.1 Motorola peripheral interface adapter, 263
12.1.2 Intel programmable peripheral interface unit, 269

12.2 **Microcomputer System Communications, 274**

12.2.1 Motorola Communication Interface, 275
12.2.2 Intel Universal Interface, 279

12.3 **Problems, 285**

CHAPTER 13 OPERATION AND CONFIGURATION OF INDEPENDENT SYSTEMS 288

13.1 **System Map, 289**

13.1.1 Motorola System Map, 289
13.1.2 Intel System Map, 291

13.2 **Configuring Microcomputer Systems, 292**

13.3 **Program Initialization, 293**

13.3.1 Motorola Initialization, 293
13.3.2 Intel Initialization, 293

APPENDIX A DEFINTION OF A BOOLEAN ALGEBRA 295

APPENDIX B NUMBER SYSTEMS 297

B.1 The Decimal System, 297
B.2 Other Radix Systems, 298
B.3 Basic Arithmetic, 299
B.4 The Binary System, 300
B.5 Binary Coded Decimal, 300
B.6 Nine's Complement Codes, 300
B.7 Computer Number Systems, 301
B.8 Gray Code, 302
B.9 Notational Convenience, 303
B.10 Instruction Codes and Formats, 303
B.11 Alphanumeric Characters, 304
B.12 Problems, 305

APPENDIX C MOTOROLA'S EXECUTABLE INSTRUCTIONS 307

C.1 Symbols, 307
C.2 Instruction Set, 308

APPENDIX D INTEL'S EXECUTABLE INSTRUCTIONS 327

D.1 Symbols, 327
D.2 Instruction Set, 328

APPENDIX E MOTOROLA DIRECTIVES 341

APPENDIX F INTEL DIRECTIVES 343

APPENDIX G MOTOROLA PROGRAMMING 344

G.1 System Start-Up, 344
G.2 Maid Module Commands, 345
G.3 Editor Module Commands, 346

G.4 Assembler Module Commands, 347

G.5 Directory Module Command, 347

APPENDIX H INTEL PROGRAMMING AID COMMANDS 348

H.1 System Start-Up, 348

H.2 Editor Module Commands, 348

H.3 Assembler Module Commands, 350

H.4 Monitor Module Commands, 350

H.5 Directory Module Command, 351

H.6 Print Module Command, 351

APPENDIX I MOTOROLA INTERFACE ADAPTERS 352

I.1 PIA Registers, 352

I.2 ACIA Registers, 352

APPENDIX J INTEL PERIPHERAL INTERFACES 356

J.1 PPIU Registers, 356

J.2 USART Registers, 356

REFERENCES 362

SUBJECT INDEX 365

INTEGRATED DEVICES IN DIGITAL CIRCUIT DESIGN

CHAPTER **1**

Combinational Design

Did you ever wonder where the word *logic* comes from? As you look at the most dominant products of the age of logic, you begin to realize that they, above all else, follow the whims of the designer. When using a minicomputer or microprocessor, there is no test of logic that will lead to your salvation. So why the term *logic*? Boole was particularly interested in the laws of thought, as can be verified by Reference 1. He developed a process by which certain propositions could be tested for truth. However, propositions imply a sentence structure outlining hypotheses, forming a conjunction, and stating a conclusion. So the concepts of true and false are from a calculus of propositions leading to today's truth tables used by manufacturers. Observation of the truth table in today's form shows that the original structure is not used, but the term *logic* remains.

Boolean algebra gives a definition of a list of operators, and these operators are stated in the form in which they are useful in the development of present-day logic. However, Boolean algebra contains one operator which is not useful in logic and omits operators which are very useful. So logic may be better described as a switching algebra. Switching algebra from the work of Shannon, Venn, and Karnaugh is developed in this chapter [2]. The map methods, which give a pictorial view of operators, are fundamental to all chapters that follow.

1.1 LOGICAL OPERATIONS

There are really three basic operators required to develop a switching algebra. All other operators are designed from some combination of these. To illustrate the concept behind the basic operators, consider a square that represents the set of all things affected by the operators. A subset of this square is represented by a circle within the square. Figure 1.1 is an illustration of such a figure with two subsets Y and Z represented as circles within the square. From this figure, called a Venn diagram, the operator *and* can be defined and assigned the notation •. The shaded area in Figure 1.2 represents the combination of subsets described by Y • Z and is our definition of the *and* operator.

The *and* operator is a basic operator because of its ability to select a minimum area. Four such minimum areas are labeled in Figure 1.3. The absolute minimum area in our square would be nil, but this is the trivial case and not useful to our purpose. In essence using the *and* operator identifies one of the basic areas.

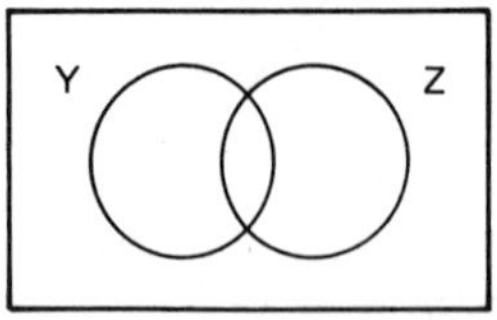

Figure 1.1 Set and subsets.

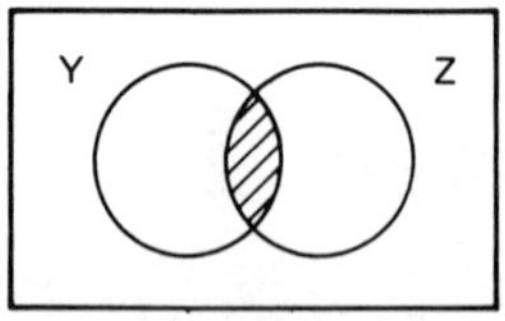

Figure 1.2 Definition of *and*, Y • Z.

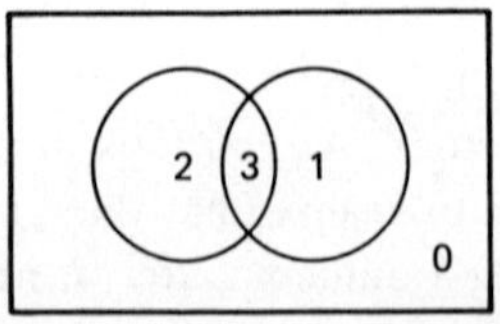

Figure 1.3 Areas of a Venn diagram.

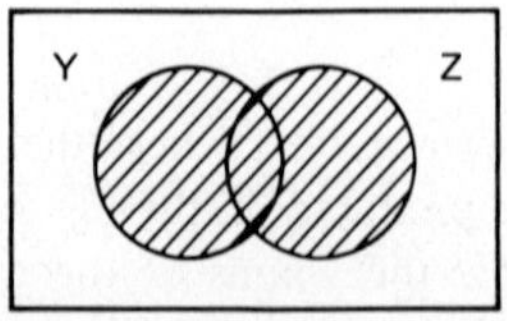

Figure 1.4 Definition of *or*, Y + Z.

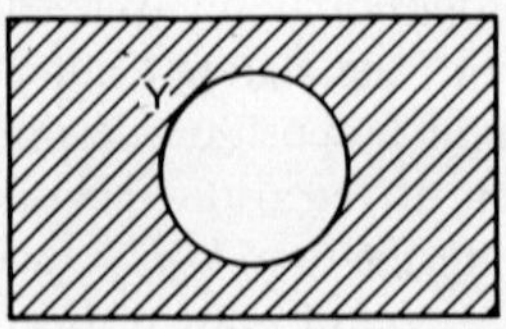

Figure 1.5 Diagram of $\overline{Y}$.

In Figure 1.4 the *or* operator is represented by the shaded area and assigned the notation +. The reader should not get confused because the plus sign is used. There are a certain number of basic operators available on any typewriter or any set of printers' mats. The fact that the plus sign is used to define other operators is strictly a matter of convenience.

The *or* operator is required because of the need to describe the maximum area within the square without describing the trivial case of the entire area. The ability to select the maximum area gives the *or* operator its usefulness and purpose.

The operators · and + are limited in their ability to select specific areas. To overcome this difficulty, the basic operator *not* is required. It is shown in Figure 1.5. If one considers a subset Y the operator *not* assigned to Y is described by the shaded area. The notation for *not* is ‾, and the shaded area in Fig. 1.5 is written $\overline{Y}$ and read as "Y *not*." This operator gives us the flexibility to make the operators *and* and *or* apply to other subsets within the square.

1.2 TRUTH TABLES

It is common practice for logic component manufacturers to use truth tables instead of Venn diagrams. In truth tables a set of conventions different from those used above is adopted. However, the truth table can be directly related to these diagrams. If a true (T or 1) is used to designate the presence of the set Y, or a

Table 1.1 Truth Table for *and*

Y	Z	∘
0	0	0
0	1	0
1	0	0
1	1	1

Table 1.2 Truth Table for *or*

Y	Z	+
0	0	0
0	1	1
1	0	1
1	1	1

Table 1.3 Label of YZ Combinations

Y	Z	Combination Label
0	0	0
0	1	1
1	0	2
1	1	3

false (F or 0) is used to represent the nonpresence of Y, and if a similar thing is done with Z, a truth table representation for the *and* operator is given in Table 1.1. The truth table for the *or* operator is given in Table 1.2. Note that there are four different combinations of Y and Z in the truth tables and that these combinations have a direct relation to the area labeling as shown in Figure 1.3. The combinations of YZ and their equivalent labels are shown in Table 1.3.

1.3 KARNAUGH MAPS AS INPUT OUTPUT EXPRESSIONS

A Venn diagram may be easy for the student to understand, and a truth table may be the standard technique for defining the operation of an IC chip by manufacturers, but neither technique is very useful to the student who is interested in developing a rapid and useful technique for logic design. The more convenient form is the Karnaugh map [2–8]. Figure 1.6 shows a conversion from a Venn diagram to a Karnaugh map. The illustration shows that the Karnaugh map is much easier to draw. Figure 1.7 shows the same conversion from a truth table. Here the Karnaugh map is much more concise and gives a clear representation of the thoughts that the author wishes to convey to the student. In this representation the student should think of the operator in a little different way. The convention that is adopted is to call the sets or variables (Y and Z) *inputs.* The logical operator forms an *output.* In the Karnaugh map form the input variables are marked on top and left-hand edge of the map. The map itself consists of a number of cells. Identifying certain cells within this map defines the logical operator or operations. Because the emphasis here is on each cell, this is called a map.

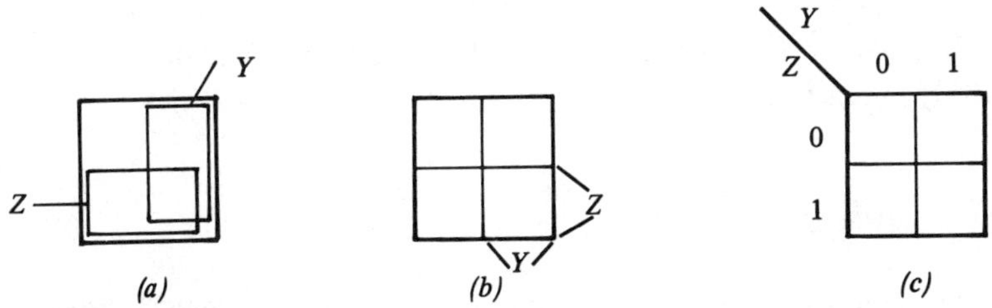

Figure 1.6 Conversion to Karnaugh map of Venn diagram. *(a)* Step 1; *(b)* Step 2; *(c)* Step 3.

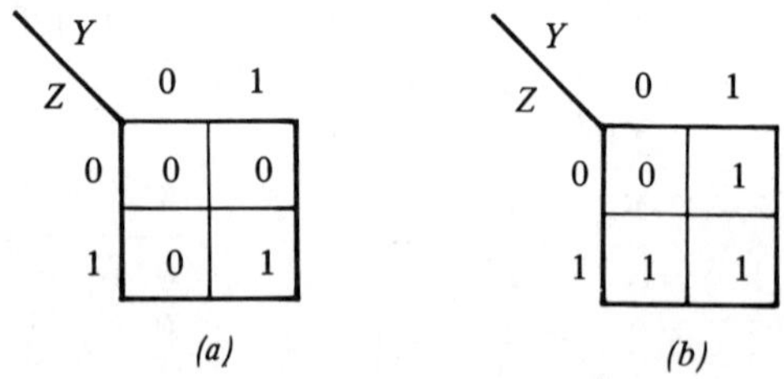

Figure 1.7 Karnaugh map of truth tables. *(a) and* operator; *(b) or* operator.

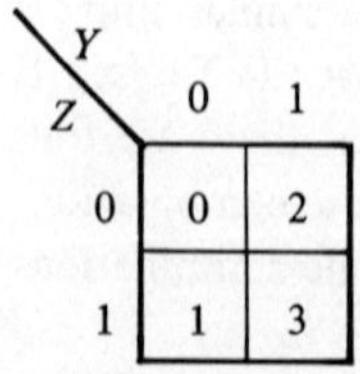

Figure 1.8 Cell designation.

1.4 MINIMUM FORMS

Each cell within this map has a particular number that identifies the combination of input variables. Thus the map shown in Figure 1.8 has the following representation. Cell 0 designates the input variables $\overline{Y}$ and $\overline{Z}$. Cell 1 designates input variables $\overline{Y}$ and Z. Cell 2 designates input variables Y and $\overline{Z}$, and cell 3 designates input variables Y and Z. Each cell identifies an *and* function of all possible input variables. These variables may occur in their normal or inverted form. A function of all possible input variables is a minterm expansion [2].

A convenient notation to represent the *and* function is $\Sigma(3)$. This notation identifies cell 3, which has inputs $Y \circ Z$ so the function f(Y, Z) is

$$f(Y, Z) = Y \circ Z = \Sigma(3)$$

This is the definition of the *and* operator. Similarly,

$$f(Y, Z) = \Sigma(0) = \overline{Y} \cdot \overline{Z}$$

$$f(Y, Z) = \Sigma(1) = \overline{Y} \cdot Z$$

$$f(Y, Z) = \Sigma(2) = Y \cdot \overline{Z}$$

It is convenient to shorten f(Y, Z) to f and $Y \circ Z$ to YZ. The shorter form is used extensively in this book.

1.5 MAXIMUM FORMS

The cell notation for the *or* operator is given in Figure 1.9. Figure 1.9(*a*) represents the function $\overline{Y} + \overline{Z}$. To agree with the convention adopted above let us call the

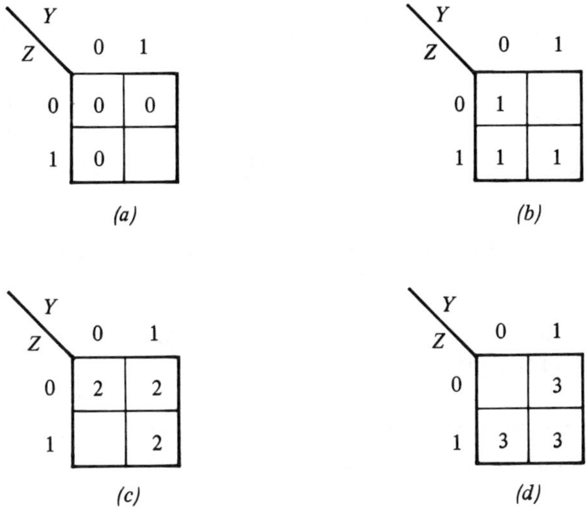

Figure 1.9 Maps of $\pi(0)$, $\pi(1)$, $\pi(2)$, $\pi(3)$. *(a)* $\bar{Y} + \bar{Z}$; *(b)* $\bar{Y} + Z$; *(c)* $Y + \bar{Z}$; *(d)* $Y + Z$.

identified cells 0 for the *or* operator. In Figure 1.9(*b*), $\overline{Y} + Z$, the designated cells are labeled 1. Similarly the functions $Y + \overline{Z}$ and $Y + Z$ are identified by cells labeled 2 and 3 in Figures 1.9(*c*) and 1.9(*d*) respectively. These are all maximum nontrivial functions, and in order to designate them as such the notation π is used. Thus

$$\pi(0) = \overline{Y} + \overline{Z}$$

$$\pi(1) = \overline{Y} + Z$$

$$\pi(2) = Y + \overline{Z}$$

$$\pi(3) = Y + Z$$

If

$$f(Y, Z) = \pi(0)$$

and

$$\bar{f}(Y, Z) = \pi(\bar{0})$$

then Figure 1.10 gives a simplified version of the π map of YZ.

Representing the function on this basis is called *maxterm representation.*

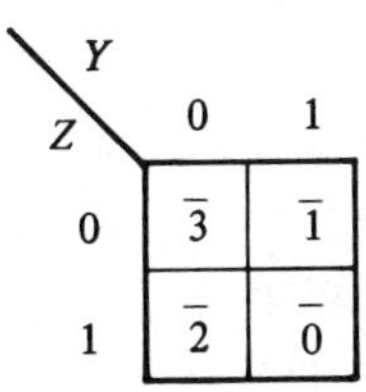

Figure 1.10 Simplified representation for π maps.

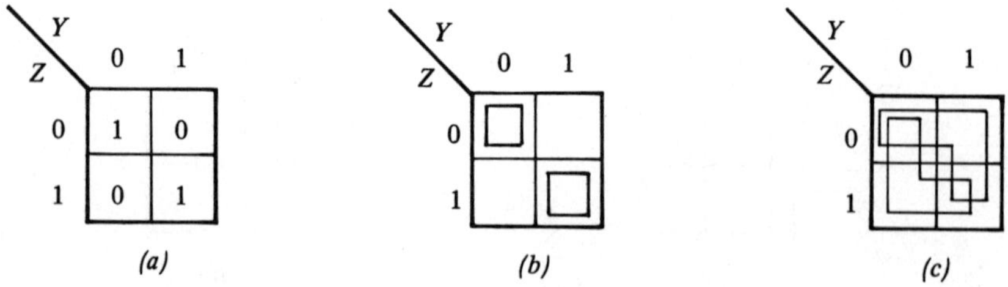

Figure 1.11 f(Y, Z). *(a)* Map; *(b)* minimum; *(c)* maximum.

Consider the function described by the map of Figure 1.11(*a*). This function is realized by one of two methods:

1. Take the maximum combination of minimum forms.
2. Take the minimum combination of maximum forms.

The minimum form is shown in Figure 1.11(*b*). The upper left cell $\Sigma(0)$ represents $\overline{Y}\overline{Z}$, and the lower right cell $\Sigma(3)$ represents YZ. Taking the maximum combination of these two cells the output function f is

$$f = \overline{Y}\overline{Z} + YZ$$

Any function with this general form is called a *sum of product* (S of P) to mean a maximum combination of minimum forms. The function f can also be written

$$f = \Sigma(0), \Sigma(3)$$

or simply

$$f = \Sigma(0, 3)$$

The maximum form is shown in Figure 1.11(*c*). The upper right maximum group of cells $\pi(2)$ represents $Y + \overline{Z}$, and the lower left maximum group of cells $\pi(1)$ represents $\overline{Y} + Z$. A minimum combination gives the output function f as

$$f = (Y + \overline{Z}) \cdot (\overline{Y} + Z)$$

Any function with this general form is called a *product of sums* (P of S) to mean a minimum combination of maximum forms.

The function can also be written

$$f = \pi(1), \pi(2)$$

or simply

$$f = \pi(1, 2)$$

Generally the S of P or minterm representation are favored in this book.

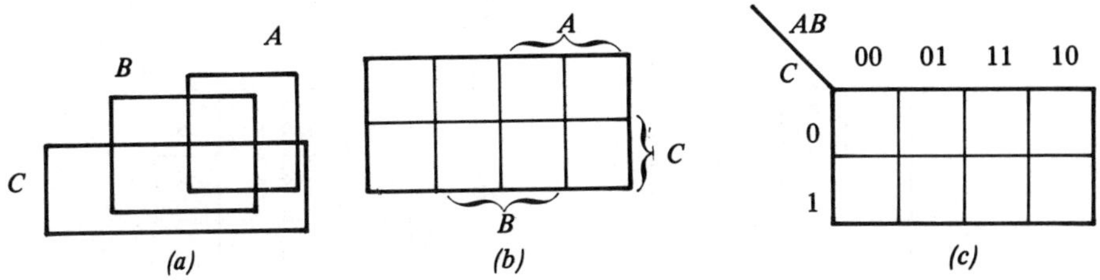

Figure 1.12 Conversion of three variables into a map. *(a)* Step 1; *(b)* Step 2; *(c)* Step 3.

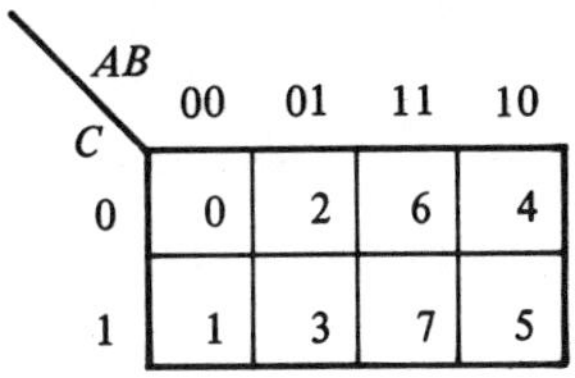

Figure 1.13 Karnaugh map showing Σ designation for each cell.

1.6 MINTERM FORMS

Expansion to three variables is easily done in accordance with Figure 1.12 for the variables A, B, and C. Figure 1.12(*a*) shows the area that each variable designates. Figure 1.12(*b*) shows the map in conventional form with the function represented by cells with the input variables along the outside borders of the figure. Figure 1.12(*c*) gives the cells of the Karnaugh map with the three variable inputs across the top and down the left edge and uses notation compatible with truth tables. For convenience the minterm representation of ABC is shown in Table 1.4.

The Σ designation is shown in Figure 1.13.

Four and more variables are expressed in Karnaugh map form by expanding in a similar way. Figure 1.14(*a*) shows cell arrangement in a Karnaugh map, and Figure 1.14(*b*) gives the equivalent minterm Σ designation for each cell.

Karnaugh maps are called *K maps* or simply *maps* in this book.

Table 1.4 Three Variable Minterms

A	B	C	Σ
0	0	0	0
0	0	1	1
0	1	0	2
0	1	1	3
1	0	0	4
1	0	1	5
1	1	0	6
1	1	1	7

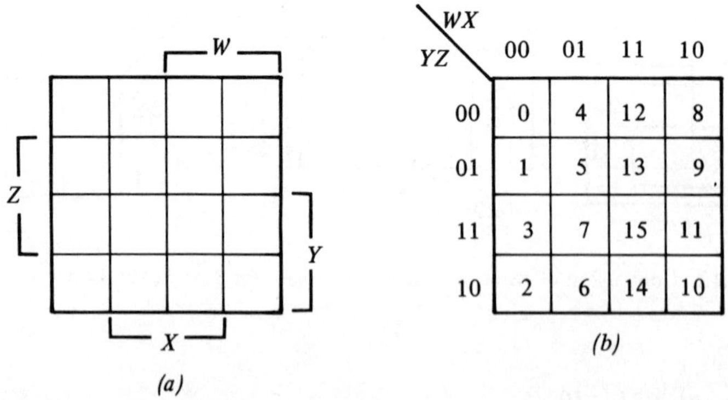

Figure 1.14 Four variable Karnaugh map. *(a)* Cell arrangement; *(b)* Σ designation.

1.7 SYMBOLS

It is convenient to express logic in diagrammatic form, and for this purpose symbols are needed. Symbols for both the basic operators and the derived logic operators are required. Derived logic operators occur for a number of reasons. One of the more common is illustrated in the manufacture of IC chips. The common feature of these chips is that their manufacture requires the inversion of the output signal, which in logical terms *nots* the output. Thus for the *and* and *or* basic operators there is an inherent form using the negated output. Figure 1.15 shows the K map form of the operators *and or* and their symbolic representation. Figure 1.16 shows the *standard* symbolic representation for the *not* operator. The little circle is taken to be the sign of inversion. Numbers (101, 102, and 103) are placed inside the gate

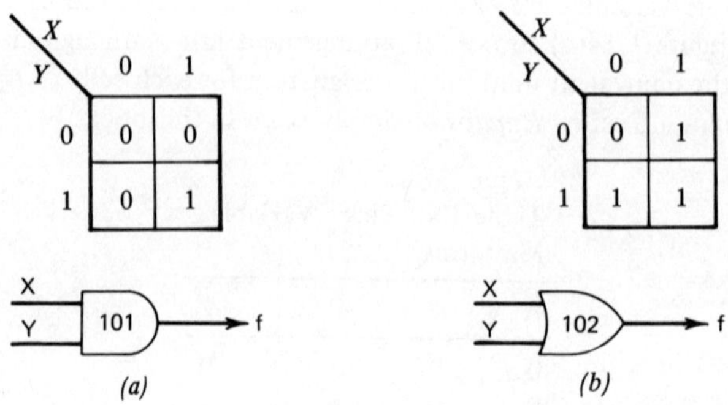

Figure 1.15 Symbols for *and* and *or.* *(a) and*; *(b) or.*

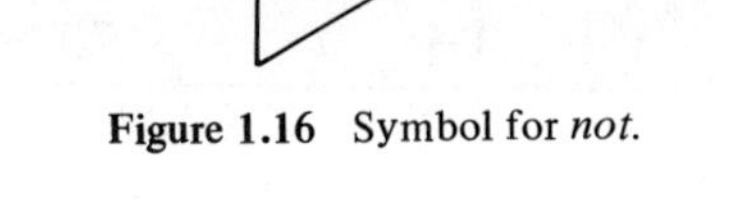

Figure 1.16 Symbol for *not.*

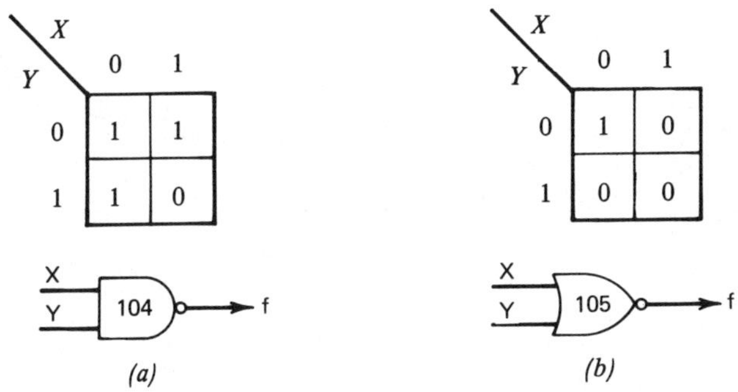

Figure 1.17 Symbols for *nand* and *nor*. *(a) nand*; *(b) nor*.

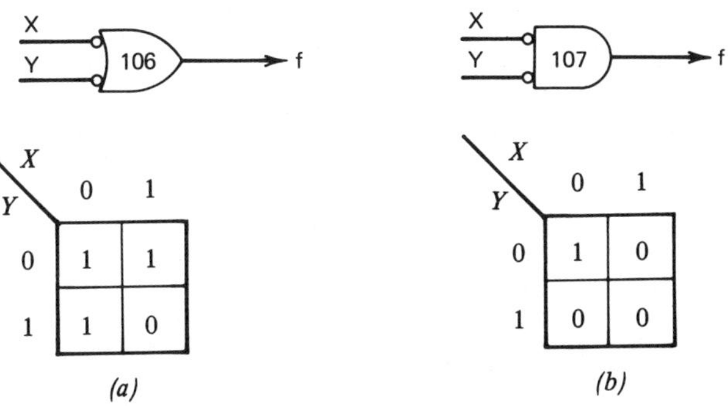

Figure 1.18 Alternate forms. *(a) nand*; *(b) nor*.

as an identifier. The first number is the chapter number. Remaining numbers are assigned in sequence. Figure 1.17(*a*) shows the inverted *and* operator, called a *nand*, and its symbolic representation. Figure 1.17(*b*) shows the Karnaugh map illustration of the inverted *or* operator, called a *nor*, and its symbolic form. Reversing this procedure and starting with a symbol, Figure 1.18(*a*) shows the *or* symbol with its inputs negated as represented by the small circles. Its Karnaugh map representation is developed immediately below. Comparison with Figure 1.17 shows that this symbol is directly equatable to the *nand* operator. Figure 1.18(*b*) shows the *and* symbol with its inputs negated. Its Karnaugh map shows that this symbol is another form of the *nor* operator.

1.8 FUNDAMENTAL RELATIONSHIPS

In this section diagrams are used to illustrate important rules. There is a one-to-one correspondence between the rule label and the individual parts of Figure 1.19. The student should note the sharp contrast obtained from parts (*a*), (*e*), and (*i*) with the conventional number system.

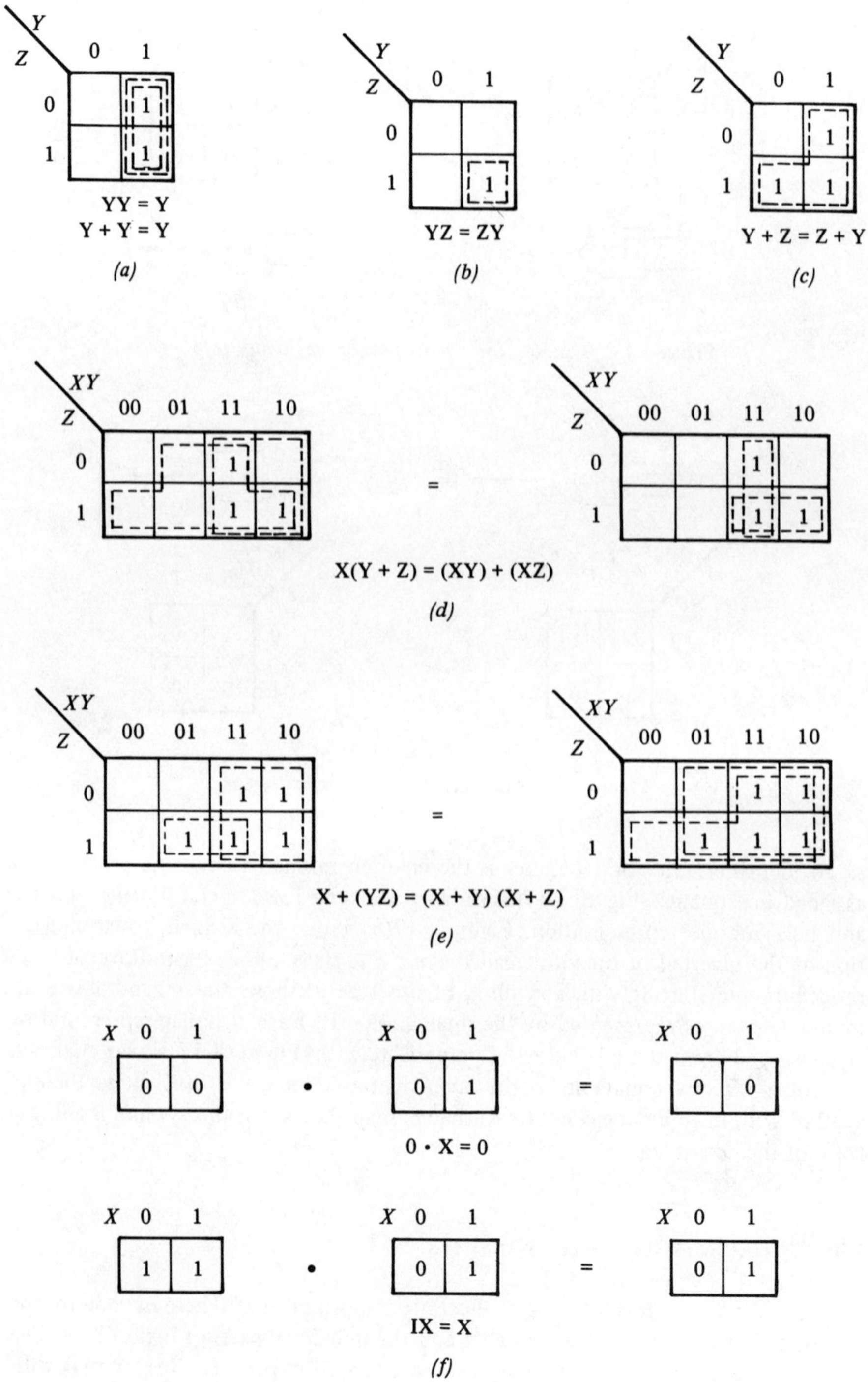

Figure 1.19 Rule illustrations.

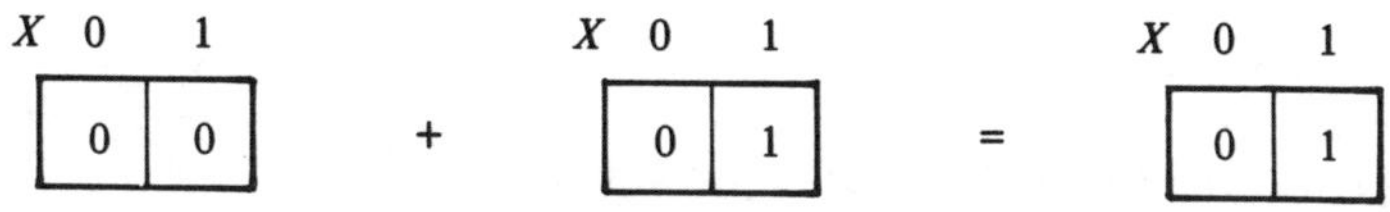

$0 + X = X$

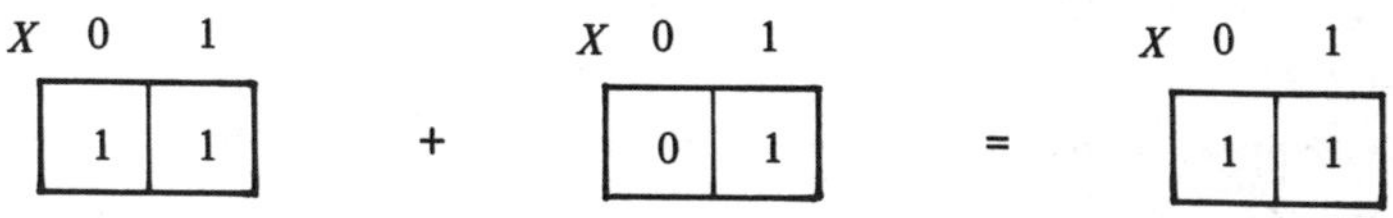

$1 + X = 1$

(g)

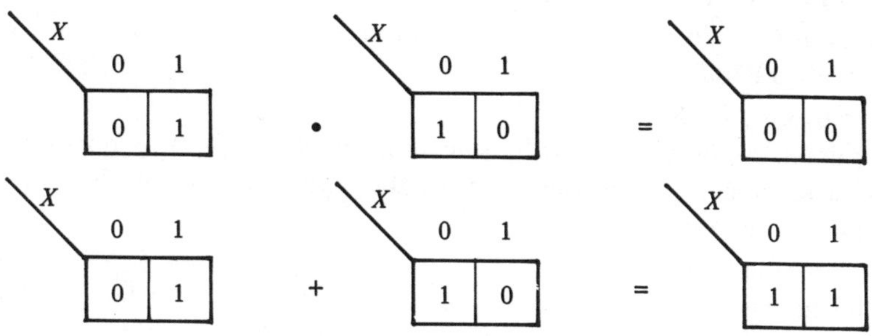

$X \cdot \overline{X} = 0$

$X + \overline{X} = 1$

(h)

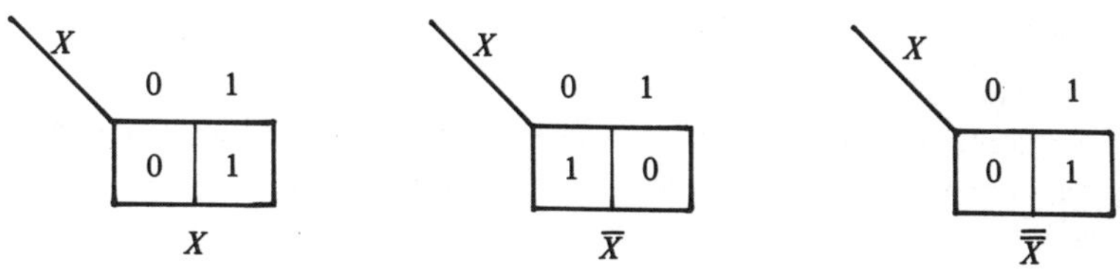

$\overline{\overline{X}} = X$

(i)

$\overline{XY} = \overline{X} + \overline{Y}$

(j)

Figure 1.19 (*Continued*)

(*a*) $YY = Y$
$Y + Y = Y$
This important result is useful in function manipulation.

(*b*) $YZ = ZY$
The order in which the variables appear under the *and* operator does not affect the output function.

(*c*) $Y + Z = Z + Y$
The *or* operator also does not require a specific ordering of the input function.

(*d*) $X(Y + Z) = (XY) + (XZ)$
This algorithm gives a distributed method to change a product of sum form into a sum of product form.

(*e*) $X + (YZ) = (X + Y) \cdot (X + Z)$
This algorithm for distributing X gives a technique for changing a S of P expression into a P of S.

(*f*) $0X = 0$ and $1X = X$
A minimum combination with something nonexistent, 0, is nonexistent, 0, and a minimum combination of a variable with the field of all variables, 1, results in the variable.

(*g*) $0 + X = X$ and $1 + X = 1$
A maximum combination taking the nil field, 0, with a variable produces the variable, and similarly, a maximum combination with the field of all variables, 1, results in the field of all variables, 1.

Cases f and g are a restatement of the properties of the basic operators when dealing with the special cases of fields containing no variables and fields consisting of all possible variables.

(*h*) $X\overline{X} = 0$ and $X + \overline{X} = 1$
In a minimum combination of a variable with its complement the cells described are mutually exclusive, and therefore the result is a field with no variables. However, a maximum combination of a variable and its complement results in a field of all variables.

(*i*) $\overline{\overline{X}} = X$
The right-hand cell of the Karnaugh map [Figure 1.19(*i*)] for this function describes the variable X. Complementing this variable moves the output function to the left-hand cell. Further complementing of the resultant function transfers the function back to the right-hand side, which is the original cell.

The above two algorithms describe special characteristics of the *not* operator.

(*j*) $\overline{XY} \equiv \overline{X} + \overline{Y}$
It can be seen from Figure 1.19(*j*) that $f = XY$ is represented by the lower right-hand cell in the map. Negating this function results in the function describing the remaining three cells. This is identically equal to $\overline{X} + \overline{Y}$. The reader may have noticed this equality from Figures 1.17(*a*) and 1.18(*a*) in the derivation for the symbol *nand*. A significant point to be noted here is that the

combined operators *and not*, $\overline{\cdot}$ are equal to *or*, and the combined operators *or not*, $\overline{+}$ are equal to *and*.

Incidentally, rule *j* is known in many books on logic as DeMorgan's theorem.

The student may find it interesting to note that in Sections 1.1 to 1.7 equations are of standard logic form. However, the rules given in this section equate variables and operators.

The equal sign = should be really read as identically equal to ≡. The expressions in this section are of the form of a calculus of propositions and can be tested in the logic sense. Since the calculus of propositions is useful only in remembering a set of identities, its illustration is limited to Figure 1.18(*j*): $\overline{XY} = \overline{X} + \overline{Y}$. In Table 1.5 the first two columns represent the propositions X and Y written in their true-false rather than 0-1 form (so that a truth test can be conducted). Column 3, XY, gives the result of the *and* operation for each row of columns 1 and 2. In columns 4 and 5 the values of X and Y respectively are complemented. Columns 4 and 5 are subjected to the *or* operation row by row, and the result $\overline{X} + \overline{Y}$ is given in column 6. Column 7, $\overline{X \cdot Y}$, is a simple negation of column 3. Column 8 is an equality test, ≡, on columns 6 and 7.

1.9 KARNAUGH MAPS AS DESIGN TOOLS

The Karnaugh map is one of the most important design tools in the development of logic systems. Certainly in this book it forms one of the main building blocks, and the student cannot progress without thoroughly mastering its use.

There are two essential points about this map. The first is that it is cell representation of all possible minterms of a function. The other essential point about this map is the unique layout of the cells. Adjacent cells differ by one variable and one variable only. Thus a transition from one cell to the next adjacent cell changes only on variable. Because of the way in which our map is developed, it should be noted that the first and last rows are adjacent as are the first and last columns.

In this section techniques are outlined to minimize the number of gates and the number of gate inputs required to realize a circuit.

Gate minimization directly reduces the number of chips required. Gate input minimization permits the use of IC chips with more gates per chip because fewer pins are required for a two-input gate than a three-input and so forth. Minimiza-

Table 1.5 Truth Test for $\overline{XY} = \overline{X} + \overline{Y}$

1	2	3	4	5	6	7	8
X	Y	•	$\overline{X}$	$\overline{Y}$	+	$\overline{\cdot}$	=
F	F	F	T	T	T	T	T
F	T	F	T	F	T	T	T
T	F	F	F	T	T	T	T
T	T	T	F	F	F	F	T

YZ \ WX	00	01	11	10
00	0	0	0	0
01	0	0	0	1
11	0	0	0	0
10	0	0	0	0

Figure 1.20 Map of $f(W, X, Y, Z) = W\bar{X}\bar{Y}Z$.

tion of gates and input in turn minimizes the connections. This makes the process important in the subsequent construction of circuits.

Figure 1.20 shows the representation of the 16 different possible combinations of the four variables W, X, Y, and Z. The expression implied by the map is complete and, if written completely, would be a function (W, X, Y, Z) in polynomial form as follows:

$$\begin{aligned} f(W, X, Y, Z) = {} & (0\bar{W}\bar{X}\bar{Y}\bar{Z}) + (0\bar{W}\bar{X}\bar{Y}Z) + (0\bar{W}\bar{X}Y\bar{Z}) + (0\bar{W}\bar{X}YZ) + (0\bar{W}X\bar{Y}\bar{Z}) \\ & + (0\bar{W}X\bar{Y}Z) + (0\bar{W}XY\bar{Z}) + (0\bar{W}XYZ) + (0W\bar{X}\bar{Y}\bar{Z}) \\ & + (1W\bar{X}\bar{Y}Z) + (0W\bar{X}Y\bar{Z}) + (0W\bar{X}YZ) + (0WX\bar{Y}\bar{Z}) \\ & + (0WX\bar{Y}Z) + (0WXY\bar{Z}) + (0WXYZ) \end{aligned}$$

However, the simple representation of this function f(W, X, Y, Z) is

$$f = \Sigma(9) = W\bar{X}\bar{Y}Z$$

Applying the techinique just developed,

$$f = \Sigma(9, 13) = W\bar{X}\bar{Y}Z + WX\bar{Y}Z$$

indicates the presence of two minterms. Using rule d developed above, this can be rewritten as

$$f = W\bar{Y}Z(X + \bar{X})$$

However, from rule h

$$X + \bar{X} = 1$$

so the function f is minimized to

$$f = W\bar{Y}Z$$

or in another form

$$\Sigma(9, 13) = W\bar{Y}Z$$

Inspection of Figure 1.21 shows that $f = \Sigma(9, 13)$ is a function where X is both 1 and 0 in the enclosed adjacent two cells. Thus it is evident that X can be neglected directly when reading from the map. Figure 1.22 gives a map where four adjacent

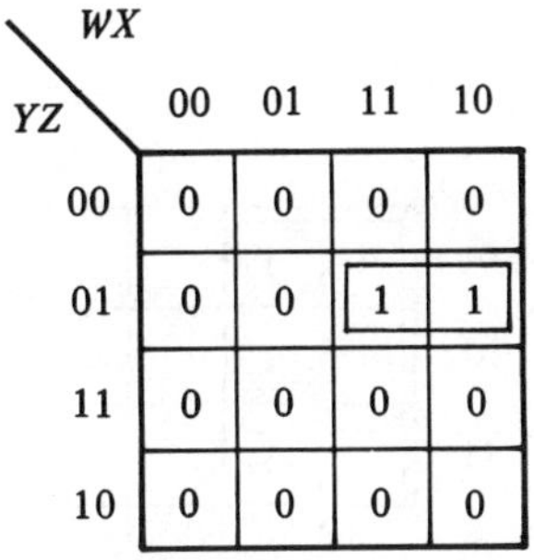

Figure 1.21 Map of $f = W\bar{Y}Z$.

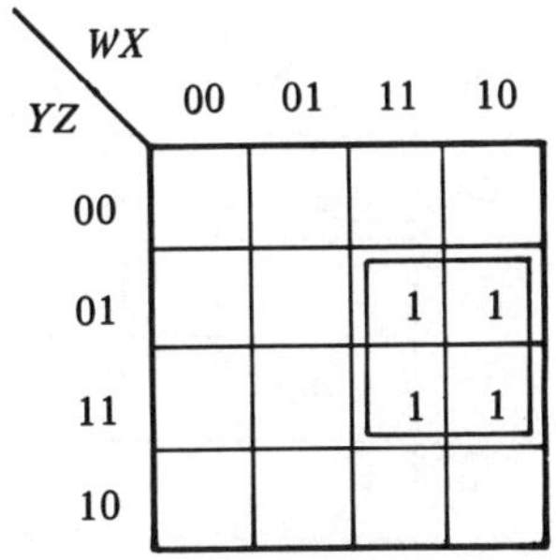

Figure 1.22 Map of $f = WZ$.

cells are enclosed. From the development above, the two cells in row 2 can be represented by

$$f = W\bar{Y}Z$$

Similarly, the two cells in row 3 can be represented by

$$f = WYZ$$

so that the function f is

$$f = W\bar{Y}Z + WYZ = WZ(Y + \bar{Y}) = WZ$$

This result can be obtained from direct inspection of the map.

The enclosed cells can be identified first in the column sense by W because X has values of both 1 and 0, and in the row sense by Z because Y has values of both 0 and 1.

Another grouping of four adjacent cells is shown in Figure 1.23. By inspection of the map, the only possible values for the columns are $\bar{W}X$. All rows are present; thus no row designation is needed, and $f = \bar{W}X$.

Figure 1.24 shows that adjacencies can occur between the first and last columns and rows. Figure 1.25 shows eight adjacent cells. The only variable that is of importance is $\bar{W}$ since all other variables X, Y, and Z have values of both 1 and 0. This grouping of cells, however, is obvious because it defines the manner in which the map is constructed in the first place.

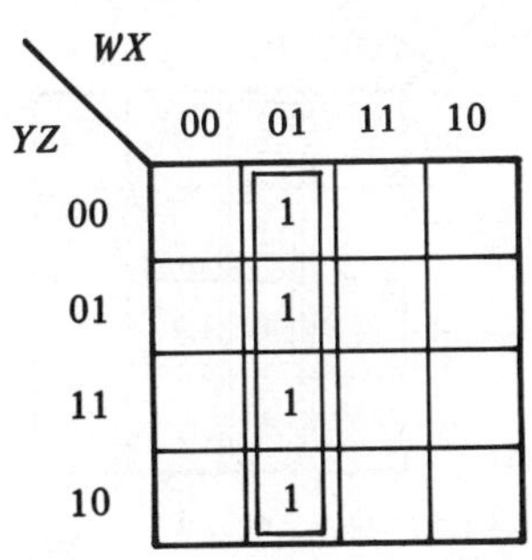

Figure 1.23 Map of $f = \bar{W}X$.

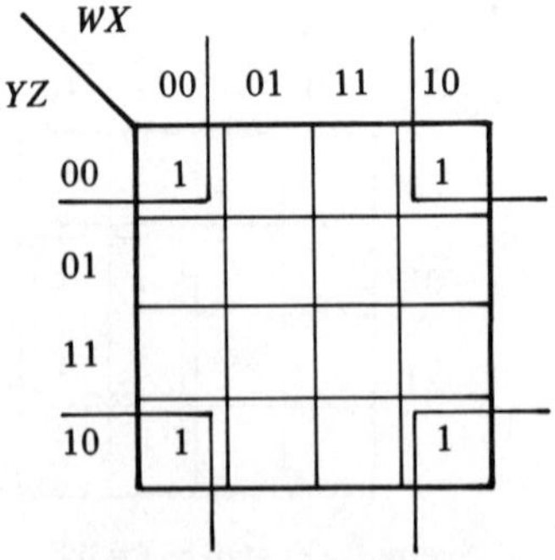

Figure 1.24 Map of f = $\overline{X}\overline{Z}$.

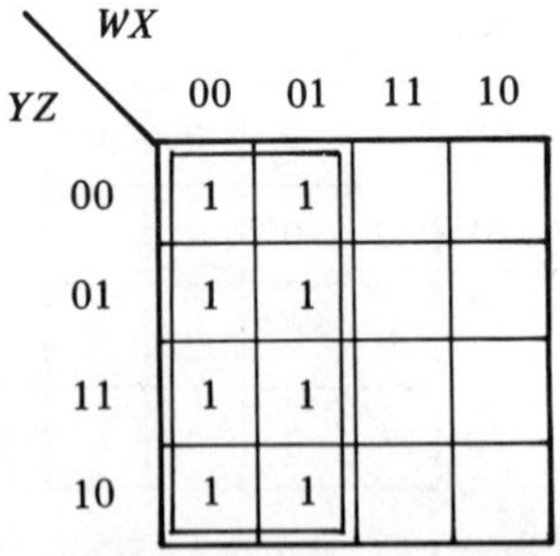

Figure 1.25 Map of f = $\overline{W}$.

Adjacencies are grouped to take in (1) rectangulars, (2) squares, (3) rows, and (4) columns, under the constraint that the number of cells enclosed must equal 2^n (2, 4, 8, 16, etc.).

The map in Figure 1.26 illustrates two sets of enclosed cells so that a minimum realization of the two functions is

$$f = YZ + \overline{W}\overline{Y}$$

Figure 1.27 illustrates a grouping in which two of the cells are in both enclosures. Making the enclosures as large as possible minimizes the form. Rule a states that a cell can be duplicated without affecting the output function. Figure 1.28 illustrates another case where enclosing the largest possible number of cells minimizes the function. Figure 1.29 shows two enclosures covering four cells. Addition of the enclosure shown by the dotted line does not minimize the function and leads to additional components in the realization. Thus the general rule for enclosing adjacent cells is that enclosures should be as large as possible and include all cells in the output function, but that the number of such groupings should be kept to a minimum.

Figure 1.30 shows the output function for a five-variable map of V, W, X, Y, Z. For maps of more than four variables, adjacencies are difficult to represent on a plane surface. Note that the cell $\overline{W}\overline{X}\overline{Y}\overline{Z}$ in the V map is adjacent to the cell $\overline{W}\overline{X}\overline{Y}\overline{Z}$

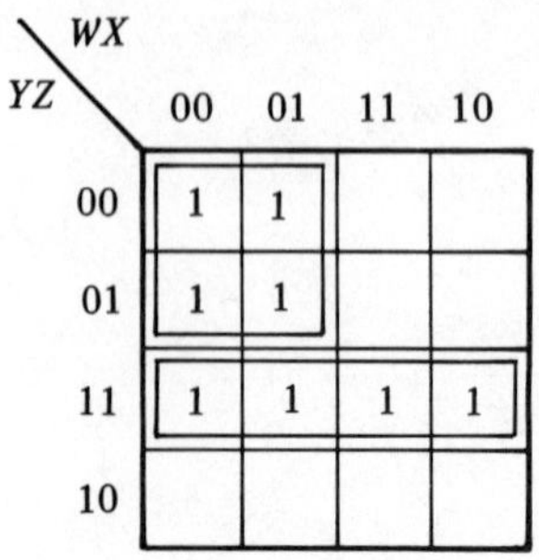

Figure 1.26 Map of f = YZ + $\overline{W}\overline{Y}$.

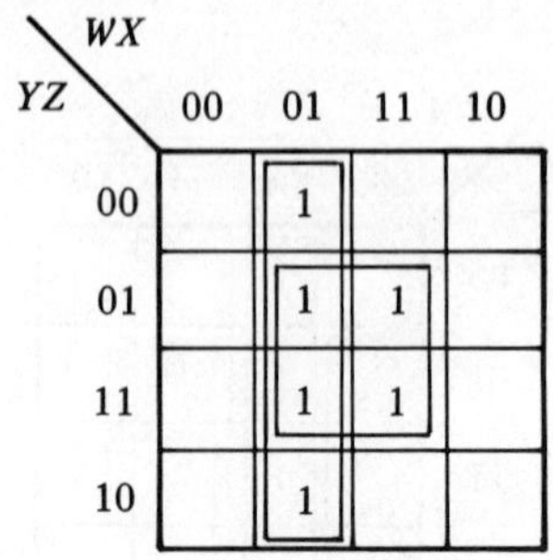

Figure 1.27 Map of f = $\overline{W}$X + XZ.

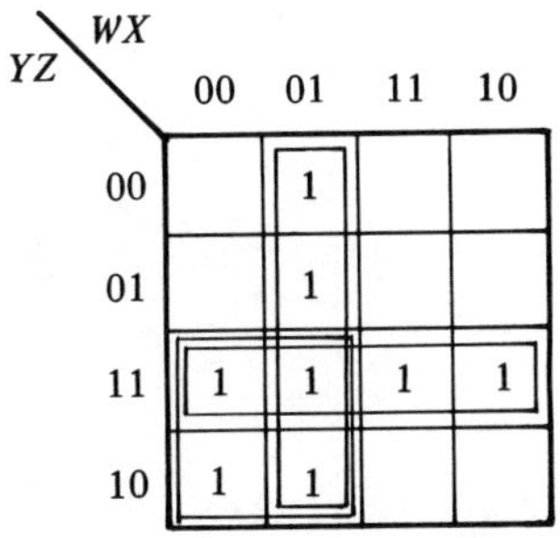

Figure 1.28 Map of $f = \bar{W}Y + YZ + \bar{W}X$.

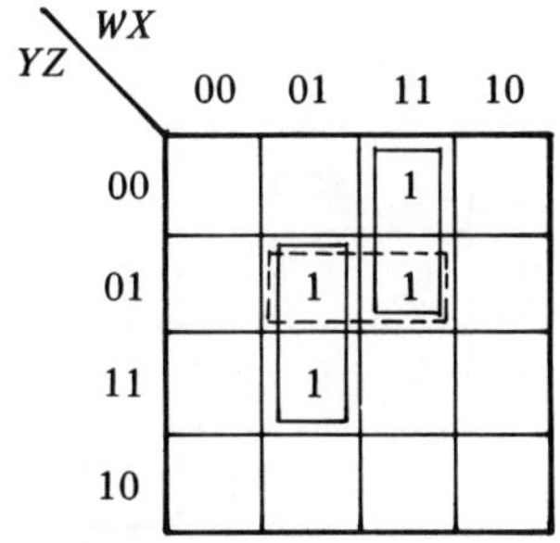

Figure 1.29 Map of $f = WX\bar{Y} + \bar{W}XZ$.

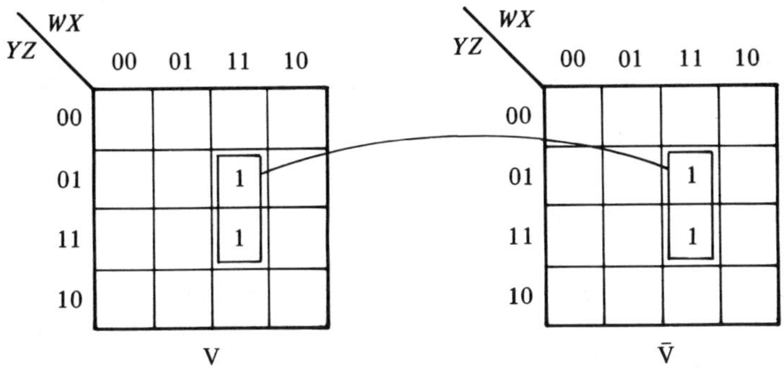

Figure 1.30 Map of f(V, W, X, Y, Z) = WXZ.

in the $\bar{V}$ map. Similarly, the two cells enclosed in the V map are adjacent to the two cells enclosed in the $\bar{V}$ map.

The map shown in Fig. 1.31 can be realized by taking three groups of eight cell adjacencies. The output function f is

$$f = \bar{X} + \bar{Y} + \bar{Z}$$

However, in some cases it is much simpler to group the empty cells. Grouping

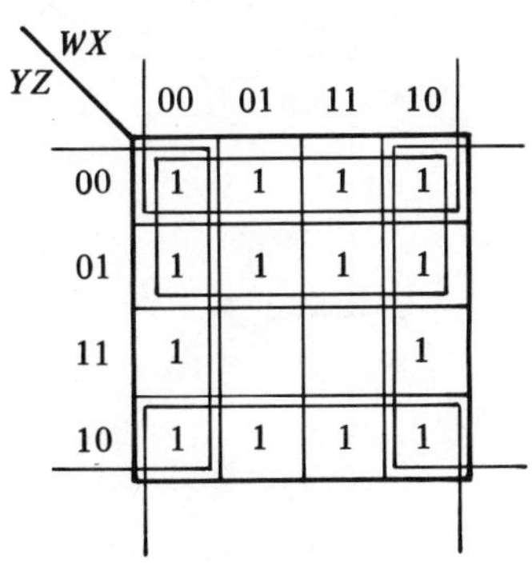

Figure 1.31 Map of $\bar{f} = XYZ$.

YZ \ WX	00	01	11	10
00	0	0	0	0
01	0	1	1	0
11	Φ	Φ	1	0
10	0	0	0	0

Figure 1.32 Map of f = XZ with don't care condition.

empty cells is an expression of $\overline{f}$. Thus

$$\overline{f} = XYZ$$

and from rules j and k,

$$f = \overline{X} + \overline{Y} + \overline{Z}$$

1.10 DON'T CARE CONDITIONS

In many practical problems there are physical constraints that result in the condition that many cells in the Karnaugh map cannot occur. These cells are regarded as *don't care cells.* Don't care cells can be included in any group to aid in the minimization procedure.

Don't care conditions can be indicated by a ϕ in the same cell and can be regarded as a 1 or 0 according to the designer's preference. Figure 1.32 illustrates such a map. By including a don't care cell in the output function, a minimization f = XY with four cells is achieved.

1.11 IMPLEMENTATION OF LOGIC FUNCTIONS

Implementation of output functions is conveyed using logic diagrams. Consider an S of P function

$$f = \overline{W}X + Y\overline{Z}$$

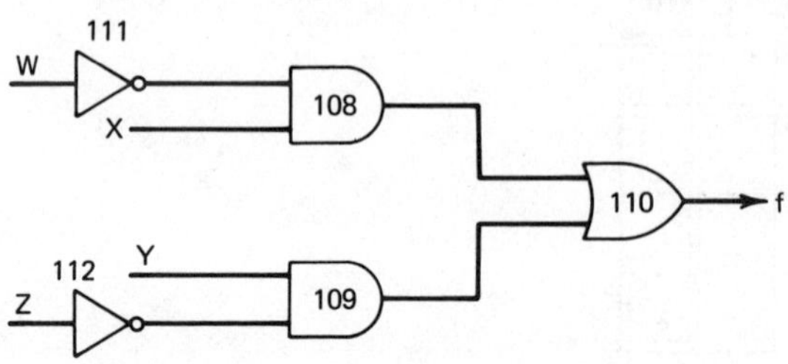

Figure 1.33 Implementation of $f = \overline{W}X + Y\overline{Z}$.

Starting on the left-hand side of Figure 1.33, the *not* operator is used to invert the sign of the variable W. The *and* gate is then used to combine variables $\bar{W}X$. The combination of the variables $Y\bar{Z}$ is realized in this manner. The outputs of the *and* gates are combined using an *or* gate. The output of the *or* gate gives the function f.

1.12 LOGIC *nand* IMPLEMENTATION

With the advent of integrated circuits ICs the use of *nand* gates became very important. To illustrate *nand* gate implementation, inverting nodes are inserted in Figure 1.33 at the output of the *and* gate and the input of the *or* gates. Rule i states that the result of the 2 inversions is nil. From Figure 1.17, an *and* gate with its output *not*ted is a *nand* gate. Similarly Figure 1.19 shows that an *or* gate with its inputs *not*ted is also a *nand* gate. Thus Figure 1.34 shows the simple implementation of the function $f = \bar{W}X + Y\bar{Z}$ using integrated IC chips of the *nand* type. It can be seen from rule *j* that $\overline{\bar{W}X} = W + \bar{X}$ and $\overline{Y\bar{Z}} = \bar{Y} + Z$.

1.13 LOGIC *nor* IMPLEMENTATION

Nor gate implementation is included here for completeness but is not considered an essential building block in this book. The S of P function,

$$f = \bar{W}X + Y\bar{Z}$$

is written in P of S form as

$$f = (\bar{W} + Y)(\bar{W} + \bar{Z})(X + Y)(X + \bar{Z})$$

This circuit is shown in *or*, *and* implementation form in Figure 1.35. Inserting pairs of inverting nodes at each end of the summing-leads (i.e. the output of the *or* gates and the input of the *and* gate) has no net effect according to rule i. However, the

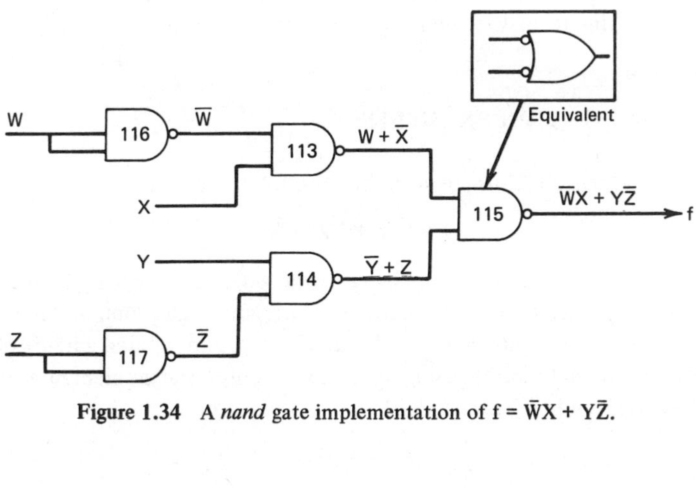

Figure 1.34 A *nand* gate implementation of $f = \bar{W}X + Y\bar{Z}$.

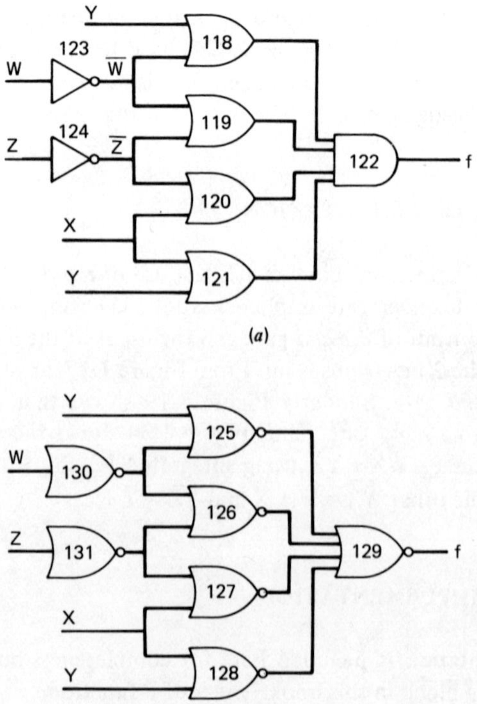

(a)

(b)

Figure 1.35 Implementation of $f = (\bar{W} + Y)(\bar{W} + \bar{Z})(X + \bar{Z})(X + Y)$: *(a) or/and* form; *(b) nor* form.

circuit can be realized using the equivalent gates. Figure 1.35(*b*) shows the *nor* gate implementation.

Nor gate implementation is usually much more easily accomplished by determining $\bar{f}$ in S of P form and then negating both sides of the function. Sometimes a simple *nand* gate implementation is achieved by determining the S of P form of $\bar{f}$ and simply inverting to realize the output function equation.

1.14 IMPLEMENTATION EXAMPLES

The solution to Figure 1.28 is given as the function f, which is

$$f = \bar{W}Y + YZ + \bar{W}X$$

Implementation of this function using *nand* gates is shown in Figure 1.36(*a*). *Nand* gates are available with two, three, four, and eight inputs. There are 4-2 input, 3-3 input, 2-4 input, and 1-8 input gates per integrated circuit chip (IC). As an example of a function using three input chips the implementation of $f = WX\bar{Y} + \bar{W}XZ$ is given in Figure 1.36(*b*).

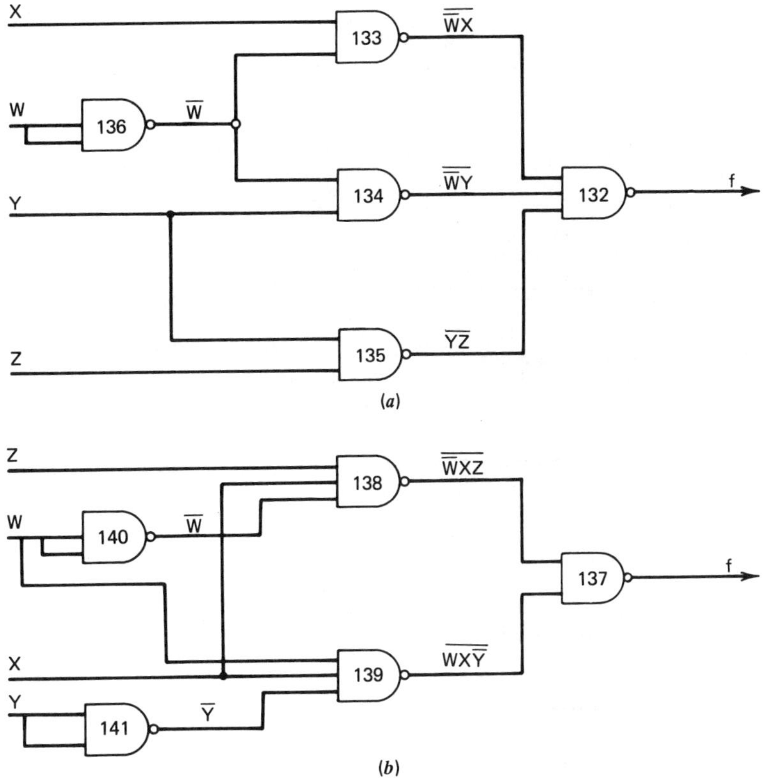

Figure 1.36 Additional *nand* gate implementation: *(a)* implementation of $f = \overline{W}Y + YZ + WX$; *(b)* implementation of $f = WX\overline{Y} + \overline{W}XZ$.

1.15 THE *exclusive or* FUNCTION

With the continued use of the *and* and *or* gates and their practical implementation using *nand* and *nor* gates, certain functions arise repeatedly. One such function is the *exclusive or* gate, which is a direct analogy to the "or" in common English usage. The definition of the function f is

$$f = A\overline{B} + \overline{A}B$$

in S of P form. The Karnaugh map of the function is shown in Figure 1.37(*a*). The circuit implementation of the minimized form of this function is shown in Figure 1.37(*b*) following the systematic method of minimizing circuits using K maps. Such a circuit can also be reduced by factoring to collect negative terms as follows.

$$f = (\overline{A} + \overline{B})(A + B) = (\overline{AB})(A + B)$$

The implementation using this form is shown in Figure 1.38. Note that one gate is saved. This circuit is so important that it is referred to by the derived operator $\oplus$

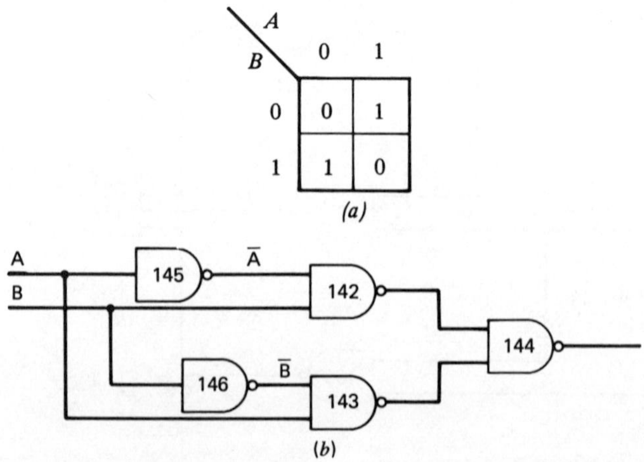

Figure 1.37 Standard *or* gate forms. *(a)* Map of *exclusive or*; *(b)* minimized form of $F = \bar{A}B + A\bar{B} = A \oplus B$.

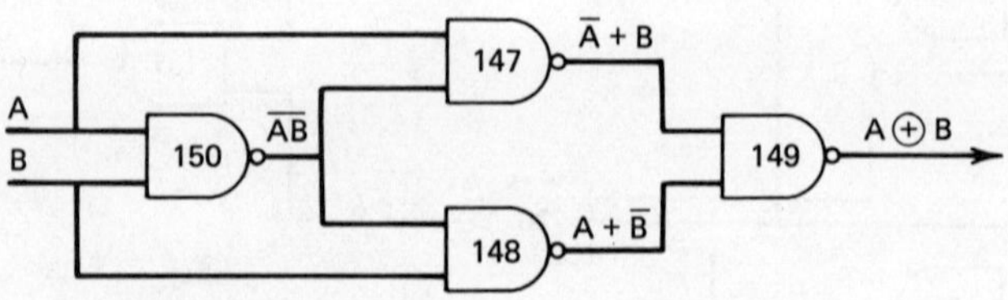

Figure 1.38 Reduced form of $f = A \oplus B$.

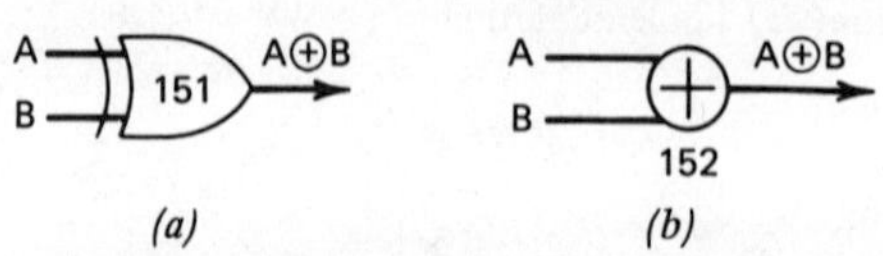

Figure 1.39 An *exclusive or* gate. *(a)* Conventional symbol; *(b)* alternative form.

and is given its own logic symbol as shown in Figure 1.39(*a*). For ease in drawing the alternate form of Figure 1.39(*b*) is used in this book.

1.16 SYMMETRICAL FORMS

Figure 1.40 shows a symmetrical form of outputs for input variables W, X, Y, and Z, which often occurs in the design of logic circuits. Figure 1.41(*a*) gives the tree-type circuit implementation and Figure 1.41(*b*) gives the unbalanced circuit implementation. A fact that should be observed is that the tree circuit has two $\oplus$ gates between input and output, and the unbalanced circuit may have as many as three gates.

If the output of a *nand* gate is at one level, then the input to the gate is said to be at the next higher level. When we consider the output function to be at level 0, the input to the output gate is at level 1. Similarly, progression from a gate output to its input increases the level by 1, so that a total of three levels is achieved in

YZ \ WX	00	01	11	10
00		1		1
01	1		1	
11		1		1
10	1		1	

Figure 1.40 K map of symmetrical form of f(W, X, Y, Z).

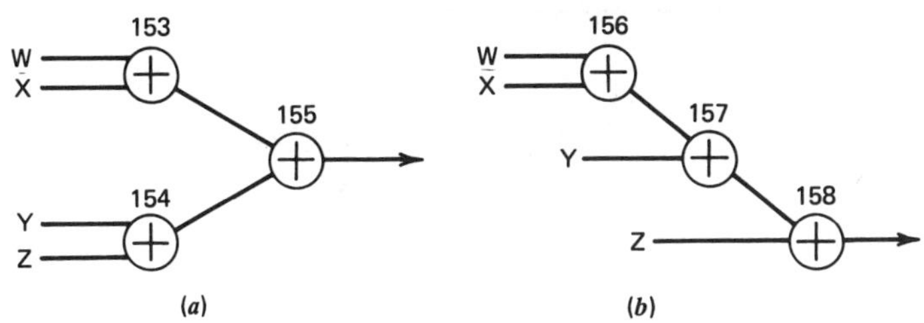

Figure 1.41 Implementation of symmetrical circuits. *(a)* Tree circuit; *(b)* unbalanced form.

Figure 1.38. Each of these levels of gating in a digital circuit represents a measure of the electronic delay. Electronic delay can be used to demonstrate proper action of special logic circuits as discussed in Chapter 5, and the level concept can be extended to the design of special-purpose combination circuits.

1.17 REDUCTION

In Section 1.15 the *exclusive or* function ($f = A\bar{B} + \bar{A}B$) is implemented using the minimized form of Figure 1.37 (5 gates) and the reduced form of Figure 1.38 (4 gates). This process is called reduction because sometimes it leads to the saving of gates. In this section a procedure for reduction is given.

Figure 1.38 is redrawn as Figure 1.42 with levels indicated. For an S of P implementation, even numbers can be equated to minimum combinations and odd levels equated to maximum combinations. Thus level 0 indicates the output function.

Each input line in level 1 should normally represent a maximum term to be *and*ed into our function in the negated form. Level 3 signifies maximum term to be eliminated from level 2. For gate reduction all inputs should be positive. So placing AB as an input at the third level of Figure 1.42 eliminates cell 3 from second-level functions.

To further illustrate this technique, the function of Figure 1.43 is reduced. By ignoring cell 7, this function could be realized from three variables at level 2, as shown in Figure 1.44. However, cell ABC must be eliminated from this function, and this is accomplished by placing the input variables for that cell in at the third level. This is shown in Figure 1.45.

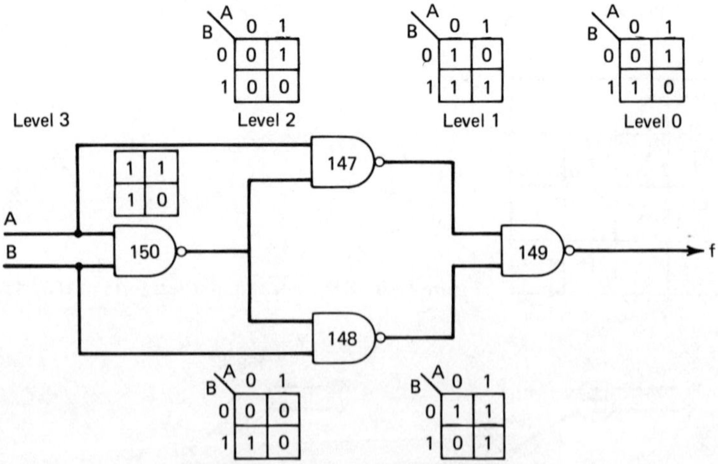

Figure 1.42 f = A ⊕ B in level form.

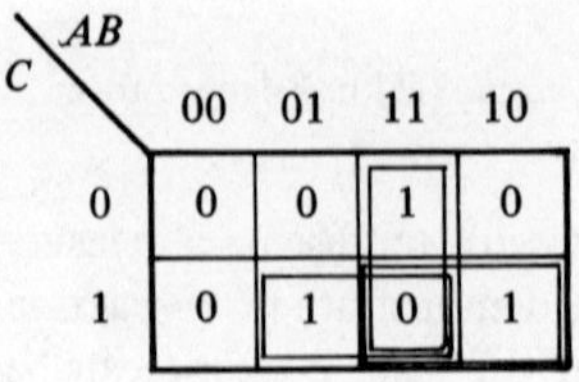

Figure 1.43 Reduction of a three-variable function.

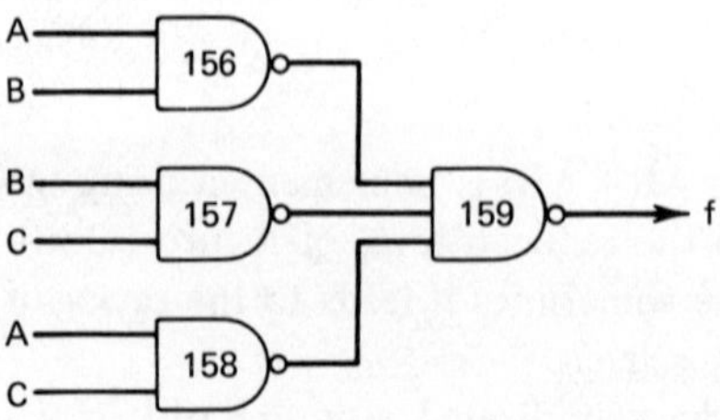

Figure 1.44 Logic diagram for f = AB + BC + AC.

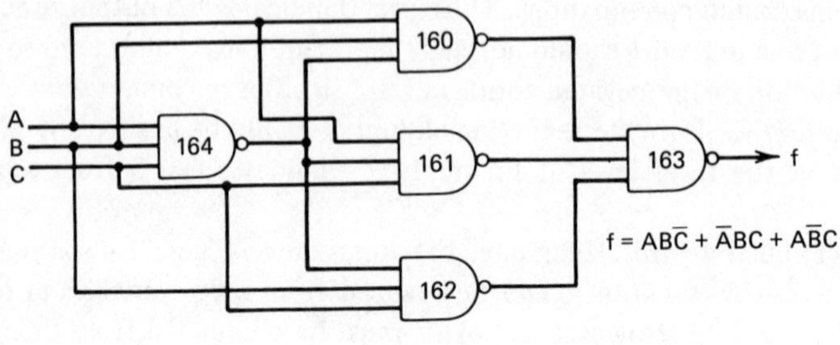

Figure 1.45 Realization of f = $AB\bar{C} + \bar{A}BC + A\bar{B}C$ using reduction.

1.18 PRACTICAL CIRCUITS

There are many cases where *nand* gates can be used to realize practical circuits. All such circuits can be described under the broad classification of function generator codes and decodes. An example of functions that can be generated is the sum and carry of binary numbers. The circuit in Figure 1.46 illustrates a binary adder.

The operations of the individual blocks 0 to 3 are

$$A = a3\ a2\ a1\ a0$$

and

$$B = b3\ b2\ b1\ b0$$

which form two binary numbers to be added.

$$K = k3\ k2\ k1\ k0$$

are the carry bits from the previous bit addition. A, B, and K are inputs. The most significant bits (MSB) are a3, b3, and k3. The least significant bits (LSB) are a0, b0 and k0. The outputs are

$$C = c3\ c2\ c1\ c0$$

$$S = s3\ s2\ s1\ s0$$

where C and S are two binary numbers representing carry and sum respectively.

For simplicity the operation of the individual blocks is illustrated in Table 1.6 where a b k = an bn kn and c s = cn sn for n = 0, 1, 2, 3.

The steps in the design of such a circuit are

1. Determine the inputs and outputs.
2. Draw a Karnaugh map for each output with the inputs defining the rows and columns of the map.
3. Write logic expressions for each output variable.
4. Draw the minimized *nand* circuit implementation.

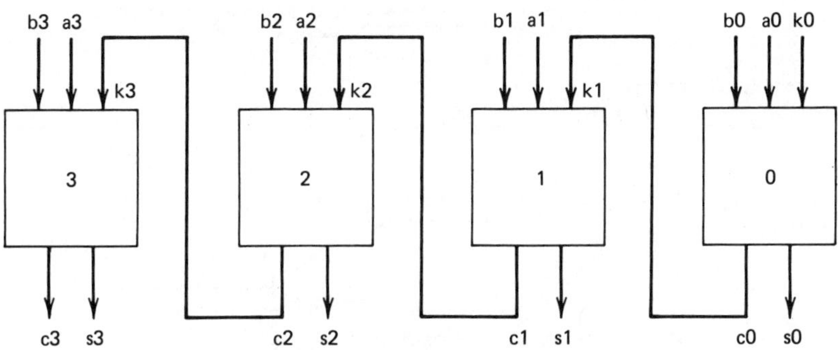

Figure 1.46 Block diagram of binary adder.

Table 1.6 Full Binary Adder

	Inputs			Outputs	
Σ	a	b	k	c	s
0	0	0	0	0	0
1	0	0	1	0	1
2	0	1	0	0	1
3	0	1	1	1	0
4	1	0	0	0	1
5	1	0	1	1	0
6	1	1	0	1	0
7	1	1	1	1	1

Execution of the steps is as follows:

Step 1. The inputs are ab and k. The outputs are c and s.

Step 2. The Karnaugh maps of c and s are shown in Figure 1.47.

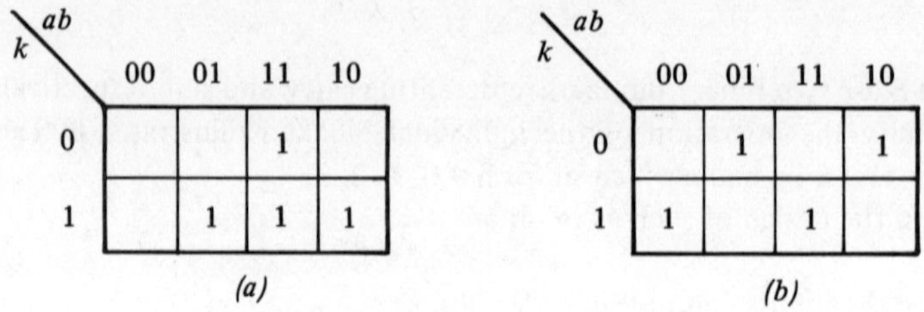

Figure 1.47 Full adder map: *(a)* c map; *(b)* s map.

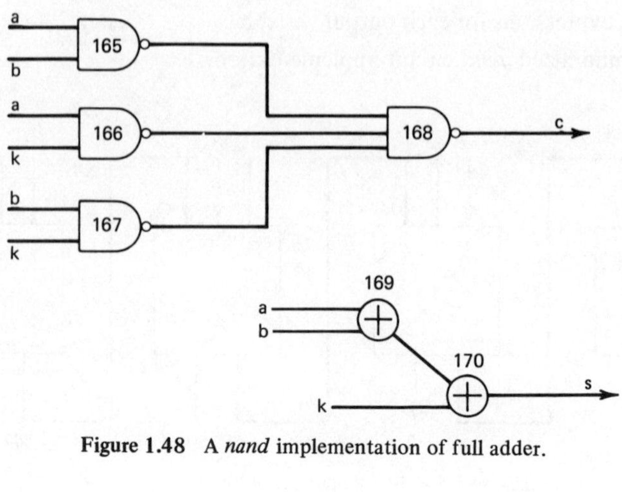

Figure 1.48 A *nand* implementation of full adder.

Step 3. The logic expressions for c and s are as follows:

$$c = ab + bk + ak$$

$$s = a \oplus b \oplus k$$

Step 4. Circuit implementation is shown in Figure 1.48.

An illustration of a circuit defined as a coder is given in Table 1.7. This is a sine function generator [3].

Figure 1.49 gives the characteristic output of Y = y3, y2, y1, y0 for inputs X = x3, x2, x1, x0 varying monotonically from 0 to 15.

For this case y = f(x). It is clear that the inputs are X and the outputs are Y. Thus determination of the inputs and outputs as required in step 1 is complete. Step 2 above calls for a Karnaugh map of each output. They are shown in Figure 1.50.

In step 3 the logic expressions are written. The expressions for the sine generator are as follows:

$$y3 = \overline{x3}\,x2 + x3\,\overline{x2} + x2\,\overline{x1} + \overline{x3}\,x1\,x0$$

$$y2 = x3\,\overline{x2} + x1\,\overline{x0} + \overline{x3}\,x2\,x0$$

$$y1 = x1\,\overline{x0} + x2\,x1 + x2\,\overline{x0} + \overline{x2}\,\overline{x1}\,x0 + x3\,\overline{x2}\,\overline{x1}$$

$$y0 = x2\,x1\,x0 + x3\,\overline{x2}\,\overline{x1} + \overline{x3}\,\overline{x1}\,x0 + x2\,\overline{x1}\,\overline{x0} + x3\,x1\,x0$$

Table 1.7 Four-Bit Sine Function Coder

Inputs				Outputs			
x3	x2	x1	x0	y3	y2	y1	y0
0	0	0	0	0	0	0	0
0	0	0	1	0	0	1	1
0	0	1	0	0	1	1	0
0	0	1	1	1	0	0	0
0	1	0	0	1	0	1	1
0	1	0	1	1	1	0	1
0	1	1	0	1	1	1	0
0	1	1	1	1	1	1	1
1	0	0	0	1	1	1	1
1	0	0	1	1	1	1	1
1	0	1	0	1	1	1	0
1	0	1	1	1	1	0	1
1	1	0	0	1	0	1	1
1	1	0	1	1	0	0	0
1	1	1	0	0	1	1	0
1	1	1	1	0	0	1	1

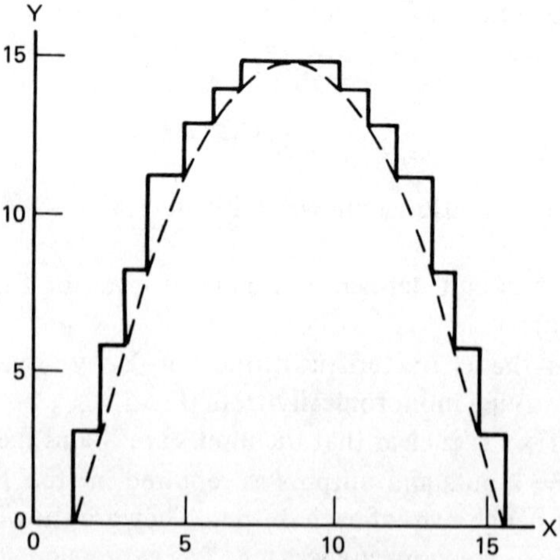

Figure 1.49 Output waveform.

x1x0 \ x3x2	00	01	11	10
00	0	1	1	1
01	0	1	1	1
11	1	1	0	1
10	0	1	0	1

(a)

x1x0 \ x3x2	00	01	11	10
00	0	0	0	1
01	0	1	0	1
11	0	1	0	1
10	1	1	1	1

(b)

x1x0 \ x3x2	00	01	11	10
00	0	1	1	1
01	1	0	0	1
11	0	1	1	0
10	1	1	1	1

(c)

x1x0 \ x3x2	00	01	11	10
00	0	1	1	1
01	1	1	0	1
11	0	1	1	1
10	0	0	0	0

(d)

Figure 1.50 Sine generator K-maps: *(a)* y3 map; *(b)* y2 map; *(c)* y1 map; *(d)* y0 map.

The circuit implementation called for in step 4 is given in Figure 1.51. For simplicity the connections between Aa, Bb, and so forth are omitted.

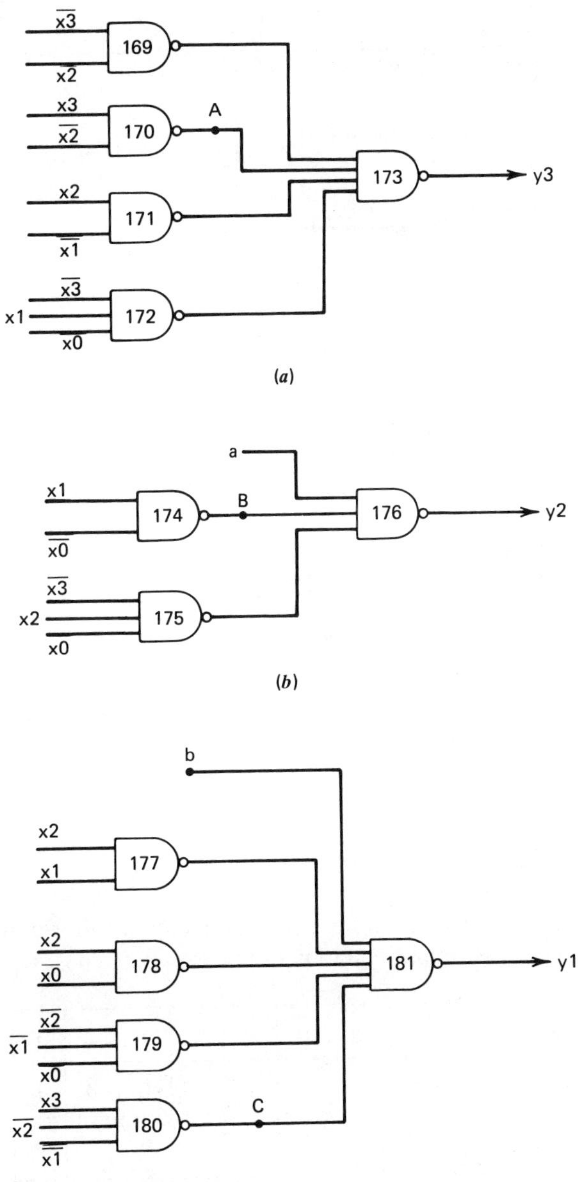

Figure 1.51 Sine generation *NAND* gate implementation: *(a)* y3 realization; *(b)* y2 realization; *(c)* y1 realization; *(d)* y0 realization.

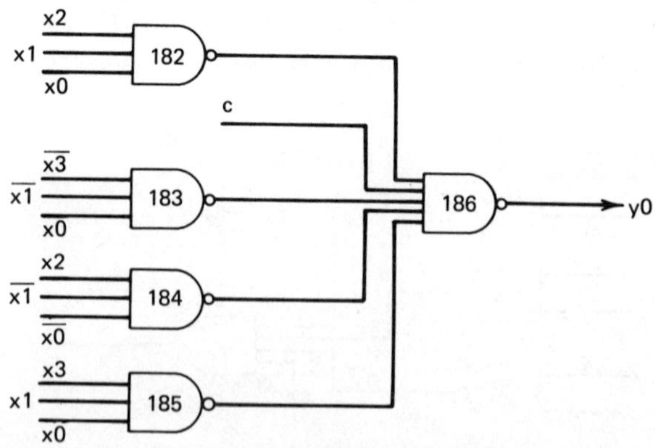

(*d*)

Figure 1.51 (*Continued*)

1.19 PROBLEMS

P1.1 Convert the following functions to minterm form using K maps.
(a) $f(W, X, Y, Z) = YZ + X(Y + Y\bar{Z})$.
(b) $f(A, B, C) = A + B\bar{C}$.
(c) $f(A, B, C, D) = AB + CD$.
(d) $f(X, Y, Z) = XYZ + Y\bar{Z}$.
(e) $f(A, B, C, D, E) = (A + C)D + \overline{ABCD}$.

P1.2 Minimize the following. Use K maps.
(a) $f(W, X, Y, Z) = \Sigma(0, 1, 2, 3, 7, 11, 15)$.
(b) $f(A, B, C) = \Sigma(3, 5, 6)$.
(c) $f(W, X, Y, Z) = \Sigma(4, 5, 8, 9, 10, 12, 13, 14)$.
(d) $f(A, B, C, D) = \pi(0, 4, 6, 7, 10, 12, 13, 14)$.

P1.3 Design a 2 × 2 multiplier.
Input ABCD
Output WXYZ

P1.4 The truth table and K maps for a full subtractor are shown in Figure P1.1. Design the *nand* gate implementation of this circuit.

Minuend	*X*	0	0	0	0	1	1	1	1
Subtrahend	*Y*	0	0	1	1	0	0	1	1
Borrow	*V*	0	1	0	1	0	1	0	1
Difference	*D*	0	1	1	0	1	0	0	1
Borrow	*B*	0	1	1	1	0	0	0	1

(*a*)

Figure P1.1 Subtracter. *(a)* K maps; *(b)* B map; *(c)* D map.

V \ XY	00	01	11	10
0		1		
1	1	1	1	

(b)

V \ XY	00	01	11	10
0		1		1
1	1		1	

(c)

Figure P1.1 (*Continued*)

P1.5 (a) Write the binary numbers of the integers 0–18 to complete the following table:

INTEGER	BINARY MSB Z4	Z3	Z2	Z1	Z0 LSB
0	0	0	0	0	0
1	0	0	0	0	1
2	0	0	0	1	0

(b) Write the BCD numbers of the integers 0–18 to complete the following table:

INTEGER	BCD MSB S4	S3	S2	S1	S0 LSB
0	0	0	0	0	0
1	0	0	0	0	1
2	0	0	0	1	0

(c) Figure P1.2 is a block diagram of a decoder circuit to convert from binary to BCD. Do a *nand* implementation of this circuit.

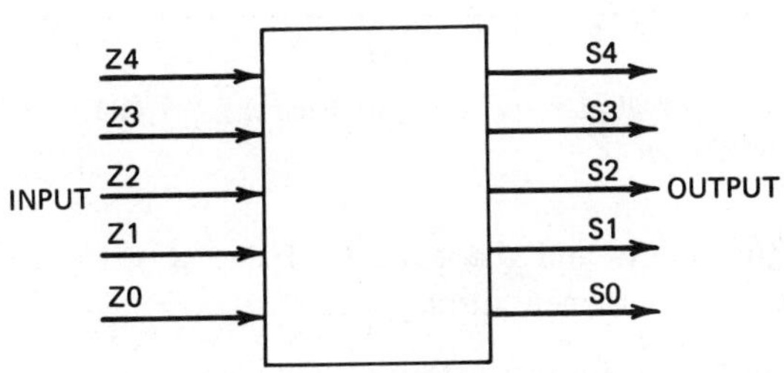

Figure P1.2 BCD decoder.

P1.6 Figure P1.3 shows the layout for a seven-segment decoder. Part (a) shows the segments in the LED display. Part (b) shows the segments illuminated to display integers. Part (c) shows the circuit you are to design to convert from binary coded decimal to seven-segment display. Part (d) gives the decimal

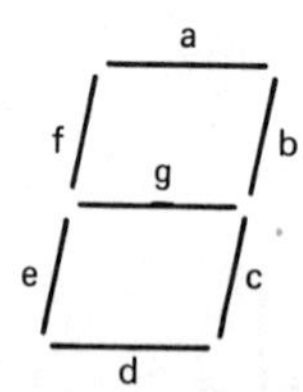

(a)

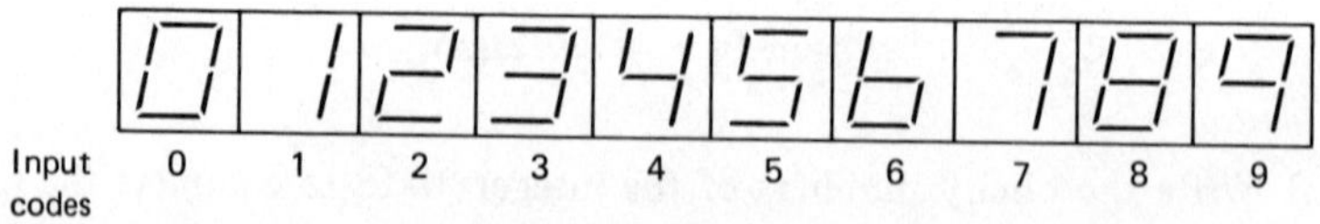

(b)

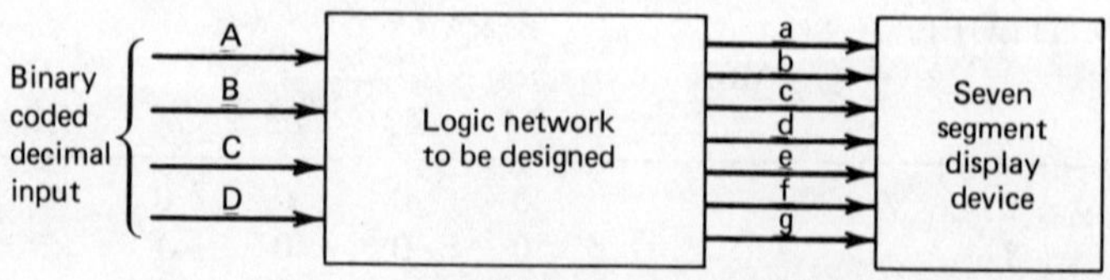

(c)

Decimal displayed	Inputs				Outputs						
	A	*B*	*C*	*D*	*a*	*b*	*c*	*d*	*e*	*f*	*g*
0	0	0	0	0	1	1	1	1	1	1	0
1	0	0	0	1	0	1	1	0	0	0	0
2	0	0	1	0	1	1	0	1	1	0	1
3	0	0	1	1	1	1	1	1	0	0	1
4	0	1	0	0	0	1	1	0	0	1	1
5	0	1	0	1	1	0	1	1	0	1	1
6	0	1	1	0	0	0	1	1	1	1	1
7	0	1	1	1	1	1	1	0	0	0	0
8	1	0	0	0	1	1	1	1	1	1	1
9	1	0	0	1	1	1	1	0	0	1	1

(d)

Figure P1.3 Seven-segment display. *(a)* LED arrangement; *(b)* coded arrangement; *(c)* logic decoder; *(d)* input output functions.

displayed, the inputs and the outputs. Using K maps design the decoder circuit for the seven-segment display.

P1.7 Use a Karnaugh map to minimize the following expressions and draw a *nand* gate implementation.

(a) $f(A, B, C) = \Sigma(1, 4, 5, 6, 7)$

(b) $f(a, b, c, d) = \Sigma(3, 4, 6, 7, 11, 12, 14, 15)$

(c) $f(X, Y, Z) = \Sigma(1, 2, 4, 7)$

(d) $f(W, X, Y, Z) = \Sigma(0, 3, 4, 7, 8, 11, 12, 15)$

P1.8 A circuit contains two 2-bit binary numbers AB and CD. Design a circuit using *nand* gates to detect the condition AB greater than CD.

P1.9 Design an encoder to convert from BCD to GRAY CODE according to the following table. Implement using *nand* gates.

Table P1.1 GRAY CODE

BCD					GRAY			
A	B	C	D	I	w	x	y	z
0	0	0	0	0	0	0	0	0
0	0	0	1	1	0	0	0	1
0	0	1	0	2	0	0	1	1
0	0	1	1	3	0	0	1	0
0	1	0	0	4	0	1	1	0
0	1	0	1	5	0	1	1	1
0	1	1	0	6	0	1	0	1
0	1	1	1	7	0	1	0	0
1	0	0	0	8	1	1	0	0
1	0	0	1	9	1	1	0	1

CHAPTER 2

Multiplexers and Demultiplexers in Combinational Circuits

The techniques developed in Chapter 1 are suited to circuit design using IC chips of *nand* and *nor* gates. To realize such a circuit a layout of chips on a circuit board is required. Lines in the logic diagram represent connections between chips. Such interconnections can be extensive. Printed circuit board layout interconnections may present a complex routing problem involving crossovers and require extensive soldering.

Medium scale integration (MSI) devices are much more compact than standard IC chips. Therefore, if one is willing to design to standard MSI configurations a saving results in the number of components and in interconnections required.

Combinational circuit design procedures using MSI are presented in this chapter.

2.1 MULTIPLEXER CONFIGURATION

Figure 2.1 shows the MSI circuit for a typical four-input, one-output (4 to 1), two-selector-line multiplexer (*mux*) [9, 10, 29]. The two selector lines are labeled X, Y. Selector lines are used to select one of four possible input gates in an equivalent manner to selecting individual cells in a Karnaugh map. Thus for XY = 00, *nand* gate 201 (with input a_0) is selected. For XY = 01, *nand* gate 202 (with input a_1) is selected. Similarly, when XY = 10, gate 203 (with a_2 as an input) is selected, and when XY = 11, gate 204 (with a_3 as an input) is selected.

A simplified symbolic form of the *mux* given in Figure 2.1 is shown in Figure 2.2. For clarity the outputs, selectors, and inputs are labeled in Figure 2.2. The reader is expected to know which is which without labels. The convention is as follows: Outputs occur on the long side of the rectangle only. There are always two outputs, one of which is negated. Thus the telltale small circle and line with arrow $\circ\!\!\longrightarrow$ are present to aid in the identification of the output.

The short side of the rectangle in a clockwise direction is used for selector lines. The line designating the least significant bit of the selector is closest to the output side. The inputs are on the remaining long side of the rectangle in a clockwise direction from the selector lines. The least significant input a_0 is closest to the selector line side.

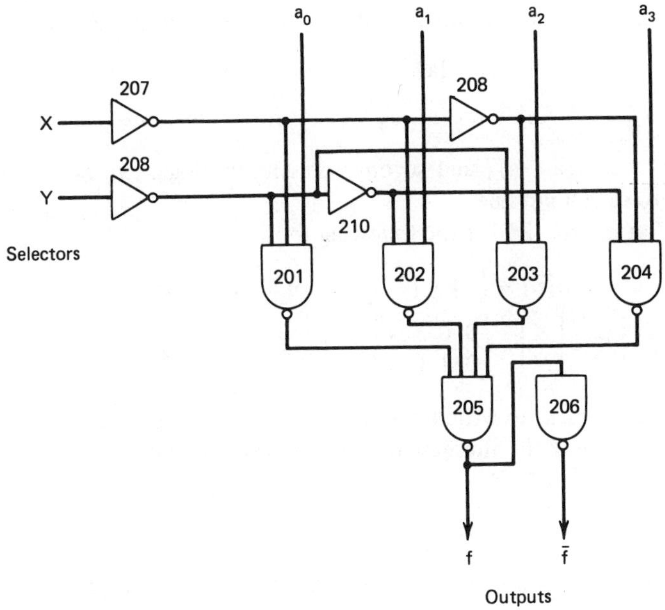

Figure 2.1 Standard *mux* circuit.

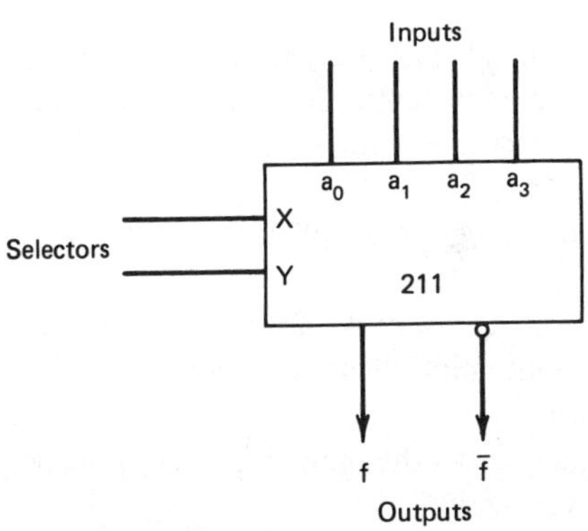

Figure 2.2 *Mux* symbol.

Muxes may be 2 to 1 (one slector line), 4 to 1 (two slector lines), 8 to 1 (three selector lines), 16 to 1 (four selector lines) and so on.

2.2 THE *mux* DESIGN NOTATION

A selection procedure expressed conveniently for application [11] to design is given in the following equation:

$$[A]\,[Z] = [f]$$

The input matrix [A] represents the value connected to input lines. Thus

$$[A] = [a_0 \ \ a_1] \qquad \text{for a 2 to 1 } mux$$

$$[A] = [a_0 \ \ a_1 \ \ a_2 \ \ a_3] \qquad \text{for a 4 to 1 } mux$$

and so on, where a_0, a_1, and so on represent the logic expression of variables connected to each input line.

The selector matrix [Z] has the following form:

$$[Z] = [X] = [0 \ \ 1] \qquad \text{for a 2 to 1 } mux$$

$$[Z] = \begin{bmatrix} X \\ Y \end{bmatrix} = \begin{bmatrix} 0 & 0 & 1 & 1 \\ 0 & 1 & 0 & 1 \end{bmatrix} \qquad \text{for a 4 to 1 } mux\text{, and so on}$$

It is more convenient to write the integer form of the binary select lines. Thus for a four to one *mux* the alternate form of the selector matrix is

$$\begin{bmatrix} X \\ Y \end{bmatrix} = [0 \ \ 1 \ \ 2 \ \ 3]$$

Actually it is most convenient to write as little as possible. Therefore the selector matrix of simply

$$\begin{bmatrix} X \\ Y \end{bmatrix}$$

represents the specific ordering of the selector variable as given above. Variables are written with most significant bits at the top and least significant bits at the bottom. The reader is cautioned that most manufacturers order their selector lines in the opposite convention.

The output [f] is obtained by application of logic multiplication to the input matrix [A] and the selector matrix [Z].

Since it can also be extended to include a number of multiplexers or the address lines on a microprocessor or minicomputer, logic multiplication is defined for the general case.

Logic multiplication makes the output [f] equal to the input [column] selected by the value of [Z]. This is an addressing function.

If [Z] = 5, [f] equals the input variable a_5 listed in column 5 (counting the first as the 0 column) of matrix [A].

2.3 APPLICATIONS TO COMBINATIONAL LOGIC

To illustrate the use of multiplexers in logic design, consider the trivial realization of the *exclusive or* circuit using a four to one multiplexer. The Karnaugh map for the *exclusive or* operator is given in Figure 2.3. From the Σ designation the output function is

$$f(A, B) = \Sigma(1, 2)$$

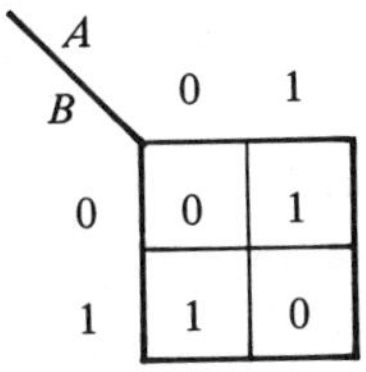

Figure 2.3 *Exclusive or* implementation.

Thus

$$[a_0 \;\; a_1 \;\; a_2 \;\; a_3] = [0 \;\; 1 \;\; 1 \;\; 0]$$

since $\Sigma(1, 2)$ is equivalent to

$$a_1 = a_2 = 1$$

This can be written in logic multiplication form as follows:

$$[0 \;\; 1 \;\; 1 \;\; 0] \begin{bmatrix} A \\ B \end{bmatrix} = [f]$$

For

$$\begin{bmatrix} A \\ B \end{bmatrix} = \begin{bmatrix} 0 \\ 0 \end{bmatrix}$$

a_0 is selected and $[f] = 0$. For

$$\begin{bmatrix} A \\ B \end{bmatrix} = \begin{bmatrix} 0 \\ 1 \end{bmatrix}$$

a_1 is selected and $[f] = 1$. For

$$\begin{bmatrix} A \\ B \end{bmatrix} = \begin{bmatrix} 1 \\ 0 \end{bmatrix}$$

a_2 is selected and $[f] = 1$. For

$$\begin{bmatrix} A \\ B \end{bmatrix} = \begin{bmatrix} 1 \\ 1 \end{bmatrix}$$

a_3 is selected and $[f] = 0$.

Consider the asymmetrical circuit given in Figure 2.4. Since

$$f(X, Y, Z) = \Sigma(1, 2, 4, 7)$$

it can be represented in logic multiplication form as

$$[0 \;\; 1 \;\; 1 \;\; 0 \;\; 1 \;\; 0 \;\; 0 \;\; 1] \begin{bmatrix} X \\ Y \\ Z \end{bmatrix} = [f]$$

It has been pointed out that this solution is trivial. The reason it is trivial is because the input lines are connected to 0's and 1's only. This represents poor utilization of the multiplexer capabilities. To achieve a more sophisticated implementa-

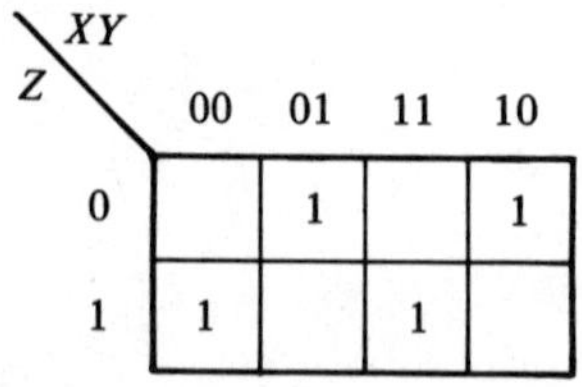

Figure 2.4 A symmetrical circuit.

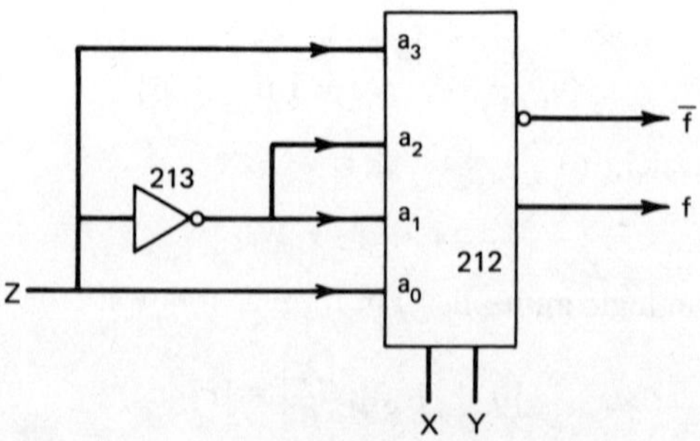

Figure 2.5 Symmetric circuit realization using 4 to 1 *mux*.

tion, let us eliminate one variable from the selector matrix. To accomplish this, the following rules from Chapter 1 can be used:

$$0 + 0 = 0$$

$$0 + Z = Z$$

$$0 + \bar{Z} = \bar{Z}$$

$$Z + \bar{Z} = 1$$

The variable form of the solution to Figure 2.4 is

$$f = \bar{X}\bar{Y}Z + \bar{X}Y\bar{Z} + X\bar{Y}\bar{Z} + XYZ$$

Rewriting in the form

$$f = Z\bar{X}\bar{Y} + \bar{Z}\bar{X}Y + \bar{Z}X\bar{Y} + ZXY$$

and using X and Y as the selector lines and Z as an input, we find $a_0 = Z$, $a_1 = \bar{Z}$, $a_2 = \bar{Z}$, and $a_3 = Z$. The logic multiplication equation is

$$[Z\ \bar{Z}\ \bar{Z}\ Z] \begin{bmatrix} X \\ Y \end{bmatrix} = [f]$$

The solution to this equation using a four to one multiplexer is shown in Figure 2.5.

2.4 DIRECT METHOD

As a further example, consider the output function f(A, B, C) described by the Karnaugh map of Figure 2.6. In this case, however, the input matrix is developed

C \ AB	00	01	11	10
0	0	0	1	0
1	0	1	1	1

Figure 2.6 K map of a carry function.

from direct inspection of the Karnaugh map rather than applications of the variable form of the logic equation. A and B are chosen as selector lines. Reading from the map for the column where

$$[A\ \ B]^t = [0\ \ 0]^t$$

the output function is 0 for all cells or values of C. Thus $a_0 = 0$. For the column where

$$[A\ \ B]^t = [0\ \ 1]^t$$

the output function is 1 if C = 1. Thus $a_1 = C$. For the column where

$$[A\ \ B]^t = [1\ \ 0]^t$$

the output function is 1 if C = 1. Thus $a_2 = C$. For

$$[A\ \ B]^t = [1\ \ 1]^t$$

the output function is 1 for all values of C. Thus, $a_3 = 1$. The four to one multiplexer realization for this is shown in Figure 2.7.

Four-input-line multiplexers are usually manufactured with two multiplexers per chip, sharing the same selector lines. This arrangement can often be used to great advantage in logic design. Eight-input-line multiplexers are very useful because they occur on a single compact chip, and can be used to realize circuits with four or more input variables. As an example, let us consider the comparator circuit shown in Figure 2.8. This is a Karnaugh map of an output f_1, which is 1 when three or more of the input variables are 1. Choosing three select lines ABC in a four-variable map means that each input line is described by two cells. These two cells constitute a K map of two cells in the remaining variable, D.

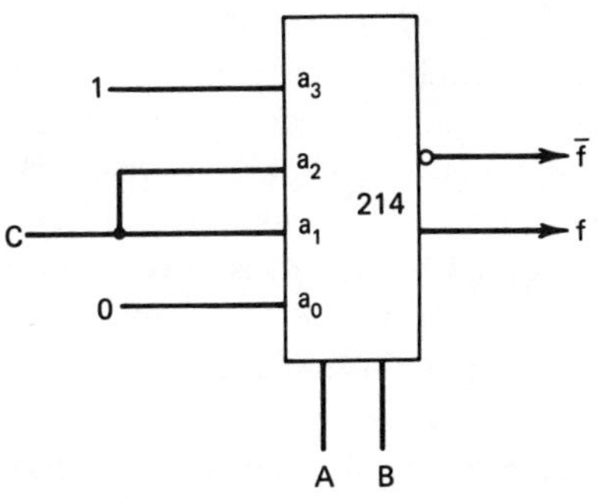

Figure 2.7 Carry map realization using *mux*.

CD \ AB	00	01	11	10
00				
01			1	
11		1	1	1
10			1	

Figure 2.8 K map of comparator function f_1.

For

$$[A\ B\ C]^t = [0\ 0\ 0]^t$$

the output function f_1 is 0 for both $\bar{D} = 0$ and $D = 0$. Thus $a_0 = 0$. Similarly for

$$[A\ B\ C]^t = [0\ 0\ 1]^t, [0\ 1\ 0]^t, \text{and } [1\ 0\ 0]^t$$

so that

$$a_1 = a_2 = a_4 = 0$$

However, for

$$[A\ B\ C]^t = [0\ 1\ 1]^t, [1\ 0\ 1]^t, \text{and } [1\ 1\ 0]^t$$

$\bar{D} = 0$ and $D = 1$ so that

$$a_3 = a_5 = a_6 = D$$

For

$$[A\ B\ C]^t = [1\ 1\ 1]^t$$

$$\bar{D} = D = 1$$

and since in a two-cell map

$$\bar{D} + D = 1$$

$$a_7 = 1$$

The eight-input multiplexer realization of this is shown in Figure 2.9.

It is common practice in this book to combine a number of output functions on the same Karnaugh map or in the same logic multiplication equation. To illustrate multiple output implementation, the foregoing circuit is redefined to that of a voting decoder. There are four judges voting yes or no on the outcome of a competition. If three or more judges vote yes, the decision is clearly yes. If two judges vote yes and two no, the competition must be rerun. If one or fewer judges vote yes, the outcome is no. The K map for such a circuit is illustrated in Figure 2.10. All cells where three or more of the variables A, B, C, D equal 1 are identified by f_1. Similarly, f_2 identifies all cells for two or more variables equal to 1. Thus,

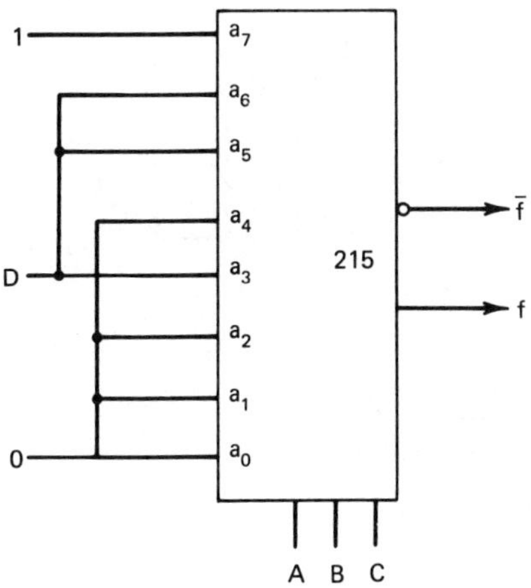

Figure 2.9 *Mux* comparator circuit.

YES = f_1, NO = $\overline{f_2}$, and RERUN = $f_2 \cdot \overline{f_1}$. The logic multiplication equation for this circuit is

$$\begin{bmatrix} 0 & 0 & 0 & D & 0 & D & D & 1 \\ 0 & D & D & 1 & D & 1 & 1 & 1 \end{bmatrix} \begin{bmatrix} A \\ B \\ C \end{bmatrix} = \begin{bmatrix} f_1 \\ f_2 \end{bmatrix}$$

Note the form of this equation. The output matrix gives the number of variables in the output function and assigns each variable a name. In the input matrix there is a row for each output function. The selector matrix defines the selector variables and the order in which they are used to designate specific columns in the input matrix.

A four-input dual realization of the voting circuit may be implemented by reduction from three select lines to two. Choosing two variables AB as the select lines in a four-variable map means that each input line is described by four cells. These four cells constitute a K map in the remaining variables CD.

For the function f_1 the two select lines AB designate four columns in Figure 2.8.

CD \ *AB*	00	01	11	10
00	00	00	01	00
01	00	01	11	01
11	01	11	11	11
10	00	01	11	01

$f_1\, f_2$

Yes = f_1, NO = $\overline{f_2}$, RERUN = $f_2 \cdot \overline{f_1}$

Figure 2.10 K map for voting decoder circuit.

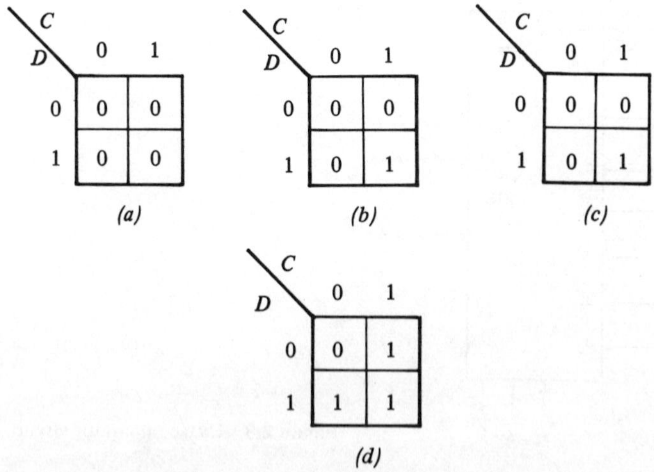

Figure 2.11 Input maps of CD for f_1. *(a)* AB = 00, $a_0 = 0$; *(b)* AB = 01, a_1 = CD; *(c)* AB = 10, a_2 = CD; *(d)* AB = 11, a_3 = C + D.

The values of the variable AB are as follows:

(*a*) AB = 00
(*b*) AB = 01
(*c*) AB = 10
(*d*) AB = 11

The corresponding maps of CD are given in Figure 2.11, parts (*a*), (*b*), (*c*), (*d*) respectively. For the function f_2 the maps of CD are shown in Figure 2.12.

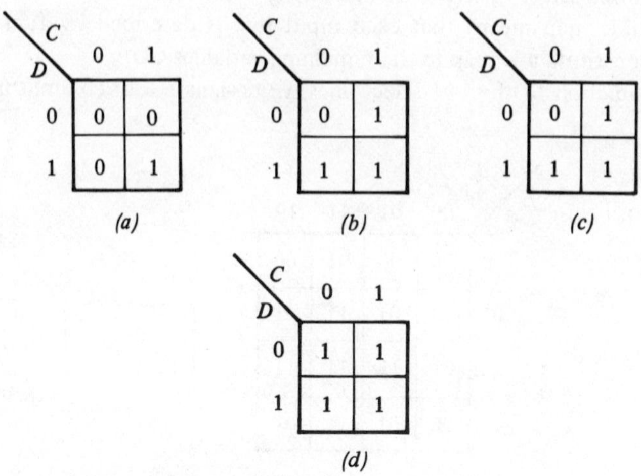

Figure 2.12 Input maps of CD for f_2. *(a)* AB = 00, a_0 = CD; *(b)* AB = 01, a_1 = C + D; *(c)* AB = 10, a_2 = C + D; *(d)* AB = 11, a_3 = 1.

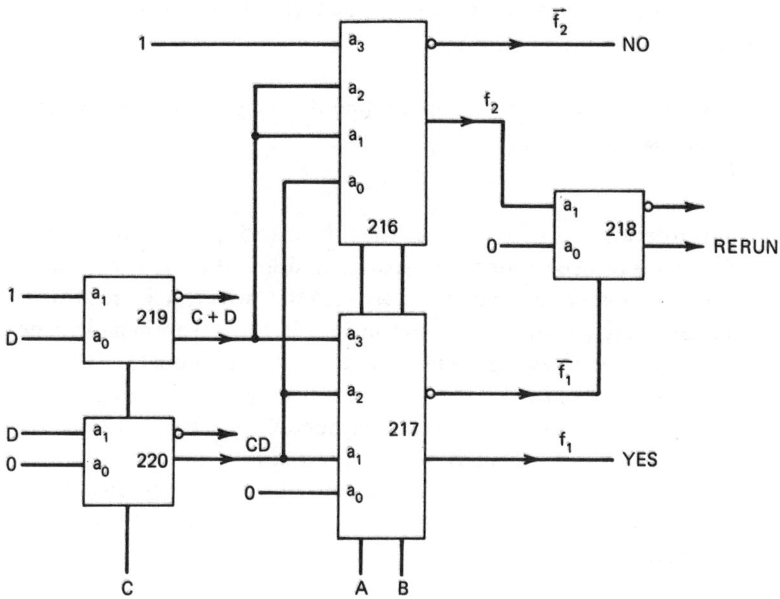

Figure 2.13 Voting decoder circuit using dual 4 to 1 and 2 to 1 *muxes.*

In designing such circuits, it is wise to look for inputs that have a common form or are a simple negated form of another input. A four to one multiplexer implementation of the vote decoder is given in Figure 2.13. For completeness, other functions in the design are included using two to one multiplexers.

Sixteen-input multiplexers are readily available and may be used for implementation of circuits with five or more input variables. However, for convenience of physical size and design, the author recommends implementation using multiplexers with eight or less inputs. Multiplexers are extremely convenient building blocks in combinational circuit realization. Their main advantages are: (*a*) reduced wiring and soldering; (*b*) reduction of capital and building costs. Their disadvantages are: (*a*) they usually have increased power consumption; and (*b*) multiplexer circuits laid out on printed circuit boards are not easily modified.

2.5 PRACTICAL CIRCUITS

The steps in the design of a multiplexer circuit are

1. Determine the inputs and outputs.
2. Write a matrix equation which includes the following:
   (a) input matrix
   (b) select line vector
   (c) output vector
3. Draw the multiplexer implementation circuit.

2.6 THE USE OF LARGE-SCALE INTEGRATION (LSI) MEMORIES AS COMBINATIONAL ELEMENTS

One step backward here and a very interesting detour presents itself. Consider the multiplication matrix

$$[A]\,[Z] = [f]$$

and a return to the trivial case; that is, when the matrix [A] consists of 0's and 1's only. It has already been noted that matrix multiplication is simply an address function. Memories such as random access (RAMs), which have read write capabilities and read only (ROMs), store information in the 0, 1 form and are provided with address lines for access. So when we place the appropriate matrix [A] and designate [Z] as the address lines on a memory element, a simple method of designing combinational circuits using LSI memory elements becomes obvious.

Such applications are particularly useful where a large number of output functions is required. It is also more worthwhile if the circuit is required to be duplicated many times. The advantages of LSI use are very minimal wiring and very compact utilization of space.

2.7 DEMULTIPLEXERS

A demultiplex (*demux*) circuit is shown in Figure 2.14. A *demux* is a standard package designed to realize the inverse function of a *mux*. Thus with one input it has many outputs.

Another way to look at a *demux* is to consider the input line as an enable switch, and a *demux* has the same circuit as a *mux* without the final output gate. Because the final output gate is missing demultiplexer outputs are generally inverted. Thus with the input at 0 all outputs except the one selected are 1's. The selected output is a 0.

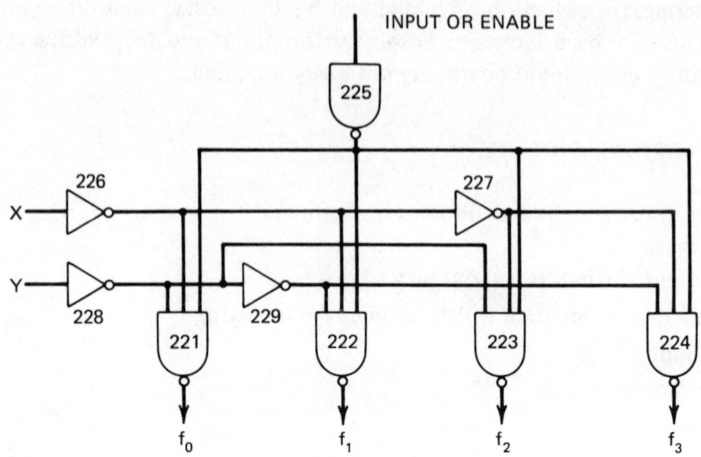

Figure 2.14 Standard *demux* circuit.

Table 2.1 Demultiplexer Output

Enable	XY	f_1	f_2	f_3	f_4
0	00	0	1	1	1
0	01	1	0	1	1
0	10	1	1	0	1
0	11	1	1	1	0

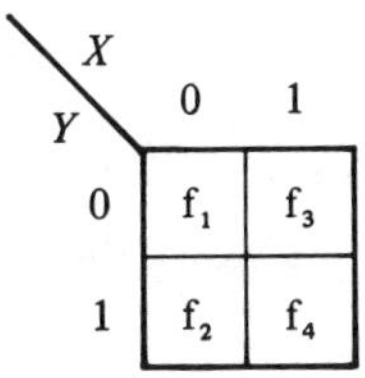

Figure 2.15 A *demux* output representation.

Demuxes are used in combinational circuits when a large number of output functions are involved. The most convenient design tool for *demux* circuits is the Karnaugh map.

When the enable or input is equal to 1 for the circuit in Figure 2.14.

$$[f_1 \ f_2 \ f_3 \ f_4] = [1 \ 1 \ 1 \ 1]$$

for all values of XY. However, when the enable is equal to 0, the select line-output relationship is shown in Table 2.1. The Karnaugh map form of Table 2.1 is shown in Figure 2.15. The cells generally contain the variable name to which they are connected rather than f_1, f_2, and so on. The fact that you may need to invert the signal is ignored in the K map representation. Of course, it cannot be ignored in circuit construction.

A convenient symbol, for the four-output *demux* shown in Figure 2.14, is shown in Figure 2.16. There the inverted outputs are indicated by the standard small circles.

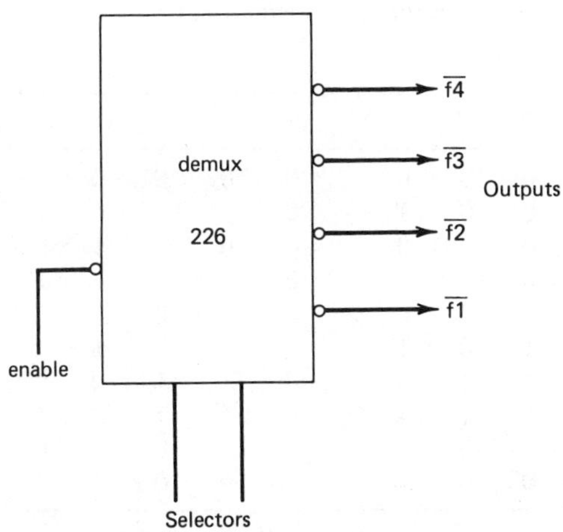

Figure 2.16 *Demux* symbol.

2.8 PROBLEMS

P2.1 Design a multiplexer implementation of the following functions, using multiplexer with one less selector lines than variables in the function.

(a) $f = B\bar{C}D + BC\bar{D}$

(b) $f = \bar{A}\bar{B}C + BC\bar{D}$

(c) $f = AC + BC$

(d) $f = BC + A\bar{C}$

(e) $f = B\bar{C}\bar{D} + A\bar{C}$

(f) $f = B + \bar{C}$

(g) $f = A\bar{B} + A\bar{C} + A\bar{D}$

P2.2 Design a multiplexer implementation of the following functions, using 4 to 1 and 2 to 1 multiplexers.

(a) $f(A, B, C, D) = \Sigma(0, 1, 7, 8, 9, 10, 14, 15)$

(b) $f(W, X, Y, Z) = \Sigma(0, 1, 2, 3, 4, 5, 10, 11, 12, 13)$

(c) $f(a, b, c, d) = \Sigma(1, 2, 4, 7, 8, 11, 13, 14)$

(d) $f(r, s, t, u) = \Sigma(0, 2, 3, 5, 10, 11, 12, 13)$

P2.3 Design a multiplexer implementation to solve the equation

$$f = x^2 + 2x$$

for all values of x from 0 to 7. Use 4 to 1 multiplexers.

P2.4 Design a multiplexer circuit that operates as a full adder or full subtracter.

P2.5 Parity codes permit error detection in a byte or word by the inclusion of one extra bit. Hamming was the first to show that the inclusion of three extra bits allows the detection and correction of a single error. Hence a detection-correction code is called a Hamming code. Table P2.1 gives the conversion from BCD to Hamming. Design and implement a multiplexer circuit to achieve this conversion.

Table P2.1 Hamming Correction Code

BCD				Hamming						
A	B	C	D	W	X	Y	1	Z	m	n
0	0	0	0	0	0	0	0	0	0	0
0	0	0	1	0	0	0	0	1	1	1
0	0	1	0	0	0	1	1	0	0	1
0	0	1	1	0	0	1	1	1	1	0
0	1	0	0	0	1	0	1	0	1	0
0	1	0	1	0	1	0	1	1	0	1
0	1	1	0	0	1	1	0	0	1	1
0	1	1	1	0	1	1	0	1	0	0
1	0	0	0	1	0	0	1	0	1	1
1	0	0	1	1	0	0	1	1	0	0

Operation of this code is discussed in Section 9.3.

P2.6 Computer hardware on some systems is designed so that the software can use base 10 arithmetic. One example is the Honeywell system. A convenient code for base 10 arithmetic is the excess 3 code. It is as follows:

Table P2.2 Excess 3 Code

BCD				Excess 3			
W	X	Y	Z	A	B	C	D
0	0	0	0	0	0	1	1
0	0	0	1	0	1	0	0
0	0	1	0	0	1	0	1
0	0	1	1	0	1	1	0
0	1	0	0	0	1	1	1
0	1	0	1	1	0	0	0
0	1	1	0	1	0	0	1
0	1	1	1	1	0	1	0
1	0	0	0	1	0	1	1
1	0	0	1	1	1	0	0

The beauty of this code is that it is a true complement code. For example,

$$\bar{0} = 9, \bar{1} = 8, \bar{2} = 7, \bar{3} = 6, \bar{4} = 5, \bar{5} = 4, \text{ and so on}$$

Such a code is convenient because subtraction is reduced to complementation, increment, and add. Design a BCD to excess 3 encoder using 8 to 1 multiplexers. Since

$$\bar{D} = z$$

you are not required to design the D multiplexer.

P2.7 Design a multiplexer circuit to realize the following function:

$$f(w, x, y, z) = wy + x(w + y\bar{z})$$

P2.8 Design a multiplexer implementation to generate a parity bit P in a 6-bit code A B C D E F.

P2.9 Design a combinational circuit using multiplexers to convert a 6-bit number a b c d e f to its 2's complement form.

P2.10 Design a multiplexer implementation of the following BCD to BiQuinary coder.

Table P2.3 Coder Bits

	BCD				BiQuinary						
Integer	A	B	C	D	M	N	V	W	X	Y	Z
0	0	0	0	0	0	1	0	0	0	0	1
1	0	0	0	1	0	1	0	0	0	1	0
2	0	0	1	0	0	1	0	0	1	0	0
3	0	0	1	1	0	1	0	1	0	0	0
4	0	1	0	0	0	1	1	0	0	0	0
5	0	1	0	1	1	0	0	0	0	0	1
6	0	1	1	0	1	0	0	0	0	1	0
7	0	1	1	1	1	0	0	0	1	0	0
8	1	0	0	0	1	0	1	1	0	0	0
9	1	0	0	1	1	0	1	0	0	0	0

CHAPTER 3

Fundamental Mode Circuits

Output functions that are solely dependent upon input variables are combinational circuits. In combinational circuits particular values of the input variables have a particular output. In sequential circuits history is also an element. A sequential circuit has a memory of past events. Therefore its output function for particular values of input variables may be different, depending upon a sequence of events. Being able to detect sequences of events is very important when designing processes or performing analysis. The real power of a computer lies in its ability to perform a large number of events in a particular sequence. This chapter is devoted to the study of classical methods of designing sequential circuits [2-8, 11-15]. In the representation of a sequential circuit, both the combinational and sequential aspects of the design are needed. These conventions are given in the following section.

3.1 FUNDAMENTAL MODE DESIGN CONVENTIONS

Figure 3.1 gives a conventional form of representing a mode diagram. So that the student may be able to construct and read such a diagram there is a strict meaning to the circle, the line, and the arrow. The arrow represents an input. The input variables are listed in the order in which they are written in the upper left-hand corner of the diagram. The values of the input variables are written above the arrow on the input line. A fundamental mode circuit is a circuit in which only one of the input variables can change its value at one time. Thus dual transitions in input variable values are not permitted in a fundamental mode circuit. The circle represents a mode or steady condition between the occurrence of input events. Modes are assigned names in number form which are placed in the circle. One practice is to assign outputs to modes and such outputs are below the circle. Outputs can be assigned to inputs, but this practice is rarely used in this book.

The inputs in Figure 3.1 are x_1x_2. Operation can be assumed to start in mode 1 (the circled 1) with output 0. When $x_1x_2 = 10$ the circuit moves to mode 2 with output 0. A further change of input $x_1x_2 = 11$ causes a change to mode 3 with output 0. Then when the inputs change to $x_1x_2 = 10$ the output becomes 1. A change of input $x_1x_2 = 11$ changes the mode to 5 and output to 0. A subsequent change of input to $x_1x_2 = 01$ results in a move to mode 6 with output 0, and when $x_1x_2 = 00$ the circuit returns to mode 1, the starting point.

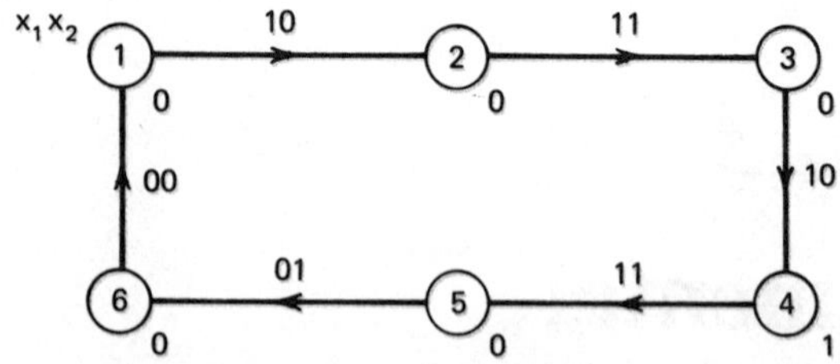

Figure 3.1 Mode diagram.

A block diagram of a sequential circuit is shown in Figure 3.2 [2]. The characteristic that gives the sequential circuit its memory capability is the state assignment, or y's. One or more modes form a state. State assignments are commonly called secondaries. Secondaries have a time delay between their excitation and the realizing of a change of state. This is represented in the schematic diagram by a delay unit on which the excitation of the secondary is described by upper case letters and the state of the secondaries is described by lower case letters.

In sequential circuits secondaries perform a feedback function. Thus, the combinational part of a sequential circuit has normal outputs and secondary excitation as outputs. The inputs to the combinational circuits are the normal circuit inputs and the present state of any secondary. The feedback that occurs leads to a logic systems variation of the concept of stability.

3.2 CONCEPT OF STABILITY

A state-by-state transition is used to design a device that moves through a set of events in a sequential fashion. To move from one state to the next state requires a change in the value of one secondary variable. The design requires that the transi-

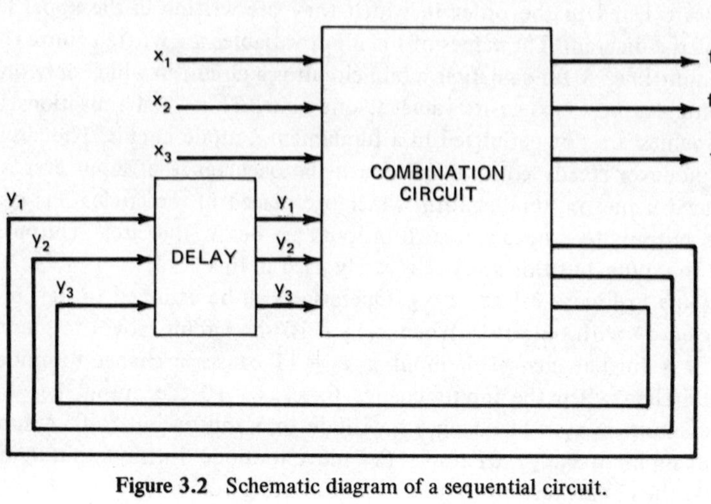

Figure 3.2 Schematic diagram of a sequential circuit.

tion be limited to change in the value of one variable. When the excitation of a secondary is the same as a present state, that is, $Y = y$, the secondary does not change. It is said to be stable. When the excitation of the secondary is not the same as the present state, that is, $Y \neq y$, the secondary is required to change, and it is said to be unstable. Thus, if $y = 0$ and $Y = 0$ there is no change required in the secondary variable y and it is stable. If $y = 1$ and $Y = 1$ the variable y again is not required to change and it is stable. If $y = 0$ and $Y = 1$ the secondary variable y must change to a 1, and the secondary is unstable. If $y = 1$ and $Y = 0$ the next state of the variable is 0 and the secondary is unstable.

In summary, a state can be said to be unstable if the value of one variable is in a state of transition. It is said to be stable if the values of all variables in the state are not changing.

3.3 BASIC OPERATION OF A SEQUENTIAL CIRCUIT

The state of the output and the secondary excitations Y's are functions of the inputs x's and the secondary state variables y's. Therefore, the secondaries enter into their own control. In a transition, the first change is in the input variables. The change in the input variables causes a change in the secondary excitations. This causes the secondary variable to change after a short delay. If the change is to a stable state, no further changes occur. However, if the present state is still unstable another change in the secondary excitation occurs, and there is a corresponding change in the state variable. Such changes continue to occur until a stable condition is reached. Since outputs are a function of inputs and secondaries, they are stable secondaries.

In the remainder of this chapter a step-by-step synthesis technique for the design of fundamental mode sequential circuits is given, and this technique is used to design a general purpose secondary circuit.

3.4 MODE DIAGRAM DESIGN

The first step in the synthesis of fundamental mode circuits is the construction of a mode diagram. The diagram is constructed from a word statement of the problem. Starting with any given set of inputs, for each change of the value of an input variable we construct a line and arrow to a new mode. Since one change is permitted in each value, the lines leaving a mode can equal the number of variables. However, the problem definition usually restricts the actual number (probably to one).

The construction is continued until all paths become closed loops.

3.5 FLOW TABLE

The second step in the synthesis is the development of a flow table. The flow table is a method of converting the diagram into a map for the development of logic expressions for secondary excitation functions and output functions. The flow table resembles a map in layout. However, as with diagrams, flow tables have a very distinct layout.

Change of an input variable initiates other changes. Each combination of values of an input variable describes a column of the table. Thus, a change in input represents a change of column in the flow table. The flow table for Figure 3.1 is shown in Figure 3.3. Each possible value of the variable x_1, x_2 is assigned a column. Each circle in Figure 3.1 is assigned a separate row in the flow table (Figure 3.3).

Stable conditions are designated by circles whereas an unstable condition is undesignated. To start ① is inserted into the flow table under input column $x_1x_2 = 00$. A change of inputs $x_1x_2 = 10$ causes a move to unstable 2 represented by a column change in this flow table. This can be considered to cause a change in excitation. A change of the excitation results in a row-to-row transition in the flow table, in this case from 2 to ②. Since outputs are generally associated with each stable mode, the assigned outputs are given under an additional column in the flow table for convenience. Similarly, when the inputs become $x_1x_2 = 11$, there is a move between columns in the flow table signified by a change from ② to unstable 3. The change in excitation causes the next event, which is a transition to ③. It is designated by a row change in the flow table.

A further change in the input variables to $x_1x_2 = 10$ results in another column change in the flow table, from ③ to 4. There is a resultant change in excitation of the secondaries that causes a row-to-row transition in the flow table from 4 to ④. The mode diagram calls for an output from ④, and this is designated in the output column. Further, $x_1x_2 = 11$ gives a change of column, ④ to 5, followed by a change of excitation that causes the transition 5 to ⑤. Next $x_1x_2 = 01$ and the change of column is denoted by the change ⑤ to 6 and a subsequent row transition 6 to ⑥ caused by excitation changes.

The next event in the sequence is a change of inputs to $x_1x_2 = 00$. The change in the flow table is designated by a column to column change ⑥ to 1. Immediately there is a change in excitation, and the result is noted in the flow table with a row-to-row transition from 1 to ①. This completes the sequence of events.

x_1x_2

00	01	11	10	f
①			2	0
		3	②	0
		③	4	0
		5	④	1
	6	⑤		0
1	⑥			0

Figure 3.3 Flow table illustration.

3.6 ELIMINATION OF REDUNDANT STABLE STATES

In the construction of the mode diagram it is possible to introduce more modes than are needed. If two modes are equivalent one of them is redundant and must be removed. Sometimes it is not obvious that some of the modes are redundant. Redundancy can be easily detected in the flow table. This check forms the next step in the synthesis.

Stable modes in the flow table are redundant if:

1. they have the same input (occur in the same column);
2. they have the same output;
3. they have a transition to the same or equivalent mode for each possible input change.

Since the example of Figure 3.3 does not have any redundant modes a new example is needed. A flow table with redundancies is given in Figure 3.4(*a*). Modes ② and ④ are equivalent. They have the same input $x_1 x_2 = 01$. They have the same output $f_1 f_2 = 10$. Transitions from ④ and ② are to 1 and 5. This is identical, column for column, thus resulting in transitions to the same rows. Therefore row ④ can be eliminated from the flow table. When a redundant row is eliminated all unstable states associated with that row must be changed to drive to the remaining redundant state as shown in Fig. 3.4(*b*).

A condition that is almost the same as redundancy is pseudoequivalence. Two rows may be pseudoequivalent if the following conditions analogous to don't care cells are present.

1. For a particular change of input, there is a transition from one row to another under a prescribed route, whereas the transition from the other row is optional.
2. The output state associated with one of the rows is described, whereas the corresponding output for the other is optional.

$x_1 x_2$

00	01	11	10	$f_1 f_2$
①	2		6	00
1	②	5		10
③	4		6	10
1	④	5		10
	4	⑤	6	01
3		5	⑥	11

(*a*)

$x_1 x_2$

00	01	11	10	$f_1 f_2$
①	2		6	00
1	②	5		10
③	2		6	10
	2	⑤	6	01
3		5	⑥	11

(*b*)

Figure 3.4 Flow table with ② and ④ equivalent. *(a)* Redundant; *(b)* row 4 eliminated.

(a)

x_1x_2: 00	01	11	10	f_1f_2
①	3		4	00
②	3	-	-	00

(b)

x_1x_2: 00	01	11	10	f_1f_2
①	3		4	00

Figure 3.5 Use of optimal transition. *(a)* Pseudoequivalence; *(b)* row reduction.

Figure 3.5 shows a pseudoequivalence where the prescribed transition for input $x_1x_2 = 10$ in the second row is optional.

3.7 MERGING ROWS

Figure 3.6 shows the merged version of the flow table from Figure 3.3. In this figure, rows ① and ② have been merged into one row as have rows ③ and ⑥ and also rows ④ and ⑤. The values of the output are not included in this table. Merging rows in a table reduces the number of implied secondaries and thus, in general, reduces the number of secondary variables required to realize a sequential circuit.

Rows are merged under the following rules:

1. Two or more rows can be merged if within the rows there are no conflicting numbers in any column. That is, two rows can be merged if each column contains either the same number or one of the rows has an option.
2. All numbers in the merge rows are written in the respective column of the merge row. If a number is circled in one of the merging rows it is circled in the merged row. The output states for each row are not considered in the merging operation. They are not repeated in any merge table.

To construct a merger diagram, the row numbers are arranged in a basically circular array, as shown in Figure 3.7. The numbers are used to identify the rows in the primitive flow table. Lines are drawn to joint rows that can be merged. Each row must be compared with every other row for possible merging. The connecting lines

x_1x_2: 00	01	11	10
①		3	②
1	⑥	③	4
	6	⑤	④

Figure 3.6 Merged flow table.

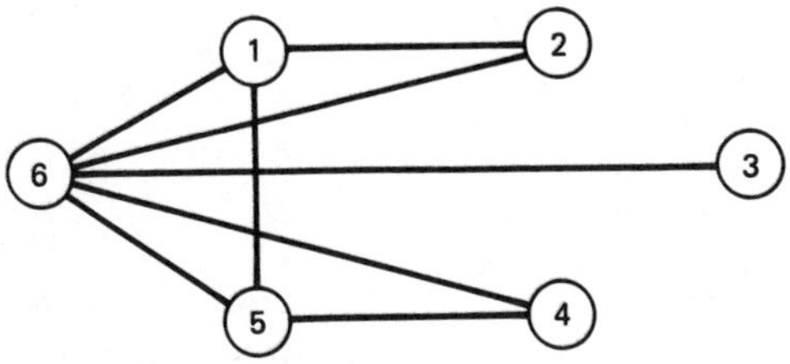

Figure 3.7 A merger diagram.

drawn in the diagram are inspected to find the best merge. Generally, the best merge results in the least number of rows after merging.

3.8 SECONDARY STATE ASSIGNMENT

The next step in the synthesis is to assign secondary states to each row in the merged table. This is preparation for construction of the excitation (Y) map. The secondaries are assigned using a transition map. A transition map has the basic form of a K map without the output function cell entries. To start this process, each row in the merge table is assigned a letter reference. These letters are to be entered into the transition map.

All possible row-to-row transitions must be determined. For transition between two rows, the row should differ in the value of only one variable. It is possible to move from unstable state to the desired stable state by way of another row provided the transition does not go through an unwanted state.

A hypothetical change is illustrated in Figure 3.8. Starting with ⑥ a change of input from $x_1x_2 = 01$ to $x_1x_2 = 00$ causes a change in excitation. The excitation causes a change ⑥ to 1 in the row labeled ⓑ. This causes a further change in excitation since state 1 is not stable in this row. The change in excitation could cause a change to the row labeled ⓒ. Again state 1 is not stable, so a further change in excitation occurs, which could cause a change to the row labeled ⓐ. Here, ① is in the stable condition and cycling ceases.

It is possible to cycle in this way through a number of rows in a merge table provided there is no transition to another state.

In Figure 3.8 row-to-row transitions occur between ⓐ and ⓑ. Excitation causes a transition from 3 in row ⓐ to ③ in row ⓑ and 1 in row ⓑ to ① in row ⓐ.

x_1x_2	00	01	11	10	
	①		3	②	ⓐ
	1	⑥	③	4	ⓑ
	1	6	⑤	④	ⓒ

Figure 3.8 Assigned letter reference and cycling diagram.

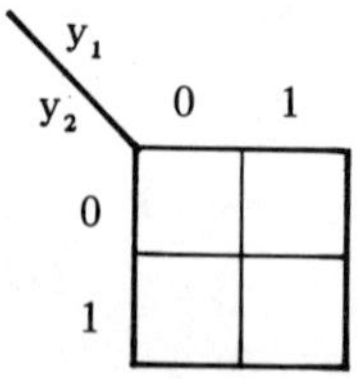

Figure 3.9 Transition map form.

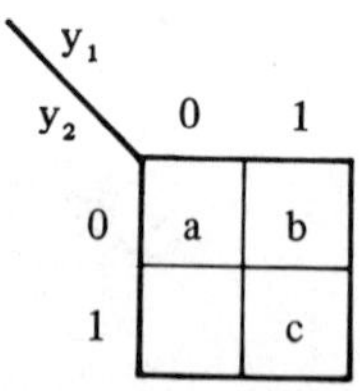

Figure 3.10 Transition map with row selection.

y_1y_2 \ x_1x_2	00	01	11	10
00	①		3	②
01				
11		6	⑤	④
10	1	⑥	③	4

Figure 3.11 Map showing secondary state assignment.

Similarly, there is a row-to-row transition between ⓑ and ⓒ denoted by 6 to ⑥ and 4 to ④.

A transition map of two variables has four cells and thus four possible row assignments. Such a map is shown in Figure 3.9. Since each row of Figure 3.8 is assigned a letter, row assignment is accomplished by choosing a position for each letter in the transition map. Placing ⓐ in cell $\Sigma(0)$ is one possibility and the choice here. Then ⓑ is placed adjacent to ⓐ in cell $\Sigma(2)$. [$\Sigma(1)$ is also a valid choice.] Then ⓒ is placed adjacent to ⓑ in cell $\Sigma(3)$. This arrangement is shown in Figure 3.10. The assigned values of y_1y_2 for the rows are a(00), b(10), and c(11). So the resulting merged flow table with secondary assignment is shown in Figure 3.11.

3.9 EXCITATION MAP (Y MAP)

The basic rule for the development of the excitation map is given in Section 3.2. It is restated here in simplified form. A stable state occurs when excitation equals row assignment $Y = y$. Figure 3.12 gives the excitation map for stable states. The entries are made as follows: For stable states ① and ②, $y_1y_2 = 00$. Thus $Y_1Y_2 = 00$. Similarly, for stable states ④ and ⑤, $y_1y_2 = 11$. Thus $Y_1Y_2 = 11$, and for stable states ③ and ⑥, $y_1y_2 = 10$. Thus $Y_1Y_2 = 10$.

Each map entry corresponding to an unstable state is the same as the entry for its destination. Thus, in general, entries with the same state name have the same value of excitation variable. For example, if stable state ⑥ has $Y_1Y_2 = 10$, state 6 has the same entry $Y_1Y_2 = 10$. The completed Y-map using this rule is shown in

$y_1y_2 \backslash x_1x_2$	00	01	11	10
00	00			00
01				
11			11	11
10		10	10	

Figure 3.12 Stable state excitation for Y_1Y_2.

$y_1y_2 \backslash x_1x_2$	00	01	11	10
00	00	Ø	10	00
01	Ø	Ø	Ø	Ø
11	Ø	10	11	11
10	00	10	10	11

Figure 3.13 Completed Y-map.

Figure 3.13. The logic expression for this map is

$$Y_1 = x_2 + y_1x_1$$

$$Y_2 = x_1\bar{x}_2y_1 + y_2x_1$$

under the assumption that blank cells are don't care cells.

3.10 DESIGN OF OUTPUT CIRCUIT

The output circuit is designed from combinational procedures. The output map f, with inputs and secondaries as variables, is drawn from Figure 3.1. The output occurs in mode ④. Placing an output in mode ④ of Figure 3.11, the resultant output map is given as Figure 3.14.

The value of mode 4 must be determined from the word statement of the problem. The logic expression for this map is

$$f = \bar{x}_2y_2$$

3.11 CIRCUIT REALIZATION

The last step before construction is to draw the *nand* gate realization of the excitation and output expression. The circuit realization of Figures 3.13 and 3.14 is given in Figure 3.15. The *nand* gates are numbered from 301 to 309 for identification.

$y_1y_2 \backslash x_1x_2$	00	01	11	10
00		Ø		
01	Ø	Ø	Ø	Ø
11	Ø			1
10				Ø

Figure 3.14 Output map.

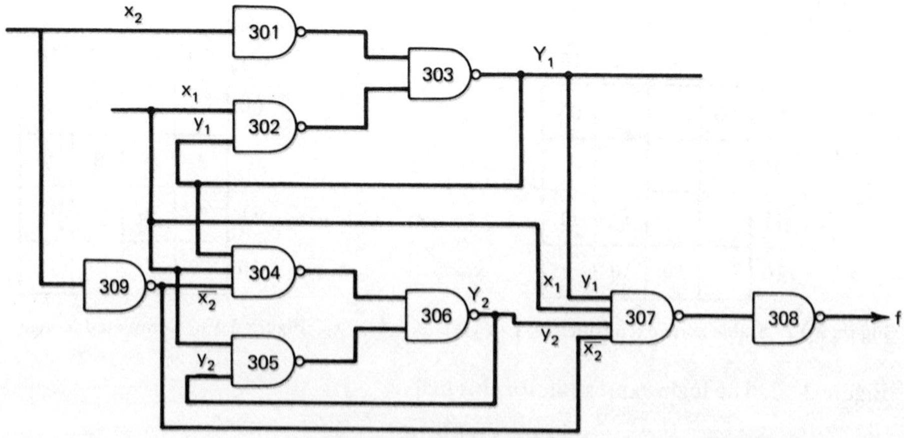

Figure 3.15 Circuit diagram.

3.12 SYNTHESIS OF FUNDAMENTAL MODE SEQUENTIAL CIRCUITS

It seems appropriate to summarize the chapter to this point so that the reader has a concise reference for sequential circuit synthesis.

In the synthesis of sequential circuits, circuit requirements are first described in a diagram. Synthesis methods are used to convert to a flow table, simplify the flow table, and realize secondary excitation and output functions. Transferring into map forms allows such functions to be read in the simple combinational sense and to realize circuits using combinational techniques. The detailed steps are the following [11]:

1. Construct a mode diagram of the circuit from a word statement of the problem.
2. Construct a primitive flow table from the diagram.
3. Test the primitive flow table for redundant states and reduce the number of rows if redundant states are found.
4. Construct a merger diagram for merging rows of the primitive flow table.
5. Assign states to the merged flow table. A transition map is used to determine the assignments.
6. Construct a secondary excitation map from the merge flow table and secondary assignment map and develop a logic expression for the secondary excitation circuits from the excitation map.
7. Construct an output map from the flow table with secondary assignment and the state diagram. Ensure the state-to-state transitions identified in the state diagram are guaranteed in this map. Determine the logic expressions for the output expressions from the output map.
8. A logic circuit is drawn from the secondary excitation and output expression.

3.13 DESIGN FEATURES

Since the procedures are summarized, it would also seem appropriate to make an evaluation of the design synthesis. *Evaluation of the design synthesis is not a required design step.* It is used here to justify a simpler approach. The realization of the circuit is given in Figure 3.15. The operating sequence is given by the excitation and changes described in Figure 3.13. To represent its operation in a more diagrammatical form, each row of Figure 3.13 is drawn as a small circle in Figure 3.16. Placed immediately adjacent to each small circle is the value of y_1y_2 associated with the row. The input variable in symbolic form is placed immediately above each arrow.

It can be seen from Figure 3.15 that the input x_2 causes an output 1 from gate 301, which in turn causes the output 1 from gate 303 or the variable Y_1 to become equal to 1. If x_1 occurs before x_2, there is no action. With input x_2 there is a transition from $y_1y_2 = 00$ to $y_1y_2 = 10$. With x_1 on, Y_1 can be sealed in by way of the feedback path through gates 302 and 303. With x_1 equal to 1 and x_2 equal to 0, the output of gate 304 is 0, and thus the output of gate 306 is 1. Y_2 is held on by way of the feedback path through gates 305 and 306. If x_2 now becomes 1, there is no resultant change of events because of these feedback paths. However, if x_1 goes to 0, Y_2 certainly goes to 0 since x_1 occurs on the input to both gates 304 and 305. If x_2 is 1, Y_1 remains 1. However, when x_2 goes to 0 Y_1 goes to 0, and the circuit returns to the initial condition.

This sequence of events describes the fundamental way the circuit operates. For each of the secondary variables y_1 and y_2 there are three specific stages in its operation. Stage 1 is the initialize operation, which is to turn the device on (*set*). Stage 2 is the hold operation. The device, once on (off), must hold itself on (off). Stage 3 is the turn-off operation (*reset*).

These stages in the operation occur for both secondaries y_1 and y_2. If we were to design more circuits, these characteristic stages would always be required. Thus it would seem sensible to design a circuit that is capable of three-stage operation. Once the circuit is designed, it could be used in the implementation of all circuits where such characteristic stages are required. This shortcut method presumably saves working through all the design steps laid out in this chapter.

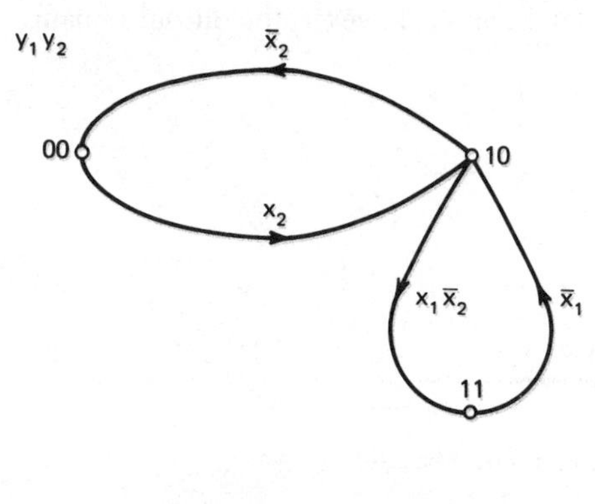

Figure 3.16 Circuit operation.

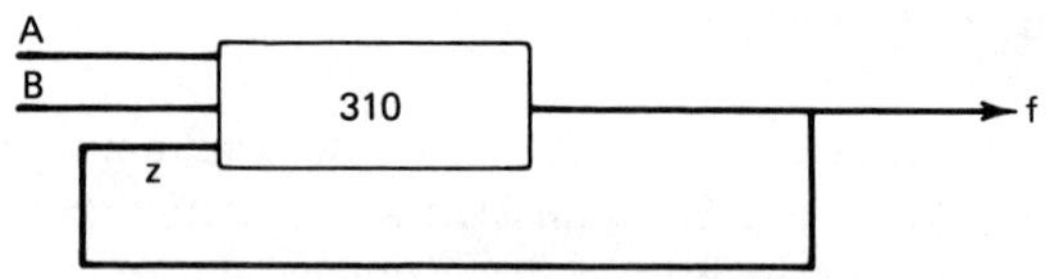

(a)

A	B	z	f
0	0	0	0
0	0	1	1
0	1	0	0
0	1	1	0
1	0	0	1
1	0	1	1

(b)

Figure 3.17 Common buffer circuit. *(a)* Block diagram; *(b)* Truth table.

3.14 BUFFER CIRCUIT DESIGN

A circuit that performs the required functions must be designed. The problem statement is: Design the circuit for the block diagram shown in Figure 3.17. When the input A occurs, the circuit output is to be on. When the input B occurs, the circuit output is to go off. There is to be a feedback path from the circuit output f to the input z such that when A or B are turned off, there is no change in output.

Figure 3.18 shows the mode diagram for the desired circuit. Starting at mode 1, turning signal A on causes the circuit to make a transition to mode 2. In mode 2 the output is 1. If A goes off, the circuit moves to mode 3. However, the output must still be on. Signal A can appear any number of times, and the device simply moves between modes 2 and 3. From mode 3, turning signal B on causes a transition to mode 4. In mode 4 the output is 0. If B disappears or there are repeated pulses of B, the circuit moves between modes 1 and 4. However, the output remains off [12].

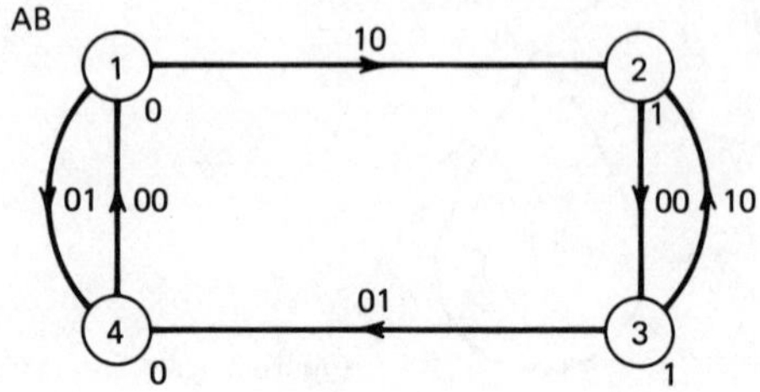

Figure 3.18 Buffer circuit mode diagram.

AB	00	01	11	10	f
	①	4		2	0
	3			②	1
	③	4		2	1
	1	④			0

Figure 3.19 Flow table for buffer circuit.

The next step in the synthesis is the design of the flow table. It is shown in Figure 3.19. Again, the primitive flow table shows that from mode ① the circuit goes to modes 2 or 4. From stable mode ② the circuit goes to mode 3. From stable mode ③ the circuit can go to modes 2 and 4; and from stable mode ④ the circuit can go to mode 1.

The next step in the synthesis is a check for redundant states. There are no redundant states.

The next step is the construction of a merge diagram for this circuit. It is shown in Figure 3.20 as developed by application of the merging test. The diagram shows that rows 1 and 4 may be merged—and rows 2 and 3. The merge flow table is shown in Figure 3.21. The transition table used to determine assignments was not used. The y assignments are chosen so that row a = 0 and b = 1. This is done so that the assignments agree with the output function, a convenience that usually does not occur.

The excitation map for the buffer circuit is shown in Figure 3.22. Stable states ① and ④ are assigned excitation Y = 0 such that Y = y for that row. Similarly, states ③ and ② are assigned Y = 1, since y = 1 for the row. The excitation Y for

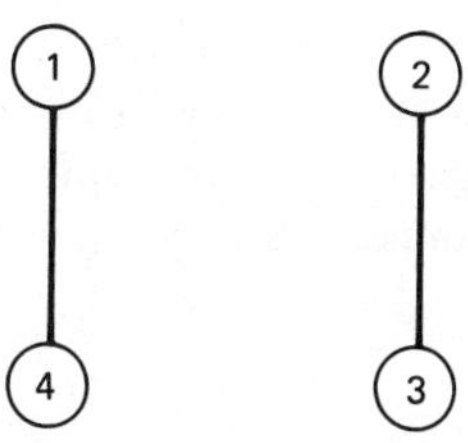

Figure 3.20 Merge diagram for buffer circuit.

AB	00	01	11	10	
0	①	④		2	a
1	③	4		②	b

y	0	1
	a	b

Figure 3.21 Merged flow table with assignment.

y \ AB	00	01	11	10
0	0	0	ϕ	1
1	1	0	ϕ	1

Figure 3.22 Excitation map and output map for buffer circuit.

the unstable state 4 is made equal to the excitation for stable ④, and Y for unstable 2 is made equal to the excitation for stable ②. The two remaining cells are given don't care designation. This circuit is unique in that the output map is exactly equal to the excitation because of the row assignments made. Thus y = z. The output function f can be realized by

$$f = A + z\bar{B}$$

Figure 3.23 shows the circuit that realizes this function. Part (*a*) shows the conventional form of the symbolic diagram realization. Part (*b*) is a reconstructed form, which is more commonly seen and presents a better presentation of the circuit. This device can be turned on or turned off. It holds the condition indicated by the last signal given. It is more commonly called a set reset flip-flop (RSFF), where A is replaced by the set signal and B is replaced by the reset signal.

Utilizing this circuit we can eliminate the drudgery from the design procedure. However, the following two points must always be kept clear when using these more sophisticated techniques:

1. The most economical use of states is desired. The general methods of eliminating redundancy and possibilities for merging can lead to component savings.
2. All events possible to occur must be included in the design. It is also wise to complete the design to include sequences for all possible states, which normally cannot occur. Small variations in manufacturing, general aging of components, lead lengths and problems in precise alignment of multipin connectors may cause the circuit to enter unused states.

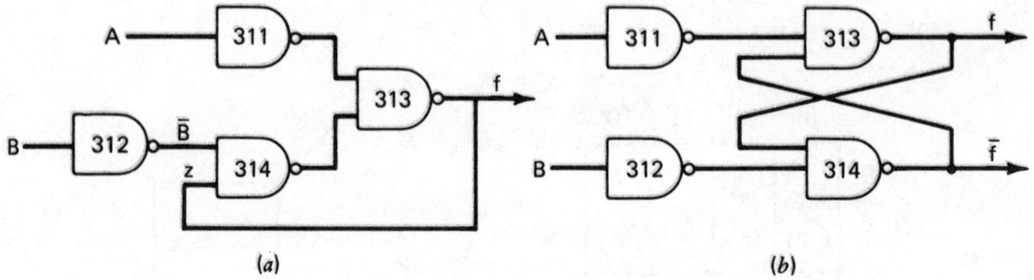

Figure 3.23 Memory circuit. *(a)* Conventional form; *(b)* reconstructed form.

3.15 PROBLEMS

P3.1 A circuit has a single input line on which pulses of varying widths occur randomly. Construct a modal diagram for this circuit with an output f on which a pulse occurs coincident with the fourth pulse of every string.

P3.2 A common feature of a computer is the step control, which enables the operator to execute a program one step at a time. To do this, we require to transmit a single pulse from the computer clock every time the push button is closed. Thus, with inputs clock and pb (push button), design a fundamental mode diagram to achieve the above function.

P3.3 The example shown in Figure P3.1 has been designed with redundant rows. Test the circuit for redundancies and eliminate the redundant rows.

ab: 00	01	11	10	$f_1 f_2$
(1)	2	-	3	00
1	(2)	4		10
5		4	(3)	11
	6	(4)	3	01
(5)	7		3	01
5	(6)	4		11
1	(7)	4		10

Figure P3.1 Primitive flow table with redundancies.

P3.4 The example of a primitive flow table is shown in Figure P3.2. Draw a merge diagram and merge rows in this table.

XY: 00	01	11	10	f
(1)	2		3	0
1	(2)	4		1
1		5	(3)	1
	6	(4)	3	1
	6	(5)	3	1
1	(6)	4		0

Figure P3.2 Primitive flow table for merging.

P3.5 Complete the excitation map started in Figure P3.3.

y_1y_2 \ x_1x_2	00	01	11	10
00	①	2	③	5
01	8	②	④	6
11	8	⑦	3	⑥
10	⑧	7	3	⑤

Figure P3.3 Incomplete excitation map.

P3.6 Construct an excitation map and realize a circuit from the following incomplete excitation map (Figure P3.4).

y_1y_2 \ *AB*	00	01	11	10
00	①	2	⑦	4
01	1	②	3	⑥
11	1	⑤	③	6
10	1	5	7	④

Figure P3.4 Flow table with assignments.

P3.7 The flow table for a circuit to indicate direction of rotation is shown in Figure P3.5. The outputs for modes ①, ②, ③, and ④ are 0 and for modes ⑤, ⑥, ⑦, and ⑧ are 1. Complete the assignment table and obtain logic equations for Y_1, Y_2, and f.

ab 00	01	11	10
②	3	5	⑥
⑦	③	4	6
7	⑧	④	1
2	8	⑤	①

Figure P3.5 Direction of rotation flow table.

P3.8 Figure P3.6 shows a fundamental mode diagram for an annunciator circuit. Figure P3.7 gives the incomplete version of the Y_1Y_2 map.

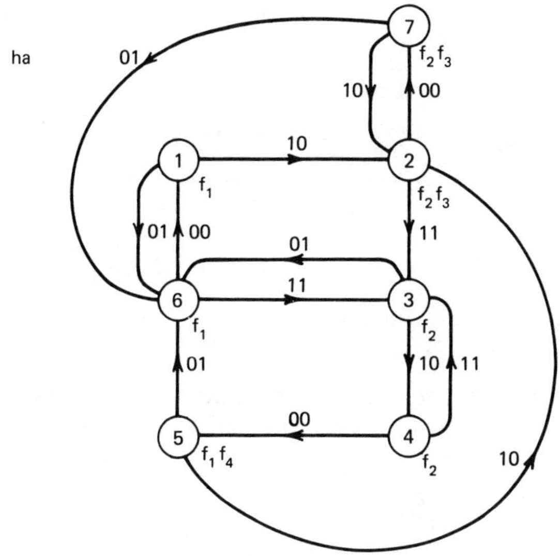

Figure P3.6 Mode diagram for annunciator circuit.

(a) Draw an excitation map for Y_1Y_2 and implement the circuit using *nand* gates.

(b) Draw the maps for f_1, f_2, f_3, and f_4.

y_1y_2 \ *ha*	00	01	11	10
00	(1)	(6)	3	2
01	(7)	6	3	(2)
11	5	6	(3)	(4)
10	(5)	6		2

$Y_1\ Y_2$

Figure P3.7 Incomplete excitation map.

P3.9 Mechanical switches among others are subject to contact bounce. An example of bouncing is inadvert duplication of characters on a printer. It is very expensive to eliminate this problem by improved switch design so a logic circuit using IC chips is used. The problem is illustrated in Figure P3.8.

As the switch moves from A to B it makes an initial contact at B and then bounces. It does not bounce high enough to contact A. A similar event occurs when the switch moves from B to A. Design a debounce circuit to produce the desired output f from inputs A and B. Use fundamental mode design technique.

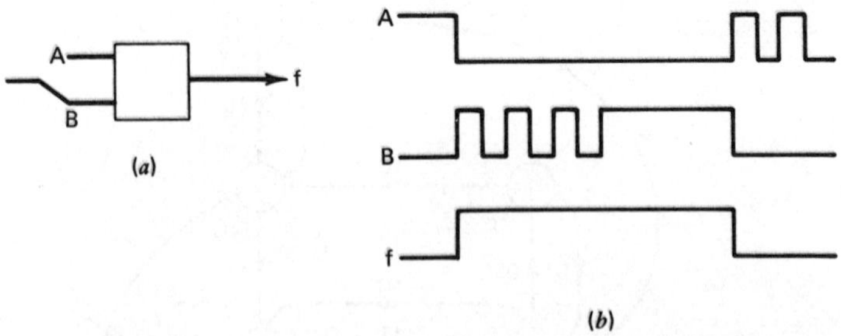

Figure P3.8 Contact bounce. *(a)* Block diagram; *(b)* timing.

P3.10 A fundamental mode circuit is to have two inputs A and B. Its timing diagram is shown in Figure P3.9. A and B cannot be equal to 1 at the same time.
(a) Draw a mode diagram.
(b) Draw a flow table.
(c) Merge rows in the flow table.

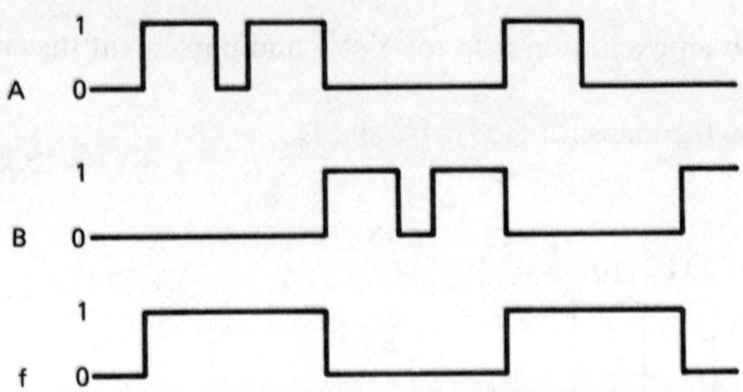

Figure P3.9 Buffer timing diagram.

P3.11 A fundamental mode circuit has two inputs AB and one output f. The circuit is to indicate the direction of rotation. The input sequence AB = 00, 01, 11, 10, or any part thereof, indicates rotation in one direction. The sequence AB = 00, 10, 11, 01, or any part thereof, indicates rotation in the opposite direction. Execute the following procedure:
(a) Draw a fundamental mode diagram for this circuit.
(b) Draw a primitive flow table for this circuit.
(c) Merge rows in the primitive flow table.

CHAPTER 4

Event-Driven Sequential Circuits

In Chapter 3 the basic concepts for sequential circuit design are developed. From the mode-to-mode change the concept of state is developed. In the concept of state, the three basic stages can be represented in a state diagram as two basic operations describing state changes.

1. The input can cause the value of a state variable to change, which is described as set or reset.
2. The state variable is held in accordance with the last input.

Progression through a sequence is accomplished by changes in the value of state variables. The essence of feedback is the hold function. It is always required in any sequential circuit, and one way to accomplish this is to utilize a specially designed hold circuit, the flip-flop.

In this chapter a method of designing event-driven sequential circuits using RSFFs is presented. The process is described from state diagram to complete circuit design. The design of state diagrams is given in Chapter 8.

4.1 STATE DIAGRAMS FOR EVENT-DRIVEN SEQUENTIAL CIRCUITS

The starting point of event-driven sequential circuit design is the state diagram. This diagram transmits all the information required in a design. The state diagram shows the sequence of events that must occur in the sequential circuit. A sequential diagram for an annunciator is used as a typical state diagram and is shown in Figure 4.1 [12].

The diagram has a very rigid format so the information it contains is absolutely clear. Deviation from this format causes confusion in the transmitted information. States in the diagram are designated by small circles and identified by the appropriate value of the secondary variables. The value of the secondary variable for each state is placed immediately above the state. The names of the secondary variables are placed in the upper left-hand corner of the diagram. Inputs are represented by lines with an arrow on the middle of the line. The input variable that causes a change in state is placed above a horizontal line over the arrow or to the right of

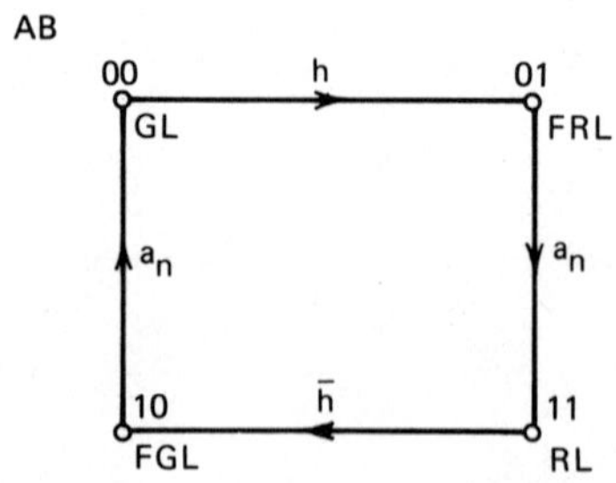

Figure 4.1 State diagram of an annunciator.

the arrow on a vertical line. Outputs in variable form are written below the line starting immediately to the right of the circle designating state.

Using this format the operation of the annunciation circuit of Figure 4.1 can be read as follows:

- Starting with AB = 00, the circuit is in the idle position, and this is indicated by the output GL (green light).
- When a fault occurs there is an input to the circuit (h = 1) and a transition to AB = 01. An operator is notified of the fault by the output FRL (flashing red light).
- The operator then acknowledges (a_n = 1) the fault causing a transition to AB = 11. The output becomes RL (steady red light).
- When the fault is cleared (h = 0) there is a transition to AB = 10. The output is FGL (flashing green light).
- The operator acknowledges the cleared fault (a_n = 1) causing the circuit to return to the starting point, AB = 00.

In an event-driven circuit, progression from one state to the next possible state causes a change in the value of one secondary variable and one variable only. A sequential circuit using RSFFs has one flip-flop assigned to each variable in the state diagram.

4.2 RSFF

The operation of the RSFF is shown in Figure 4.2. Figure 4-2(*a*) gives a new form of the Karnaugh map. In this form the Karnaugh map is used to indicate a change in the state of the RSFF. Since a change in the state is the occurrence of an event, this is called an action map [3, 12].

In Figure 4.2(*a*) the action of the RSFF is shown in map form. The present value of the RSFF is shown as a lower case q designating the value of each row. The value of the excitation (or future state) for Q is shown at the top of each column. Thus the map designates the value of R and S, the input variables. For the case or cell

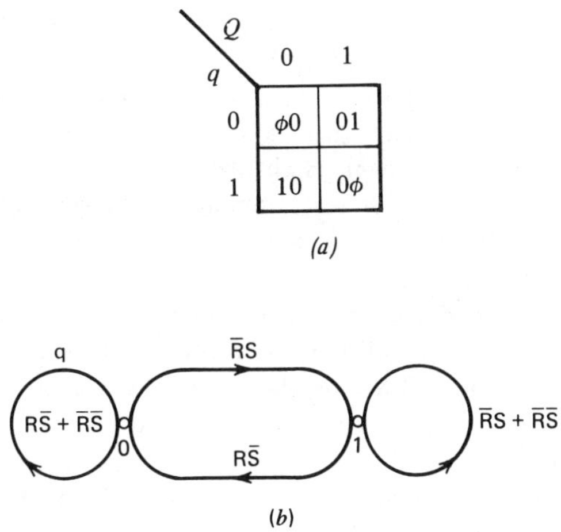

Figure 4.2 Action representation for an RSFF. *(a)* RS action map; *(b)* state representation.

when qQ = 00, RS = ϕ0. The flip-flop is not to be set, so the set input must be 0; the reset is optional. For the cell where qQ = 01, RS = 01. In this transition the flip-flop is set. For the operation to occur, R must be 0 and S must be 1. For the cell where qQ = 10, RS = 10. This is the case where the RSFF is cleared. For this to occur it is essential that R is 1 and S is 0. For the cell where qQ = 11, RS = 0ϕ. This is the case where the RSFF is set and must remain so. The reset terminal on the flip-flop must be 0, but the value of the set terminal is optional. The values of S and R for each of the state variables can be determined from this action map. All cases where a state variable changes from 0 to 1 must be described in the S function of that variable, and changes from 1 to 0 must be described in the R function. Changes in state where the variable is an unchanged 1 may be included in the S function, and an unchanged 0 may be included in the R function if a gate saving results.

In Figure 4.2(*b*) the action of the RSFF is shown in state diagram form. For the present state q and future state Q, both 0, that is qQ = 00, the input variables are $R\bar{S} + \bar{R}\bar{S}$. This is an equivalent way of saying that the value of R is optional and S must be 0. For qQ = 01, there is a transition from present state 0 to future state 1, and the input that causes this change is $\bar{R}S$ or the equivalent form RS = 01. For the case where qQ = 11, there is no transition in states. The allowed inputs are $\bar{R}S + \bar{R}\bar{S}$. Thus the value of R must be 0, and the value of S is optional. For the case where qQ = 10, there is a transition from state 1 to state 0. This is denoted by input $R\bar{S}$, which is equivalent to the RS = 10 condition.

Notice that the value of RS = 11 is not described. For RS = 11 the operation of the flip-flop is undefined, and the designer must guarantee that the situation does not occur.

The equations for the operation of the flip-flops are written with the function designated by the operation followed by the variable name. For example, S_A signifies setting variable A. R_B would designate resetting variable B.

The concept of action is applied throughout this book. In all discussions and diagrams it is the action associated with a variable that is used. For example, in Figure 4.1 the value of state variable B changes from 0 to 1 when h = 1 or when h goes high. However the value of state variable B changes from 1 to 0 when h goes low or $\bar{h} = 1$. Thus h implies that there is action where h is high (active high), and $\bar{h}$ implies that there is action when h is low (active low).

In manufacturers' literature the same action convention is used. That is, h implies action. However, say a variable exists in the form $\bar{x}$ only (x is not used). The form $\bar{x}$ implies that there is action. However, the action is when $\bar{x} = 1$ (x = 0). It is common practice to show a block diagram with an inverter (o—) for the input $\bar{x}$ when this convention is used.

The form $R/\bar{W}$ implies that W is the complement of R, and the form is intended to be more descriptive. In general the R action is taken when $R/\bar{W}$ is high. The W action is taken when $R/\bar{W}$ is low. $R/\bar{W}$ can be expressed as being low in three ways, which are $\overline{R/\bar{W}}$, $\bar{R}/W$, or $R/\bar{W} = 0$.

4.3 DESIGN FLEXIBILITY OF RSFFS

The use of a flip-flop with set and reset terminals broadens the possibilities in design enormously. By using fundamental mode design given in Chapter 3 the number of input variables is limited, and the value of no more than one variable can change at a time. However, the case of RSFFs is considerably different, as shown by the previous section. There are few restrictions on the input variables that may be used. Incidentally, lack of restrictions on the input variable and the implied consequence that the number of input variables may be great is one of the reasons why input variables are not designated as fixed variables in the state diagrams.

The horizons for the development of circuits to achieve solutions with a great deal of sophistication are much wider. The RSFF gives a way of utilizing secondaries effectively and thus an effective way of utilizing components in the design of sequential circuits.

Although RSFFs are conveniently represented in their *nand* implementation form, it is also convenient to show them in block form to draw attention to their presence and the operation of the circuit. Figure 4.3 shows a block form representation of an RSFF.

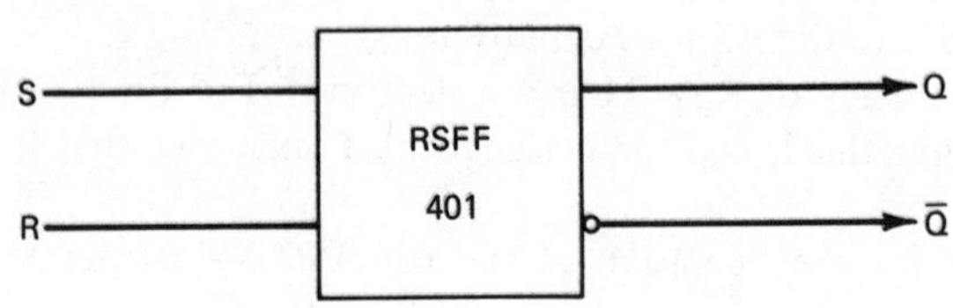

Figure 4.3 RSFF in block form.

4.4 STEPS IN THE DESIGN OF AN EVENT-DRIVEN CIRCUIT USING RSFFS

Synthesis of event-driven sequential circuits using RSFFs is achieved by the following steps:

1. Draw a closed loop to enclose all of the states where the value of a specific variable equals 1. Since transition in and out of the loops requires action input lines intersecting loops, designate this action. Arrows entering the loop designate the set excitation that turns the variable on, and arrows exiting from the loop designate reset excitation that turns the variable off.
2. Write an S of P expression for the set excitation and another for the reset excitation. Each product must contain the input variable(s) and the nonchanging state variables associated with the transition. Inclusion of the state variable guarantees that the circuit has progressed through the proper steps in the sequence. The occurrence of the input variables causes the excitation that results in the subsequent action.
3. Draw a symbolic diagram of the expressions derived above.

The method of looping variables as described in step 1 of the synthesis is illustrated in Figure 4.4 as applied to the state diagram of Figure 4.1.

Step 2 is executed as follows: From Figure 4.4(*a*) it can be seen that the set excitation for variable A, S_A, determined from the input arrow crossing into the loop, is

$$S_A = Ba_n$$

where B is the value of the nonchanging state variable and a_n is the input.

The reset excitation for variable A, R_A, determined from the input arrow exiting the loop, is

$$R_A = \bar{B}a_n$$

where $\bar{B}$ is the value of the nonchanging state variable and a_n is the input.

Similarly for Figure 4.4(*b*) the excitation S_B as determined from the input arrow

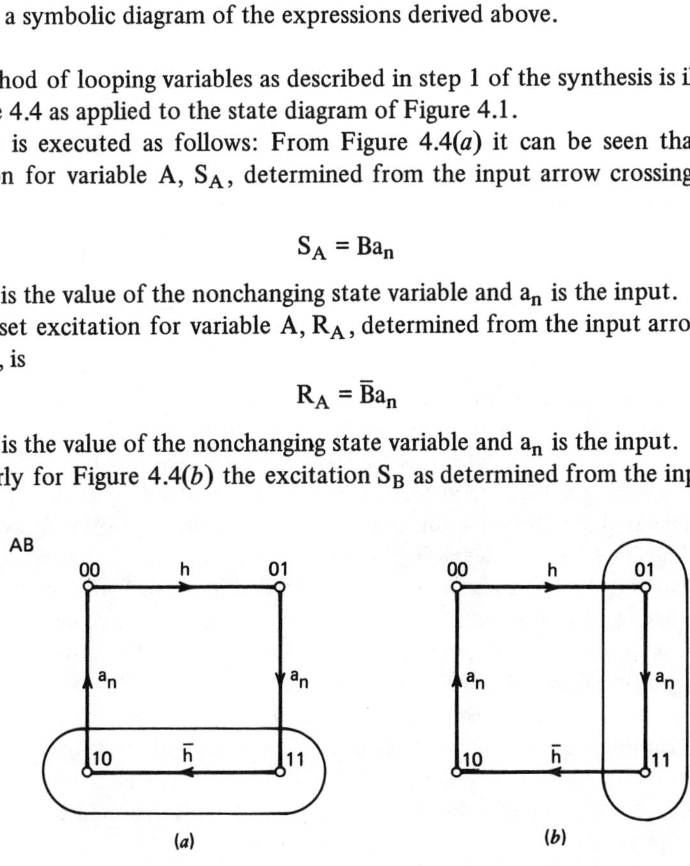

Figure 4.4 Closed loops enclosing state variables. *(a)* A loop; *(b)* B loop.

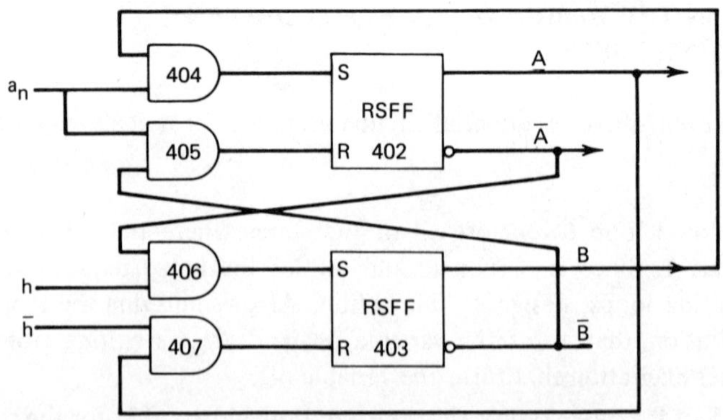

Figure 4.5 Circuit realization of an annunciator.

crossing into the loop is

$$S_B = \bar{A}h$$

and the R_B excitation as determined from the input arrow exiting the loop is

$$R_B = A\bar{h}$$

Step 3, the symbolic diagram for the realization of the circuit, is given in Figure 4.5. Since the desired inputs to R and S may be realized without using two additional *nand* gates in series, a standard *and* gate is drawn to indicate flexibility in actual implementation.

4.5 OPERATION OF AN AUTOMATIC START-UP SHUT-DOWN CIRCUIT

Figure 4.6 shows the state diagram for a circuit to start up and shut down a turbine automatically. This circuit is chosen to illustrate the design techniques using RSFFs and is not necessarily the circuit most suited to turbine control.

In the idle condition, the state variables ABCD = 0000. The circuit makes a transition to state ABCD = 0001 upon receipt of a *start* command. This *start* command may be generated by the operator or may be the output of another logic circuit. In state ABCD = 0001, a signal OPEN AUX is sent to the turbine to open the auxiliary valve. Auxiliary valves are usually provided in parallel with main valves so that the pressure can be equalized on both sides of the large valve before opening. Failure to provide equalizing pressure may cause damage to the valve. Opening of the auxiliary valve allows steam to flow.

When the speed of the turbine has reached 5%, a pickup on the turbine detects the condition and sends a signal *speed* 5% to the automatic start-up circuit. This signal generates a transition to state ABCD = 0011. In this state, an output signal START OIL PUMP is generated to start the oil pump. Oil is required in the bearings of the turbine.

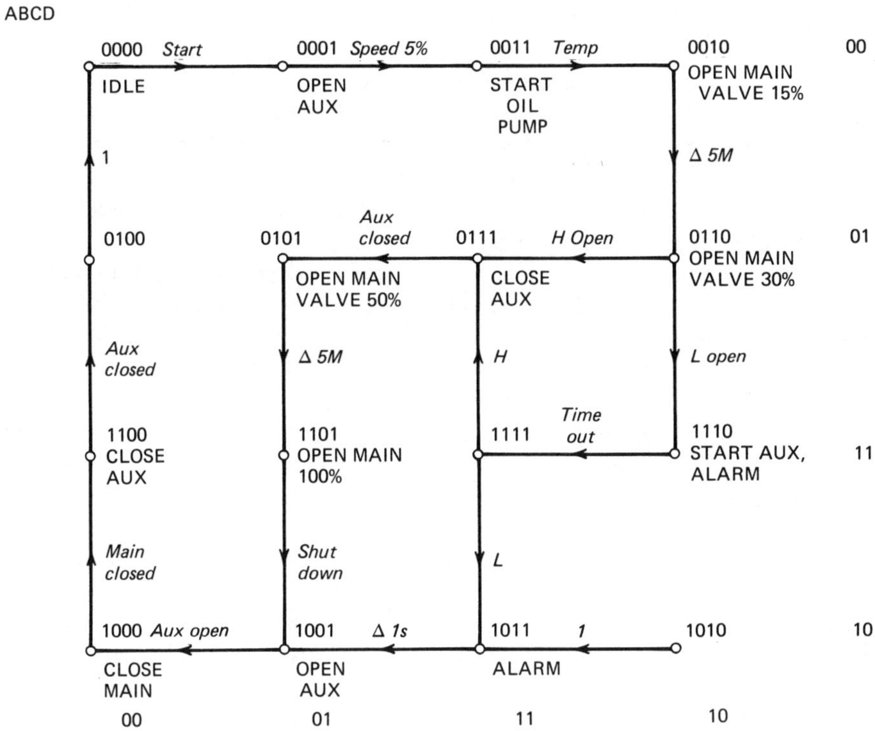

Figure 4.6 Turbine automatic start-up shut-down state diagram.

It is important that the temperature within the turbine be raised slowly so that warping of blades does not occur. When the proper temperature is reached, a signal *temp* is transmitted from the turbine to the start-up circuit, causing a transition from state ABCD = 0011 to 0010. With the temperature criteria satisfied, the main valve can be opened to 15% (OPEN MAIN VALVE 15%) to increase turbine speed. The output signal also starts a timer, and after 5-minutes lapse time the timer transmits a signal $\Delta\,5M$ to the automatic start-up shut-down circuit. This input causes a state change from ABCD = 0010 to 0110. An output signal OPEN MAIN VALVE 30% is generated.

Opening of the main valve must be controlled in distinct steps because acceleration through certain speed ranges must be done rapidly. This is because the turbine has certain resonant speeds that cannot be endured for long periods of time. When the main valve is opened to 30% (OPEN MAIN VALVE 30%) a signal *open* is transmitted from the turbine to the circuit. The open signal is *and*ed with a signal from an oil pressure monitor on the turbine bearings. If the oil pressure is high H the normal start-up procedure can continue. However, if the pressure is low L at this stage, there is a transition from ABCD = 0110 to 1110. In this state an auxiliary oil pump is started, and an early alarm START AUX, ALARM is sent to the operator at a main console.

After an appropriate delay *time out*, the state changes from ABCD = 1110 to 1111. Here again the oil pressure signal is monitored. If it is now high *H*, there is a transition to the normal starting sequence. However, if it is still low *L*, there is a state transition from ABCD = 1111 to 1011. In this case an alarm is sent to the main console operator (ALARM). State 1010 is included in this design to guarantee that the circuit does not inadvertantly settle in this state. A transition is made from state ABCD = 1011 to 1001 after a 1-second delay Δ *1s* to place the circuit at the first stage of the shut-down sequence.

Normal operation is continued with a transition from ABCD = 0110 or 1111 to 0111. An output signal is sent to the turbine control to close the auxiliary (CLOSE AUX). When the auxiliary is closed, its limit switch transmits a signal *aux closed*, which causes a transition in the state diagram from ABCD = 0111 to 0101. In this state, an output signal OPEN MAIN VALVE 50% causes the main valve to be opened to the 50% full-speed position and also initializes a time-out circuit. After a 5-minute time delay, the time-out circuit sends a signal Δ *5M* back to the circuit, and a transition from ABCD = 0101 to 1101 occurs. In this state the main valve is opened to 100%, and a signal OPEN MAIN 100% indicating that the start-up procedure is over is transmitted to the control room or the logic circuit for loading the turbine.

When a *shut-down* command is issued, the circuit makes a transition from state ABCD = 1101 to 1001. In this state an output command OPEN AUX is issued to open the auxiliary valve. If this valve is not closed, the command does not generate an action because repeated set commands do not change the state of a flip-flop. However, if the valve was closed in start-up sequence, it is opened. When the auxiliary valve is fully opened, it generates a command *aux opened* from its limit switch, causing a transition from state ABCD = 1001 to 1000. In this state a signal is sent to the turbine to close the main valve (CLOSE MAIN). When this task is complete, a limit switch on the main valve transmits a signal *main closed* to the automatic start-up shut-down circuit and causes a transition in states from ABCD = 1000 to 1100. This state causes a signal to be issued to close the auxiliary valve (CLOSE AUX). When it is closed, the limit switch on the valve generates a signal *aux closed* to the automatic circuit and causes a transition in states from ABCD = 1100 to 0100. The circuit is now returned to the normal state with an automatic transition from ABCD = 0100 to 0000. However, this automatic transition may be replaced with a time-out circuit, so that the main valve can be closed in stages. This slow down gives the steam boilers and other energy components more time to adjust to changing temperatures and pressures.

4.6 STATE VARIABLE EQUATIONS FOR THE START-UP SHUT-DOWN CIRCUIT

This circuit can be designed using the techniques described in this chapter. The first step is to redraw the state diagram for each variable with a loop circling the on conditions for the state variable. Figure 4.7 shows the diagram redrawn four times. In part (*a*) states where the value of the variable A is equal to 1 are looped. Similarly for part (*b*), the states where the variable B equals 1 are enclosed in a loop. Part (*c*) gives the same construction with the loop enclosing the states where variable C equals 1, and (*d*) shows the loop that encloses the states where variable D equals 1.

From the intersections of loops and input lines, equations for the excitation or set and reset of variables can be written. Inspection of Figure 4.7(*a*) shows that there are two transitions into the loop for the variable A. A is set in a transition from ABCD = 0101 to 1101. Note that in this transition only variable A is changing. Similarly, in the transition from ABCD = 0110 to 1110, A is set, and it is the only variable subject to change. Therefore, including the nonchanging state variables with the input variable change, the equation for the set excitation of A can be written:

$$S_A = \Delta\, 5M\; B\bar{C}D + L\ open\; BC\bar{D}$$

Similarly, inspection of Figure 4.7(*a*) shows that there are two transitions out of the loop. These are the two reset inputs. They are from states ABCD = 1100 to 0100 and states ABCD = 1111 to 0111. A is reset, and all other values of the state variables remain constant. Thus the reset function for state variable A can be written as

$$R_A = aux\ closed\; B\bar{C}\bar{D} + H\; BCD$$

In Figure 4.7(*b*), there are two transitions into the loop for B and three transitions out of the loop. From inspection of the figure, the conditions for setting B are

$$S_B = \Delta\, 5M\; \bar{A}C\bar{D} + main\ closed\; A\bar{C}\bar{D}$$

The condition for resetting B or transitions out of the loop can also be seen from the diagram. The expression for reset is

$$R_B = L\; ACD + shut\ down\; A\bar{C}D + \bar{A}\bar{C}\bar{D}$$

Note the automatic transition from ABCD = 0100 to 0000.

Inspection of Figure 4.7(*c*) gives the action status for the C variable. There is one transition into the C loop; therefore there is one term in the set C expression. It is

$$S_C = speed\ 5\%\; \bar{A}\bar{B}D$$

The reset excitation from this figure is given by two transitions out of the loop. Thus the expression for resetting of the C variable is

$$R_C = aux\ closed\; \bar{A}BD + \Delta 1s\; A\bar{B}D$$

The conditions for action on the inputs to the flip-flop for the D variable can be read from Figure 4.7(*d*). From this figure, there are four transitions into the loop. The expression for the set excitation is

$$S_D = start\; \bar{A}\bar{B}\bar{C} + H\ open\; \bar{A}BC + time\ out\; ABC + A\bar{B}C$$

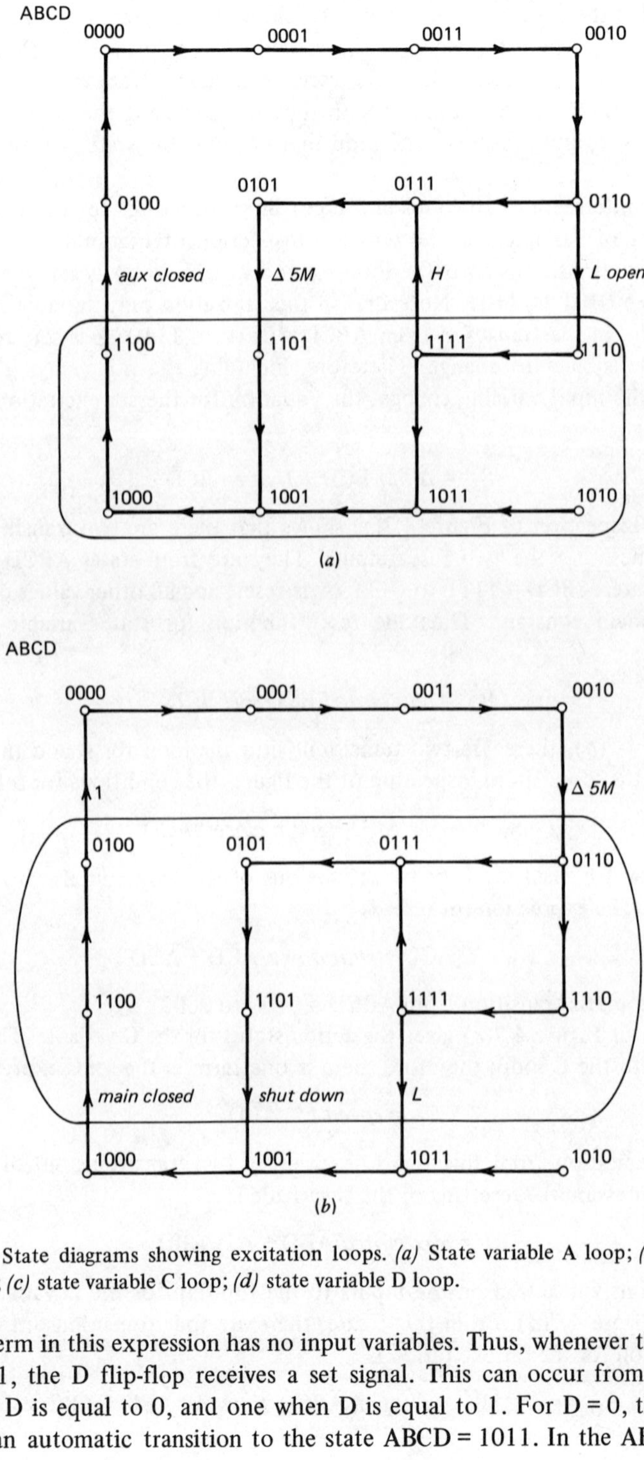

Figure 4.7 State diagrams showing excitation loops. *(a)* State variable A loop; *(b)* state variable B loop; *(c)* state variable C loop; *(d)* state variable D loop.

The last term in this expression has no input variables. Thus, whenever the values of ABC = 101, the D flip-flop receives a set signal. This can occur from two states; one when D is equal to 0, and one when D is equal to 1. For D = 0, there will always be an automatic transition to the state ABCD = 1011. In the ABCD = 1011

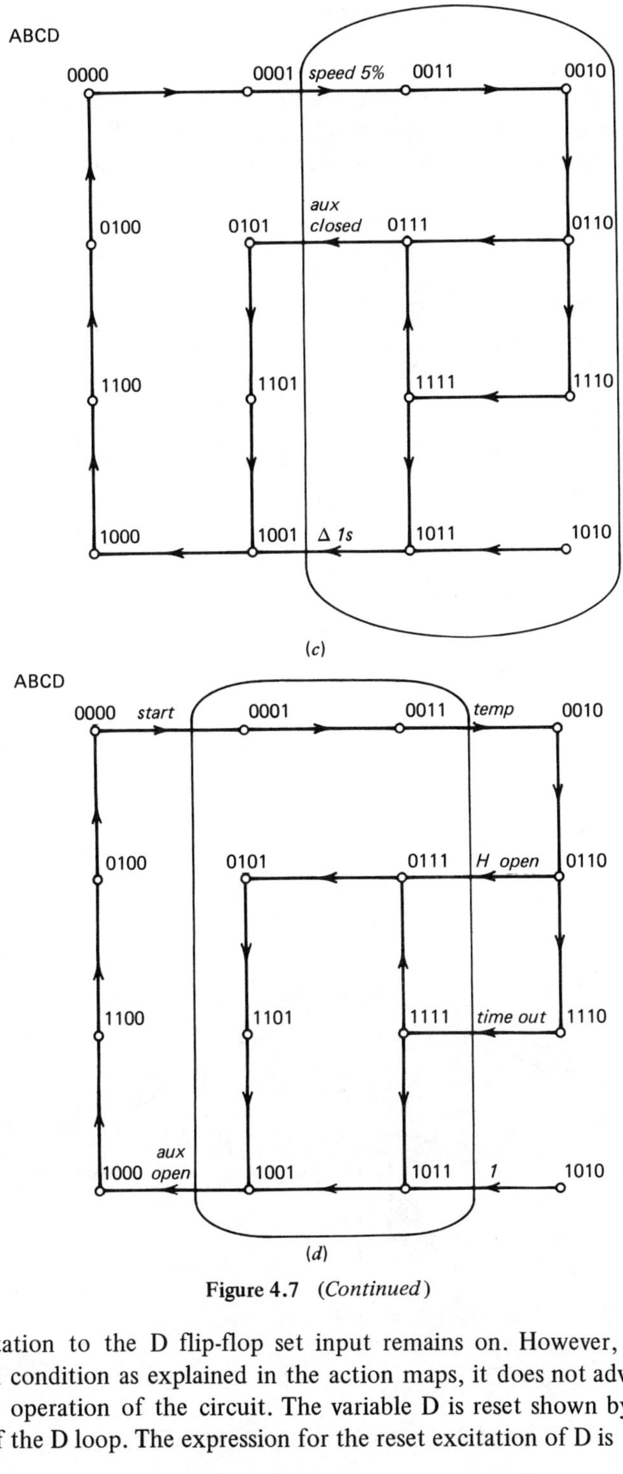

Figure 4.7 *(Continued)*

state, excitation to the D flip-flop set input remains on. However, since this is an optional condition as explained in the action maps, it does not adversely affect the desired operation of the circuit. The variable D is reset shown by two transitions out of the D loop. The expression for the reset excitation of D is

$$R_D = temp\ \overline{A}\overline{B}C + aux\ open\ A\overline{B}\overline{C}$$

The symbolic diagram for the realization of this circuit is given in Figure 4.8. *Again, some liberty is taken with the description of the gates, indicating the possible gate saving in actual implementation.* Connections of RSFF outputs to *nand* gate inputs are implied but not included to simplify the diagram.

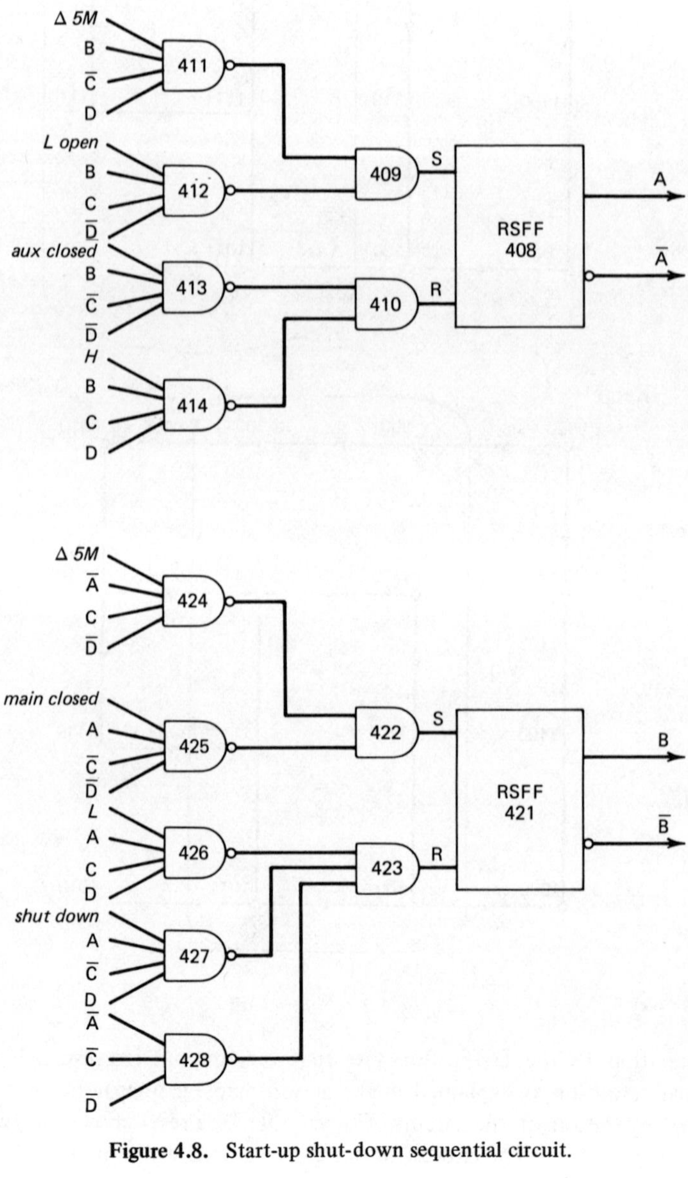

Figure 4.8. Start-up shut-down sequential circuit.

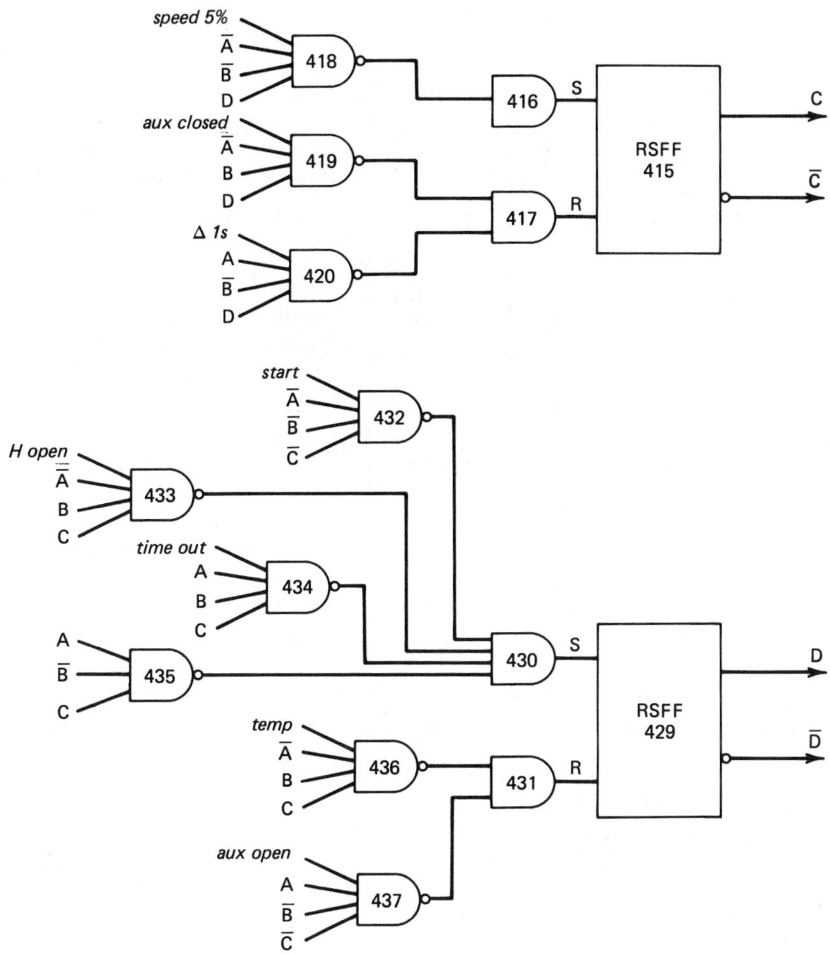

Figure 4.8 *(Continued)*

4.7 THE OUTPUT FUNCTION

In Figure 4.6 all outputs were associated with state variables only, so the output functions can be written very simply. Progressing through all the states the output variable expressions are the following:

OPEN AUX = $\bar{B}\bar{C}D$
START OIL PUMP = $\bar{A}\bar{B}CD$
OPEN MAIN VALVE 15% = $\bar{A}\bar{B}C\bar{D}$
OPEN MAIN VALVE 30% = $\bar{A}BC\bar{D}$
START AUX OIL PUMP = $ABC\bar{D}$

CD \ AB	00	01	11	10
00	IDLE		CLOSE AUX	CLOSE MAIN
01	OPEN AUX	OPEN MAIN VALVE 50%	OPEN MAIN VALVE 100%	OPEN AUX
11	START OIL PUMP	CLOSE AUX		ALARM
10	OPEN MAIN VALVE 15%	OPEN MAIN VALVE 30%	AUX ALARM	

Figure 4.9 Karnaugh representation of output from the start-up shut-down circuit.

$$\text{CLOSE AUX} = \bar{A}BCD + ABC\bar{D}\bar{D}$$

$$\text{ALARM} = A\bar{B}CD + ABC\bar{D}$$

$$\text{OPEN MAIN VALVE 50\%} = \bar{A}B\bar{C}D$$

$$\text{OPEN MAIN VALVE 100\%} = AB\bar{C}D$$

$$\text{CLOSE MAIN} = A\bar{B}\bar{C}\bar{D}$$

All these functions may be realized from a *nand* combination. However, it seems that a 4 to 16 line demultiplexer would be a great aid in realizing this circuit. The select lines would be ABCD; the input would be an enable, normally wired to 0. The above representation is given in Karnaugh map form as shown in Figure 4.9. A *demux* implementation is shown in Section 5.5.

Medium-scale integrated circuits can be used to realize much of this circuit, and techniques to accomplish this are discussed in the next chapter.

4.8 PROBLEMS

P4.1 Figure P4.1 shows a state diagram for a JK flip-flop described in section 6.2.

(a) Construct this flip-flop using RSFFs.

(b) Draw the implementation diagram.

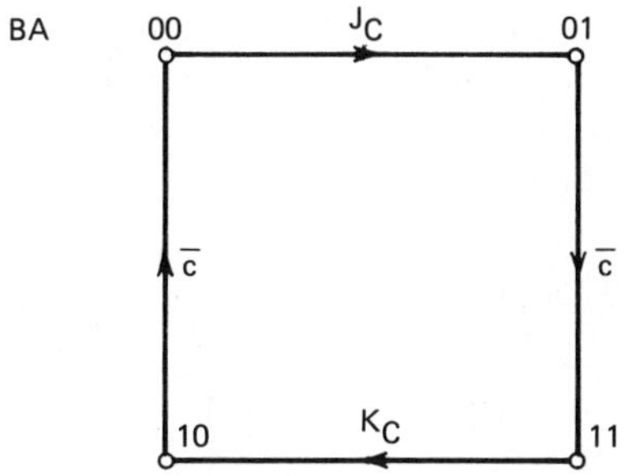

Figure P4.1 JKFF circuit.

P4.2 Design ye olde one shot using RSFFs. This circuit is used for such applications as single stepping a computer. Figure P4.2 shows timing and state diagrams. Each time the push button is depressed one (and only one) clock pulse is transmitted. Realize this circuit.

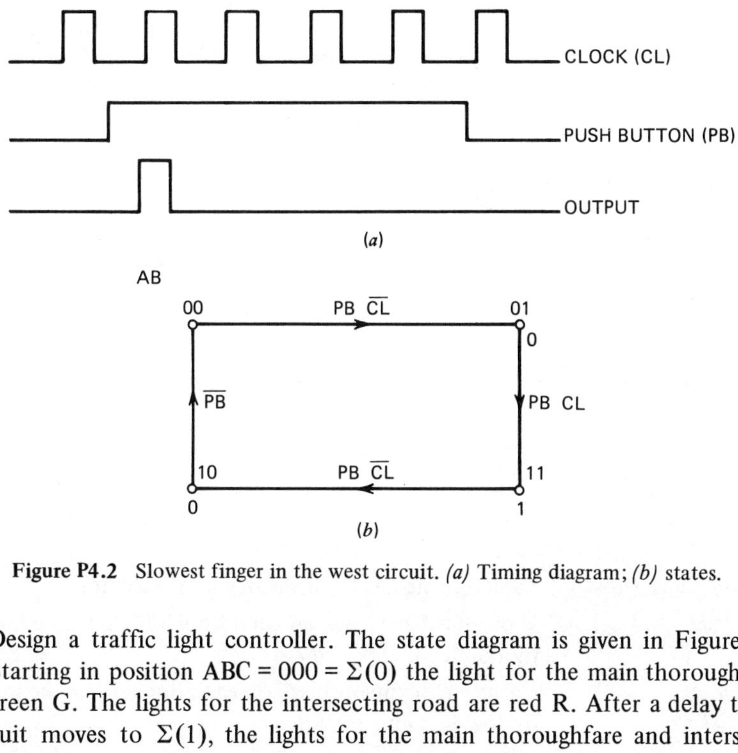

Figure P4.2 Slowest finger in the west circuit. *(a)* Timing diagram; *(b)* states.

P4.3 Design a traffic light controller. The state diagram is given in Figure P4.3. Starting in position ABC = 000 = $\Sigma(0)$ the light for the main thoroughfare is green G. The lights for the intersecting road are red R. After a delay the circuit moves to $\Sigma(1)$, the lights for the main thoroughfare and intersecting road change to amber A and red respectively. The conditions for output are as follows:

$\Sigma(3)$ red and red
$\Sigma(7)$ red and green
$\Sigma(5)$ red and amber
$\Sigma(4)$ red and red

for the main thoroughfare and intersecting road respectively. If the circuit should ever inadvertently reach Σ(2) or Σ(6) (say on start-up) an automatic transition and an $\bar{x}$ force it to Σ(7). There is no required output from these two states. Use RSFFs.

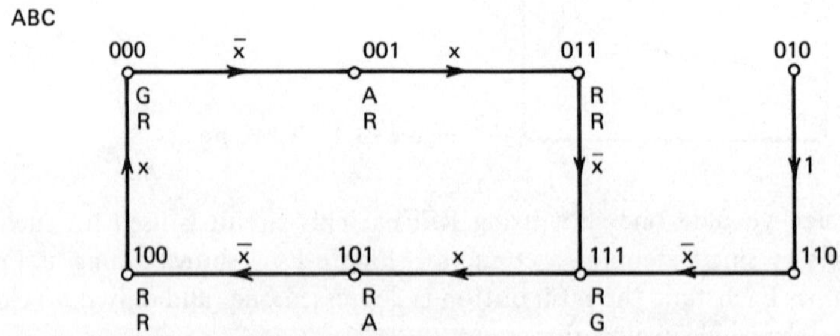

Figure P4.3 Light controller.

P4.4 The circuit shown in Figure P4.4 is used to divide the clock frequency. It divides by 2. A six-state circuit would divide by 3, eight-state by 4, ten-state by 5, and so on. Also such circuits can be used in cascade. Design this circuit.

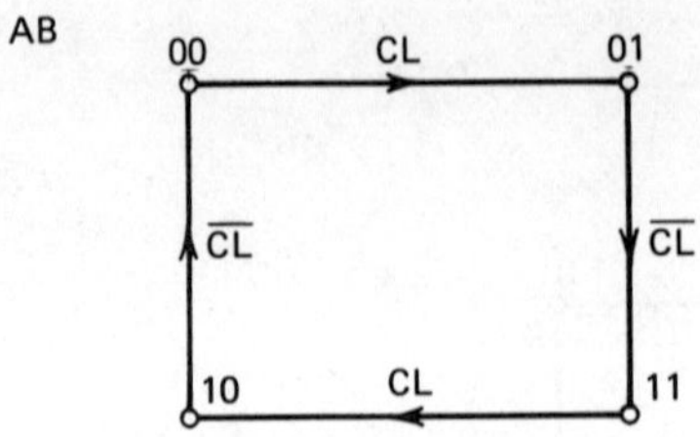

Figure P4.4 Frequency divider.

P4.5 The circuit in Figure P4.5 could be used for control of an elevator. Input signals are the following:

Call decode	Received from a decoding circuit that logs all current calls.
Up and *Down*	Direction signals from the decoding circuit.
LD and *LU*	From limit switches on each floor which can be inhabited by the decoding circuit.
Level	From an alignment plate on each floor. It can be inhibited by the decoding circuit.
Stopped time out	Intended to hold the door open.

Output signals are the following:

CLOSE DOOR	
FORWARD and REVERSE	To control the elevator motors.
SLOW	To elevator motor.

BRAKE	Elevator brakes.
OPEN DOOR	

Design and implement this circuit.

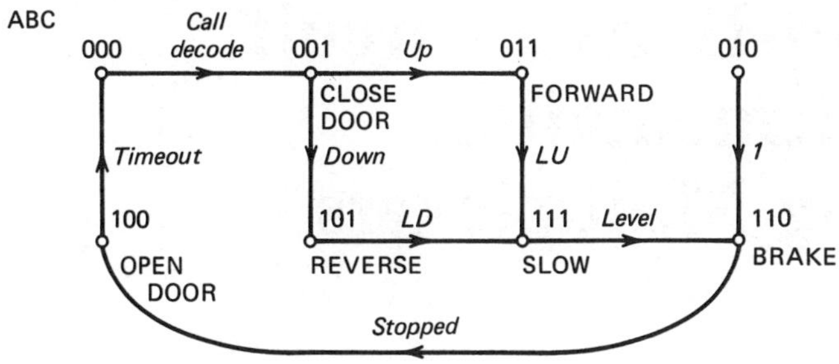

Figure P4.5 Elevator control.

P4.6 The circuit of Figure P4.6 is for the control of older type coffee machines. Inputs signals are the following:

Coin	Correct change inserted
PBC	Cream request push button
PBS	Sugar request push button
Timeout and ΔS	are delays

Output signals are the following:

CUP	Drop cup
COFFEE	Insert coffee powder
CREAM	Insert powdered milk
SUGAR	Drop sugar
HOT WATER	Pour water

Design and implement the circuit to realize the coffee machine control.

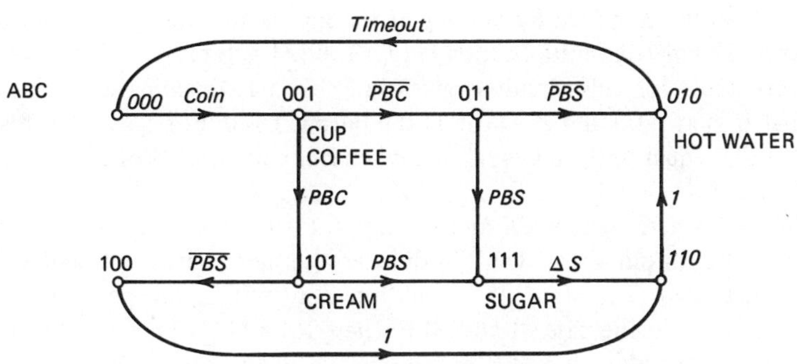

Figure P4.6 Coffee machine.

CHAPTER 5

Event-Driven Circuit Implementation Using Multiplexers

In Chapter 4 a basic method for the implementation of event-driven sequential circuits using RSFFs is given. The basic building component, an RSFF, responds to the change of an input variable and uses feedback to hold the variable state to retain the last input. In Chapter 3 the basic concepts illustrate that the ability to handle feedback is fundamental to sequential design.

In Chapter 2 methods of using multiplexer or medium scale integration devices (MSI), which lead to much more compact design, are developed. Feedback is easily implemented in a multiplexer by connecting the output signal to a select line. In this chapter a method of implementation of event-driven sequential circuits using multiplexers is given. Since MSI components give more compact circuit realization than standard IC chips, their use almost always results in a considerable saving in components and required interconnection.

5.1 MULTIPLEXERS AS A MEMORY DEVICE

A two to one multiplexer is shown in Figure 5.1. It is connected so that feedback function is provided by connecting the output to the select line. The operation of this circuit can be examined by looking at the matrix multiplication form for the multiplexer. The matrix multiplication is $[a_0 \;\; a_1]\;[Z] = [f]$.

The truth table for this circuit is given in Table 5.1. Examination of this table shows that if $a_0 a_1 = 00$ and $Z = 0$ or 1, the output f is 0. For $a_0 a_1 = 01$, f is equal to 0 when Z is equal to 0, and equal to 1 when Z is equal to 1. For $a_0 a_1 = 11$, $f = 1$ for all values of Z.

For the case where $a_0 a_1 = 10$, and for $Z = 0$, $f = 1$. This causes $Z = 1$ which, in turn causes the output $f = 0$. The multiplexer alternates between 0 and 1, so the output is unknown. Therefore, the condition $a_0 a_1 = 10$ is not permitted. This is exactly analogous to the case of an RSFF where RS = [11]. Thus the 2 to 1 multiplexer circuit resembles an RSFF except that a_1 (reset input) is inverted. If an inverting gate is placed on the input to a_1, this circuit has the same truth table as an RSFF.

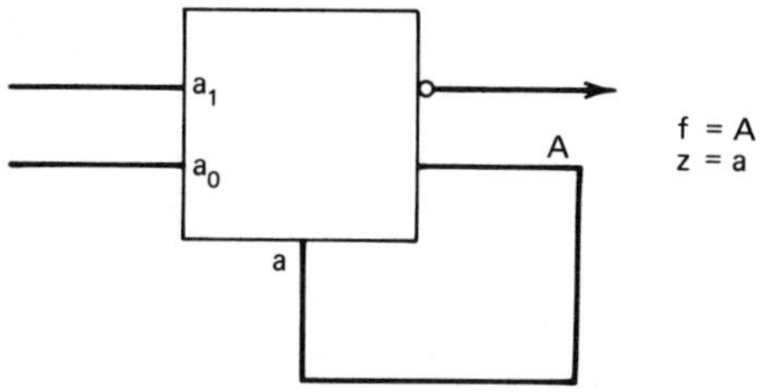

Figure 5.1 A two to one multiplexer as a memory element.

Table 5.1 Truth Table for a Two to One Multiplexer with Feedback

a_0	a_1	Z	f	
0	0	0	0	
0	0	1	0	
0	1	0	0	
0	1	1	1	
1	1	0	1	
1	1	1	1	
1	0	0	1	UNDEFINED
1	0	1	0	

5.2 SEQUENTIAL CIRCUIT MULTIPLEXER NOTATION AND CONVENTIONS

To this point in the book, it has been the convention to designate outputs by f and select lines by Z. However, it is the custom in sequential circuits to designate the state variables as uppercase letters. When using multiplexers in event-driven circuit design, the outputs are designated by uppercase letters and the select lines are designated by lowercase letters. Therefore, when feedback is utilized the same variable is used on the output (uppercase) and the select line (lowercase). This agrees with the convention adopted in Chapter 3 resulting from the definition of stability. In that convention, uppercase describes excitation; and lowercase describes variables.

In the foregoing section, a circuit to illustrate how a two to one multiplexer could be used as an RSFF is given. It is fairly obvious that a number of flip-flops can be constructed using this technique. However, this would give a very limited expansion. A more subtle and stronger technique is presented here. Using the notation given above, a matrix multiplication for two variables is given by the equation.

$$\begin{bmatrix} A \end{bmatrix} \begin{bmatrix} x \\ y \end{bmatrix} = \begin{bmatrix} X \\ Y \end{bmatrix}$$

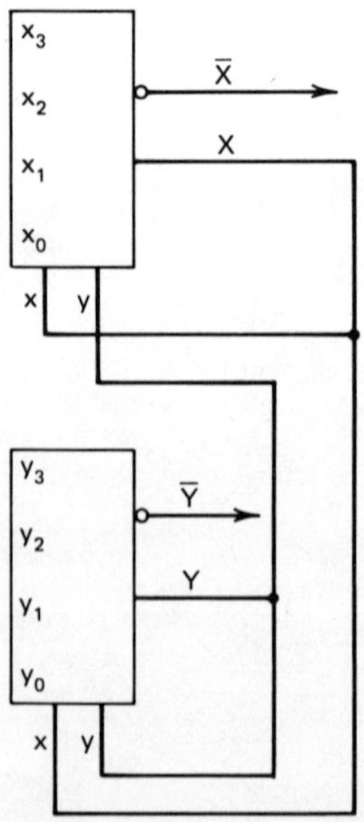

Figure 5.2 Two-variable multiplexer feedback system.

This can be expanded to the three-variable case as

$$[A]\begin{bmatrix} x \\ y \\ z \end{bmatrix} = \begin{bmatrix} X \\ Y \\ Z \end{bmatrix}$$

Two-variable and three-variable multiplexer feedback systems using this designation are shown in Figures 5.2 and 5.3 respectively. *All* multiplexers within one feedback system must be of the same type to equalize delays.

5.3 STATE DIAGRAMS IN MULTIPLEXER NOTATION

As with other techniques, the starting point for multiplexer implementation is the state diagram. The state diagram for the multiplexer uses exactly the same convention as outlined in Section 4.1.

A form of the state diagram that is completely generalized is given in Figure 5.4. In this diagram, the input function that causes a transition from XYZ = 000 to 001 is called a_0, where a_0 represents a generalized input function that causes the transition. Similarly, the input that causes the transition from XYZ = 001 to 011 is a_1, where a_1 represents the generalized input function that causes this transition. And similarly a_3, a_2, a_6, a_7, a_5, and a_4 cause the value of a state variable to change. These inputs give a general notation to represent a specific input.

Figure 5.4 can be represented by a matrix multiplication equation with three

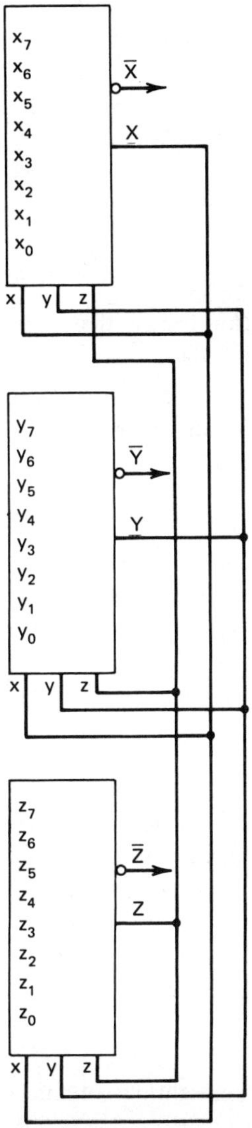

Figure 5.3 Three-variable multiplexer feedback system.

select variables and three outputs. The equation is of the form

$$\begin{bmatrix} A \end{bmatrix} \begin{bmatrix} x \\ y \\ z \end{bmatrix} = \begin{bmatrix} X \\ Y \\ Z \end{bmatrix}$$

The problem is to determine the matrix [A]. Each column in the matrix [A] can be completed by the following rules:

1. For each column, find the equivalent state in the state diagram.

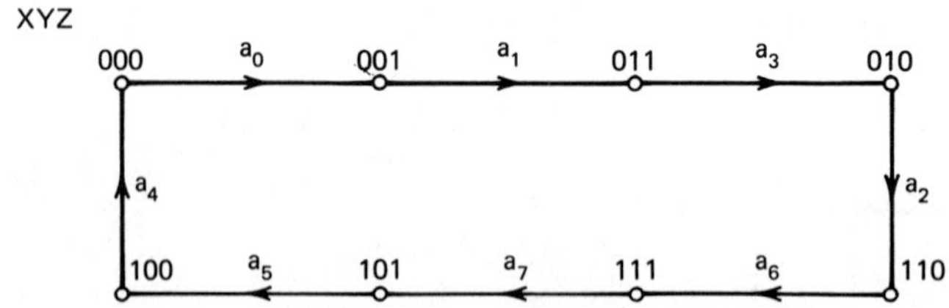

Figure 5.4 State diagram with generalized inputs.

2. Determine which variables are not changing in the transition leaving that state. Enter each of the values of these variables in the appropriate row.
3. The variable that is being changed in the transition leaving that state is replaced with the value of the input that causes the change if the change is from 0 to 1, and its complement if the change is from 1 to 0.

Column 0 is located in the state diagram where the variables xyz = 000. The transition out of this state is from xyz = 000 to 001 in Figure 5.4. Thus, x and y are unchanged and equal to 0. Enter the value 0 in rows 1 and 2 of column 0 in the [A] matrix. The value of z is changing from 0 to 1. Thus the value of the z variable must be replaced by variable a_0. The equation is complete to the following stage:

$$\begin{bmatrix} 0 & & \\ 0 & & \\ a_0 & & \end{bmatrix} \begin{bmatrix} x \\ y \\ z \end{bmatrix} = \begin{bmatrix} X \\ Y \\ Z \end{bmatrix}$$

The transition from xyz = 001 to 011 is the transition for column 1. The variable x is unchanged at 0, and the variable z is unchanged at 1. However, the value of y is changing from 0 to 1 so the value of y is replaced by the variable a_1. The next column can be added to the matrix as follows:

$$\begin{bmatrix} 0 & 0 & & \\ 0 & a_1 & & \\ a_0 & 1 & & \end{bmatrix} \begin{bmatrix} x \\ y \\ z \end{bmatrix} = \begin{bmatrix} X \\ Y \\ Z \end{bmatrix}$$

To determine the input for column 2, the state variable representing column 2 (xyz = 010) must be located in Figure 5.4. The transition is from xyz = 010 to 110 for this column. In the transition the value of y is unchanged and equal to 1; the value of z is unchanged and equal to 0; the value of x is changing from 0 to 1 and is therefore replaced by the input variable that causes the transition a_2. For column 3, that is the transition from xyz = 011 to 010, the value of x is unchanged at 0; the value of y is unchanged at 1; but the value of z is changing from 1 to 0. This is a reverse transition. So to make the operation equivalent to a reset, as explained in Section 5.1, the input must be complemented. Therefore, z is replaced by the complemented form of the input variable that causes the change $\bar{a}_3$.

For column 4, there is a change in the state variables from xyz = 100 to 000. y and z are unchanged at 0, but x is changing from 1 to 0 so the value of x is replaced by the complemented input variable that causes the change $\bar{a}_4$. For column, 5, xyz = 101, there is a transition to 100. The value of x is unchanged and equal to 1; the value of y is unchanged and equal to 0; the value of z is changing from 1 to 0; therefore z is replaced by the complemented input variable that causes the change $\bar{a}_5$. Similarly, for column 6, xyz = 110, there is a change to 111. The values of x and y are unchanged at 1; the value of z is changing from 0 to 1 and is replaced by a_6. And finally, for column 7, xyz = 111, there is a change to 101. The values of x and z are unchanged and equal to 1; the value of y is changing from 1 to 0, and $\bar{a}_7$ causes this change. The complete equation is written as

$$\begin{bmatrix} 0 & 0 & a_2 & 0 & \bar{a}_4 & 1 & 1 & 1 \\ 0 & a_1 & 1 & 1 & 0 & 0 & 1 & \bar{a}_7 \\ a_0 & 1 & 0 & \bar{a}_3 & 0 & \bar{a}_5 & a_6 & 1 \end{bmatrix} \begin{bmatrix} x \\ y \\ z \end{bmatrix} = \begin{bmatrix} X \\ Y \\ Z \end{bmatrix}$$

Note that the variable x is controlled by row 1 of matrix [A]. It is set by a_2 of column 2. The value 1 of columns 5, 6, and 7 holds x on. $\bar{a}_4$ resets x and the value 0 of columns 0, 1, and 3 holds x off.

The last step in the sequence is to replace the variables a_0, a_1, a_2, $\bar{a}_3$, $\bar{a}_4$, $\bar{a}_5$, a_6, $\bar{a}_7$ by the inputs as designated in the state to state transition. An example is worked in the next section.

5.4 REALIZATION OF AN ANNUNCIATOR CIRCUIT USING MULTIPLEXERS

The annunciator circuit given in Figure 4.1 is repeated here as Figure 5.5 for convenience. This example was realized using RSFFs in Chapter 4. Here it is to be realized using multiplexers.

The state diagram is to be changed to multiplexer notation form, and is

$$\begin{bmatrix} \lambda \end{bmatrix} \begin{bmatrix} a \\ b \end{bmatrix} = \begin{bmatrix} A \\ B \end{bmatrix}$$

In this case, the input matrix is given the notation λ to distinguish it from the selector and output lines. Figure 5.5 shows that the circuit is to make the transition such that the sequence is AB = 00, 01, 11, 10. Column 0 in the λ matrix has a state designation AB = 00. The transition is from AB = 00 to 01, since the A variable

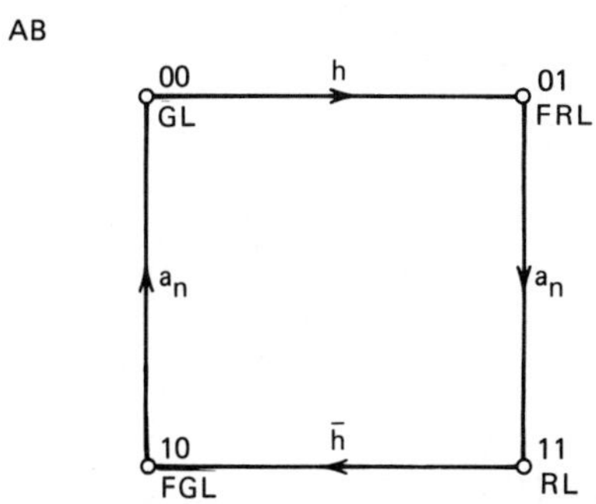

Figure 5.5 Annunciator state diagram for multiplexer design.

remains constant, and the B variable has a transition from 0 to 1. This transition can be written in matrix notation as

$$\begin{bmatrix} 0 \\ \lambda_0 \end{bmatrix}$$

For the next state, AB = 01, there is a transition from 01 to 11. Thus the A variable must change, and the B variable remains constant. So the entry for column 1 of the λ matrix is

$$\begin{bmatrix} \lambda_1 \\ 1 \end{bmatrix}$$

For column 2 the value of the state variables is AB = 10. The transition from this state is from AB = 10 to 00. Thus the A variable is changing from 1 to 0, and B is unchanged so

$$\begin{bmatrix} \bar{\lambda}_2 \\ 0 \end{bmatrix}$$

where the negation sign represents the fact that the A transition is down.

The final column in the input matrix is column 3 where the associated state is AB = 11. The transition here is from AB = 11 to 10. The variable A is unchanged, and the variable B is changing from 1 to 0, another down transition. Thus the input is

$$\begin{bmatrix} 1 \\ \bar{\lambda}_3 \end{bmatrix}$$

The completed matrix multiplication equation for this circuit can be simply written as

$$\begin{bmatrix} 0 & \lambda_1 & \bar{\lambda}_2 & 1 \\ \lambda_0 & 1 & 0 & \bar{\lambda}_3 \end{bmatrix} \begin{bmatrix} a \\ b \end{bmatrix} = \begin{bmatrix} A \\ B \end{bmatrix}$$

This equation can be completed by substituting the inputs that cause the transition into the general representation as follows:

$$\lambda_0 = h$$

$$\lambda_1 = a_n$$

$$\lambda_2 = a_n$$

$$\lambda_3 = \bar{h}$$

Therefore

$$\bar{\lambda}_2 = \bar{a}_n$$

$$\bar{\lambda}_3 = h$$

The equation is rewritten in specific form as

$$\begin{bmatrix} 0 & a_n & \bar{a}_n & 1 \\ h & 1 & 0 & h \end{bmatrix} \begin{bmatrix} a \\ b \end{bmatrix} = \begin{bmatrix} A \\ B \end{bmatrix}$$

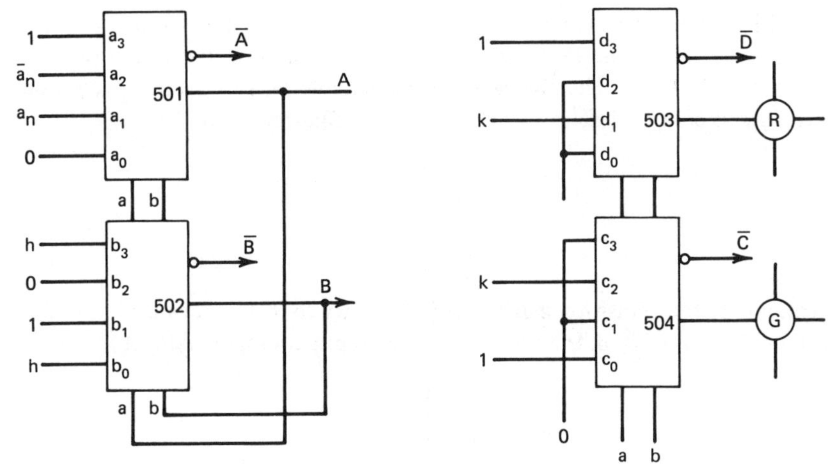

Figure 5.6 Multiplexer realization of annunciator.

The annunciator is realized using the multiplexer circuit shown in Figure 5.6. The outputs AB are fed back to the select lines ab. The inputs for the A multiplexer are read from the first row of the input matrix. Thus

$$a_0 = 0$$

$$a_1 = a_n$$

$$a_2 = \bar{a}_n$$

$$a_3 = 1$$

for the A multiplexer. Similarly,

$$b_0 = h$$

$$b_1 = 1$$

$$b_2 = 0$$

$$b_3 = h$$

for the B multiplexer. This annunciator circuit is built using a single chip incorporating dual four to one multiplexers. A dual chip has two select lines only. If a signal k alternating between 0 and 1 (clock) is available, the output circuit is designed using a dual four to one multiplexer, 503 and 504. The output circuit design follows the method shown in Chapter 2.

5.5 INPUT MATRIX FOR THE AUTOMATIC START-UP SHUT-DOWN CIRCUIT

The turbine automatic start-up shut-down circuit input matrix of Figure 4.6 can be easily designed using multiplexers. The first task is to determine the input matrix $[\lambda]$. Since the reader by now recognizes the pattern, some of the explanation is eliminated in this example. Starting with column 0, the transition is from ABCD = 0000 to 0001. The change is in the variable D, so the entry λ_0 occurs in the last row of column 1. Thus, column 0 of the λ matrix equals

$$\begin{bmatrix} 0 \\ 0 \\ 0 \\ \lambda_0 \end{bmatrix}$$

For column 1, the transition is from ABCD = 0001 to 0011, a change in variable C. Thus there is a λ_1 entry in the third row of the input column making it

$$\begin{bmatrix} 0 \\ 0 \\ \lambda_1 \\ 1 \end{bmatrix}$$

For column 2, the transition is from ABCD = 0010 to 0110. This is a change in variable B or the second row in the column. So the column becomes

$$\begin{bmatrix} 0 \\ \lambda_2 \\ 1 \\ 0 \end{bmatrix}$$

For column 3 the transition is from ABCD = 0011 to 0010. There is a change in D, but in the negative sense (that is, from 1 to 0). Thus the input for column 3 is

$$\begin{bmatrix} 0 \\ 0 \\ 1 \\ \bar{\lambda}_3 \end{bmatrix}$$

The input change for column 4 is 0100 to 0000. This is a down-moving transition in the second row of the input matrix. Thus the column is

$$\begin{bmatrix} 0 \\ \bar{\lambda}_4 \\ 0 \\ 0 \end{bmatrix}$$

The transition for column 5 is from 0101 to 1101. This is a positive-going transition in the first row, row A. Therefore, this column input is

$$\begin{bmatrix} \lambda_5 \\ 1 \\ 0 \\ 1 \end{bmatrix}$$

For column 6 a double set of transitions occurs. The resulting change in variables is from ABCD = 0110 to 0111 and 1110. The transition from 0110 to 1110 is a positive transition and a change in variable A. Therefore, the entry in row 1 is λ_{6A}. The transition from 0110 to 0111 is also a positive transition and is equal to λ_{6D}. The column equals

$$\begin{bmatrix} \lambda_{6A} \\ 1 \\ 1 \\ \lambda_{6D} \end{bmatrix}$$

By similar arguments, the complete matrix can be written as follows:

$$\begin{bmatrix} 0 & 0 & 0 & 0 & 0 & \lambda_5 & \lambda_{6A} & 0 & 1 & 1 & 1 & 1 & \bar{\lambda}_{12} & 1 & 1 & \bar{\lambda}_{15A} \\ 0 & 0 & \lambda_2 & 0 & \bar{\lambda}_4 & 1 & 1 & 1 & \lambda_8 & 0 & 0 & 0 & 1 & \bar{\lambda}_{13} & 1 & \bar{\lambda}_{15B} \\ 0 & \lambda_1 & 1 & 1 & 0 & 0 & 1 & \bar{\lambda}_7 & 0 & 0 & 1 & \bar{\lambda}_{11} & 0 & 0 & 1 & 1 \\ \lambda_0 & 1 & 0 & \bar{\lambda}_3 & 0 & 1 & \lambda_{6D} & 1 & 0 & \bar{\lambda}_9 & \lambda_{10} & 1 & 0 & 1 & \lambda_{14} & 1 \end{bmatrix}$$

Inspection of the diagram shows that

λ_0 = *Start*

λ_1 = *Speed 5%*

λ_2 = Δ *5M*

λ_3 = *Temp*

λ_4 = *1*

λ_5 = Δ *5M*

λ_{6A} = *L open*

λ_{6D} = *H open*

λ_7 = *Aux closed*

λ_8 = *Main closed*

λ_9 = *Aux open*

λ_{10} = *1*

λ_{11} = Δ *1s*

λ_{12} = *Aux closed*

λ_{13} = *Shut down*

λ_{14} = *Time out*

λ_{15A} = *H*

λ_{15B} = *L*

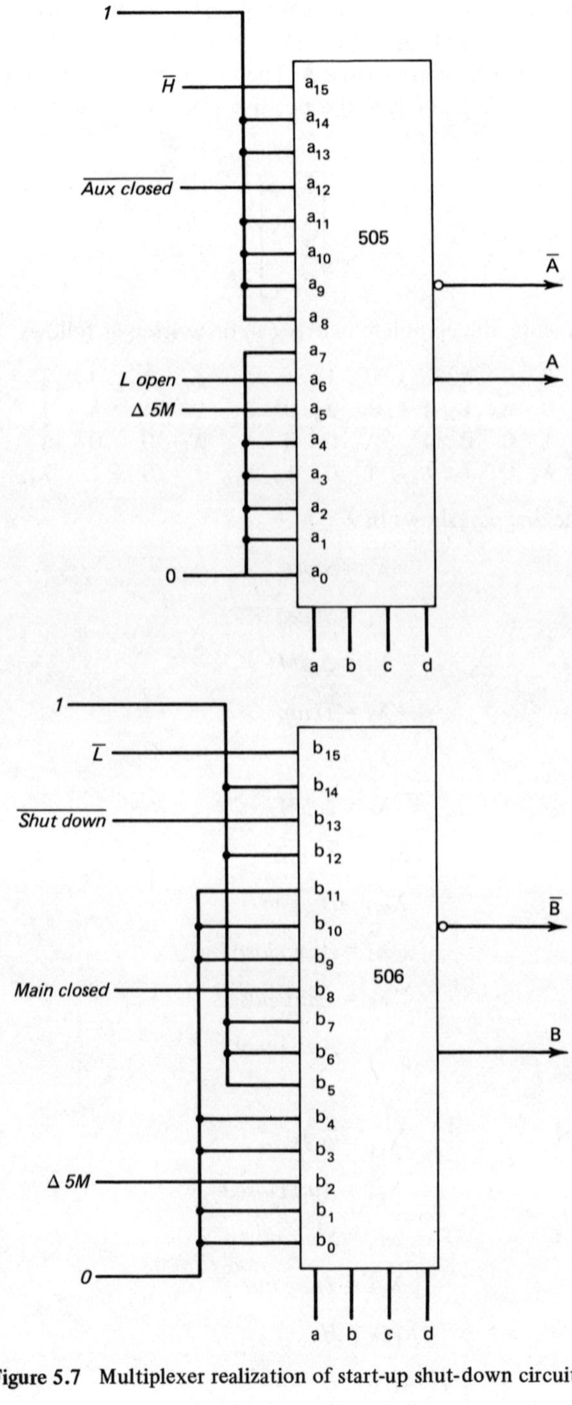

Figure 5.7 Multiplexer realization of start-up shut-down circuit.

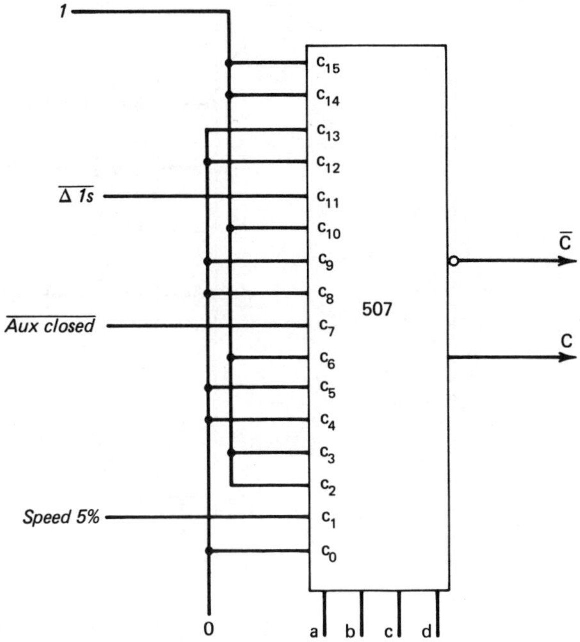

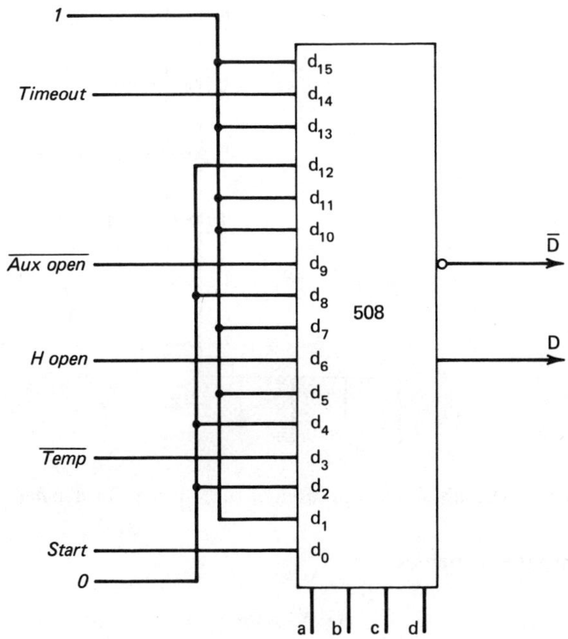

Figure 5.7 (*Continued*)

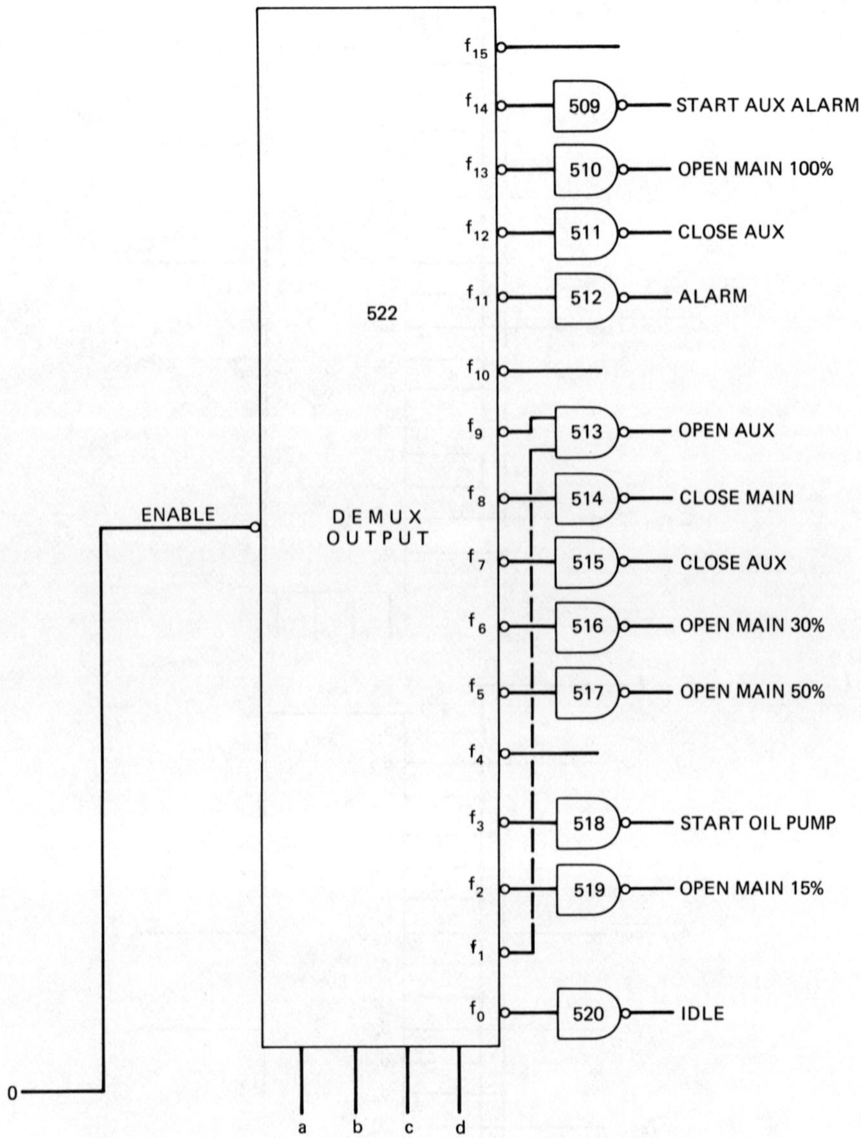

Figure 5.8 Demultiplexer realization of start-up shut-down circuit outputs.

Thus for the negated variables

$$\bar{\lambda}_3 = \overline{Temp}$$

$$\bar{\lambda}_4 = 0$$

$$\bar{\lambda}_7 = \overline{Aux\ closed}$$

$$\bar{\lambda}_9 = \overline{Aux\ open}$$

$$\bar{\lambda}_{11} = \overline{\Delta\ 1s}$$

$$\bar{\lambda}_{12} = \overline{Aux\ closed}$$

$$\bar{\lambda}_{13} = \overline{Shut\ down}$$

$$\bar{\lambda}_{15A} = \bar{H}$$

$$\bar{\lambda}_{15B} = \bar{L}$$

Figure 5.7 shows the realization of this circuit using multiplexers. For completeness, a demultiplexer designed in Figure 4.9 is realized in Figure 5.8. The feedback loops from the output of multiplexer A to the select lines of all four multiplexers and the demultiplexer are not explicitly shown in the diagram, but this connection is always implied. The circuit in Figure 5.7 can be seen to require few chips and uses relatively simple wiring to make relatively minimum demands in regard to wiring.

5.6 APPLICATION GUIDELINES

The design technique given in this chapter is new. It was developed to simplify the realization of a wide range of sequential circuits. Application of the method is straight-forward for normal event driven circuits. This is because the select-line-not gates, internal to the multiplexer structure and thus in the feedback path, provide the delay needed for stability. An increased stability margin is required when events occur at maximum frequency to overcome the detremental affect of glitches and circuit ringing. Obviously inserting additional gates in the feedback path to increase the delay and attaching resistors or capacitors to increase damping widens the stability margin. It is strongly recommended that all mux outputs be loaded with a damping capacitor (say 0.01 μf). In general clocked circuits do not need extra damping although the flexibility of design with clocked components can be increased by the simple addition of a loading capacitor on the clock line. Clocking a circuit increases the delay in the feedback path and deters ringing by waveshaping. Design techniques using clocks are given in Chapters 6 and 7.

5.7 PROBLEMS

P5.1 Design the traffic light controller of Figure P5.1 using multiplexers.

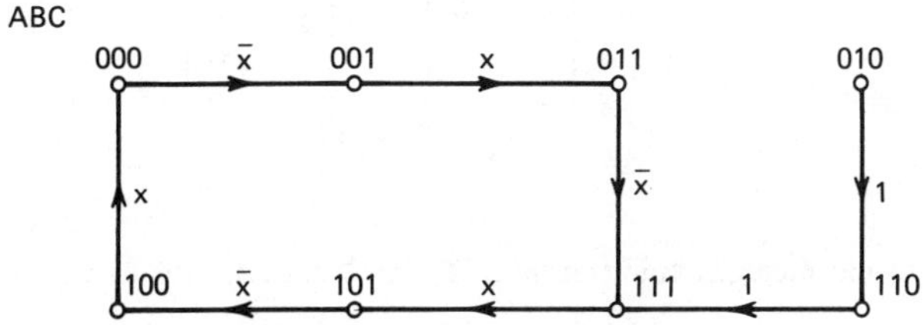

Figure P5.1 Traffic light controller.

P5.2 Seven lamps are arranged in a pattern to correspond with the spot position of a set of dice. When a key k is pressed, the circuit goes through a cycle of six states repeatedly. When the key is released the digit corresponding to the state of the circuit at the moment of release is to be displayed on the lamp. Design a multiplexer implementation of this circuit. Figure P5.2(*a*) shows the I/O signals; Figure P5.2(*b*) gives the state diagram; Figure P5.2(*c*) gives the arrangement of the spot position on the die; and Figure P5.2(*d*) gives the position of each lamp in the die.

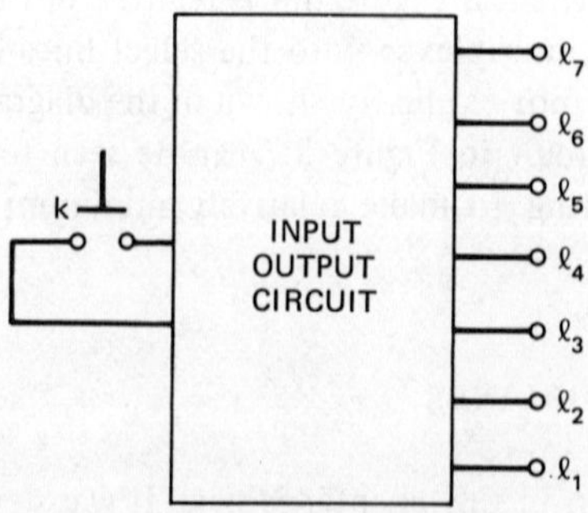

(a)

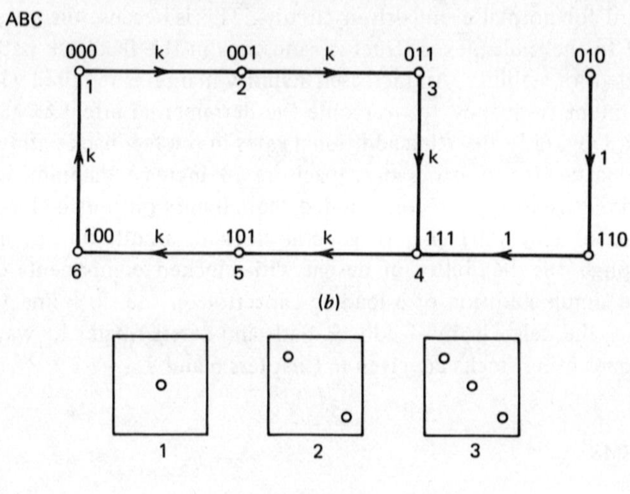

(b)

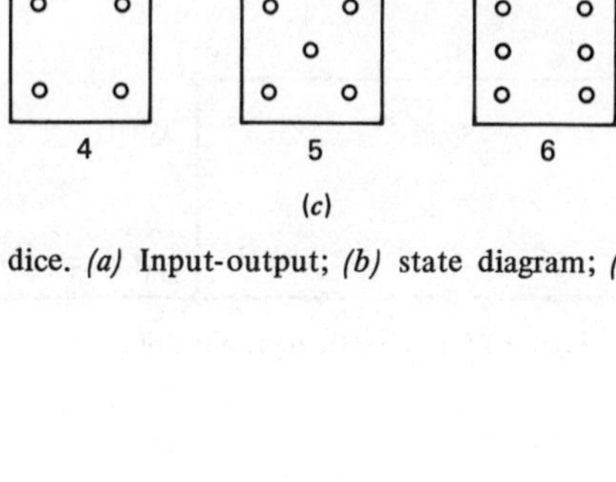

(c)

Figure P5.2 Electronic dice. *(a)* Input-output; *(b)* state diagram; *(c)* dice display; *(d)* lamp layout.

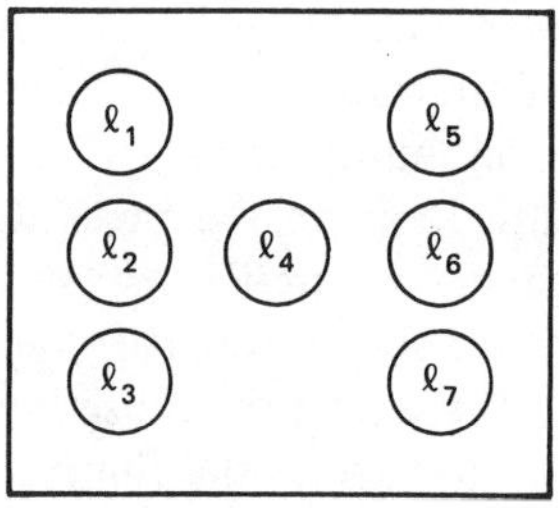

(*d*)

Figure P5.2 (*Continued*)

P5.3 Design an interface for a minicomputer. Usually such interfaces are designed to run with a synchronizing clock. However, a special problem makes it desirable to interface a slow-speed and higher-speed device together: the high speed central processing unit (CPU) is very cheap so its resources can be wasted, whereas the low speed peripheral is extremely expensive and should operate as the master.

Figure P5.3 shows an illustrative circuit. The device normally sits in the IDLE state. When the proper *control word* (*CW*) is decoded, there is a state transition and a FLAG is sent to the CPU. The CPU issues a *go* command causing a state transition. If the device is *busy* a loop is set up for continuous interrogation. When the device becomes free ($\overline{busy}$) the transmission begins on receipt of a START signal. This is outputted, and when the *transmit* signal is received data is loaded through an input (*load*) or output (REQ) cycle until the last character (*LC*) is detected and the device is returned to the IDLE state.

Design this circuit using multiplexer techniques.

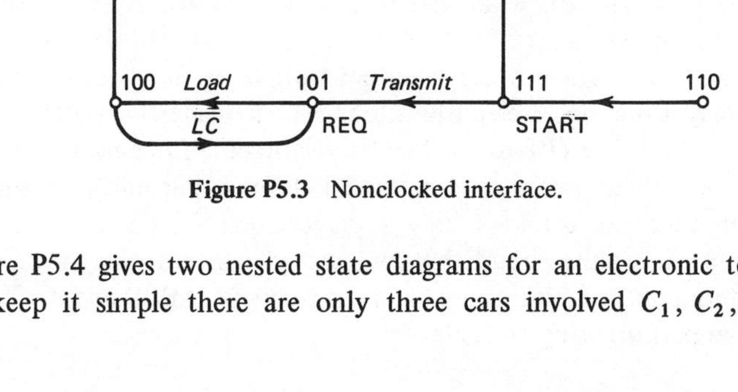

Figure P5.3 Nonclocked interface.

P5.4 Figure P5.4 gives two nested state diagrams for an electronic toy car race. To keep it simple there are only three cars involved C_1, C_2, C_3. Figure

P5.4(a) shows the status request state diagram. An "end" push button is used to obtain the results of the race. A "start" push button is used to energize the strobe (reset) lines on multiplexers ABC in the main circuit. This eliminates the need for return loops in the main state diagram. You are not required to design this circuit. Figure P5.4(b) shows the state diagram for the main car race circuit. At the start of the race the circuit is in the idle position. ABC = 000. If car C_1 is first across the finish line there is a transition to state ABC = 001. If car C_2 is first across the finish line there is a transition from ABC = 000 to 100, and so on.

Design this circuit using multiplexer techniques.

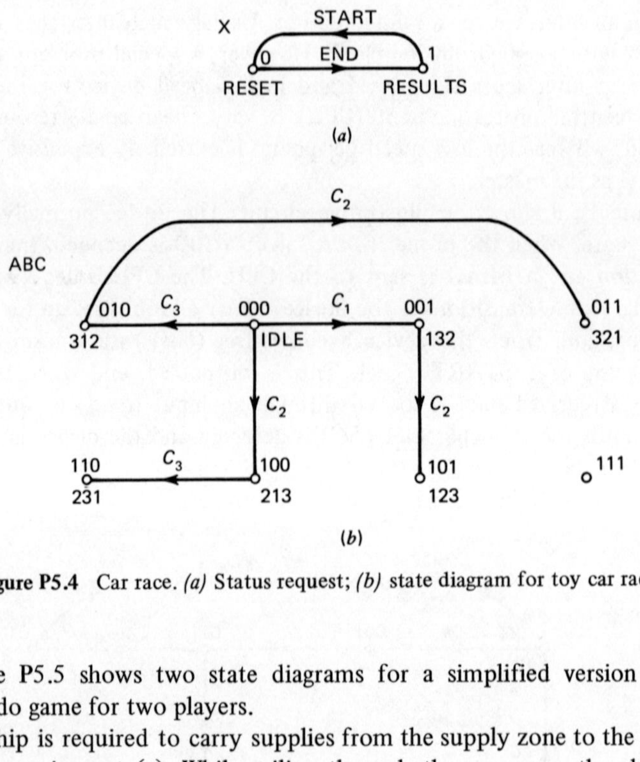

Figure P5.4 Car race. *(a)* Status request; *(b)* state diagram for toy car race.

P5.5 Figure P5.5 shows two state diagrams for a simplified version of a ship-torpedo game for two players.

A ship is required to carry supplies from the supply zone to the home zone as shown in part (a). While sailing through the war zone the ship may be sunk by a torpedo.

The ship may sail one unit step per turn along the direction of the arrow when player one operates pushbuttons u or v. Player two fires a torpedo by selecting the target (L1 to L10) and activating the fire push button. Since the war zone is three unit steps long, he has three opportunities to sink the ship. The ship's location is known only to player one.

Part (b) shows the circuit that indicates player turn. Not shown are circuits for determining if a hit is scored and the resets for ABCD and e.

Design part (a) using multiplexers.

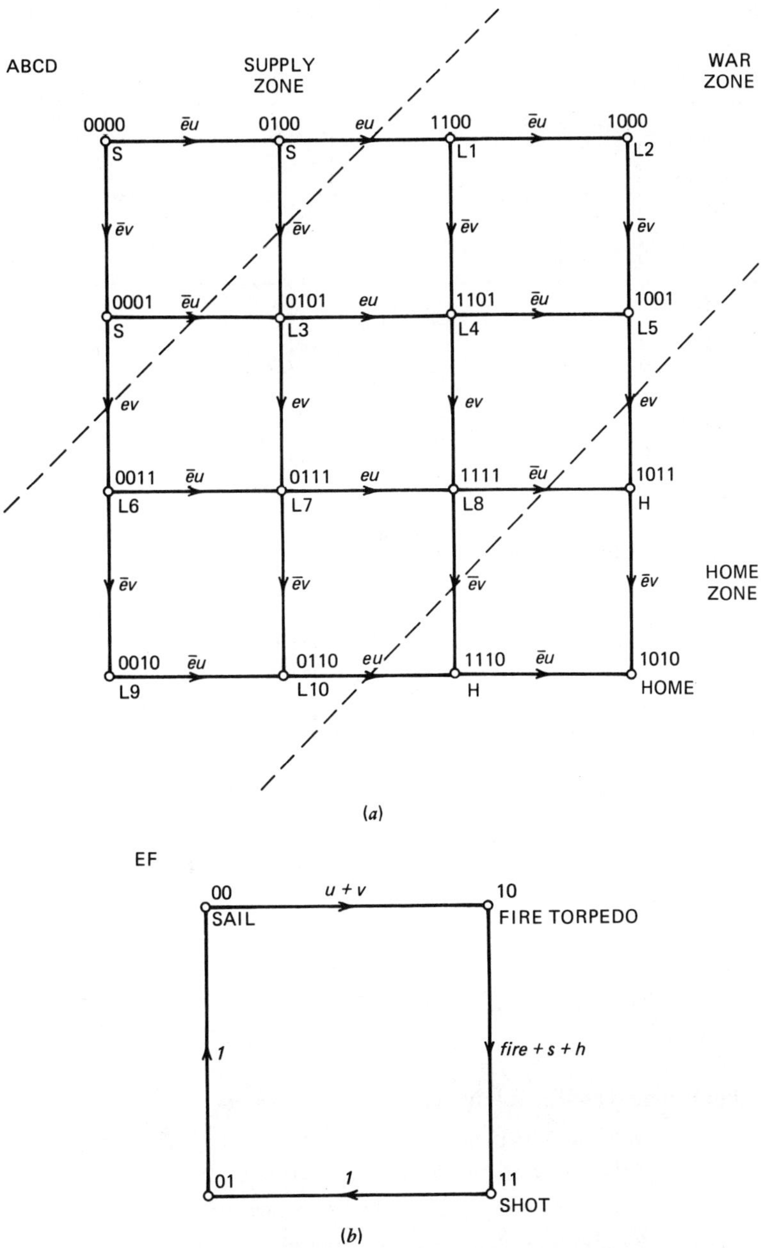

Figure P5.5 Ship-torpedo game. *(a)* Sailing routes; *(b)* fire and sail executive.

P5.6 Figure P5.6 shows the state diagram part of a data acquisition system (DAS). The DAS is complete with sample/holds, analog multiplexer, analog-to-digital converter, buffers, and instruction registers. The DAS is designed for very high-speed conversion. It also operates at device speed so a clock is not used.

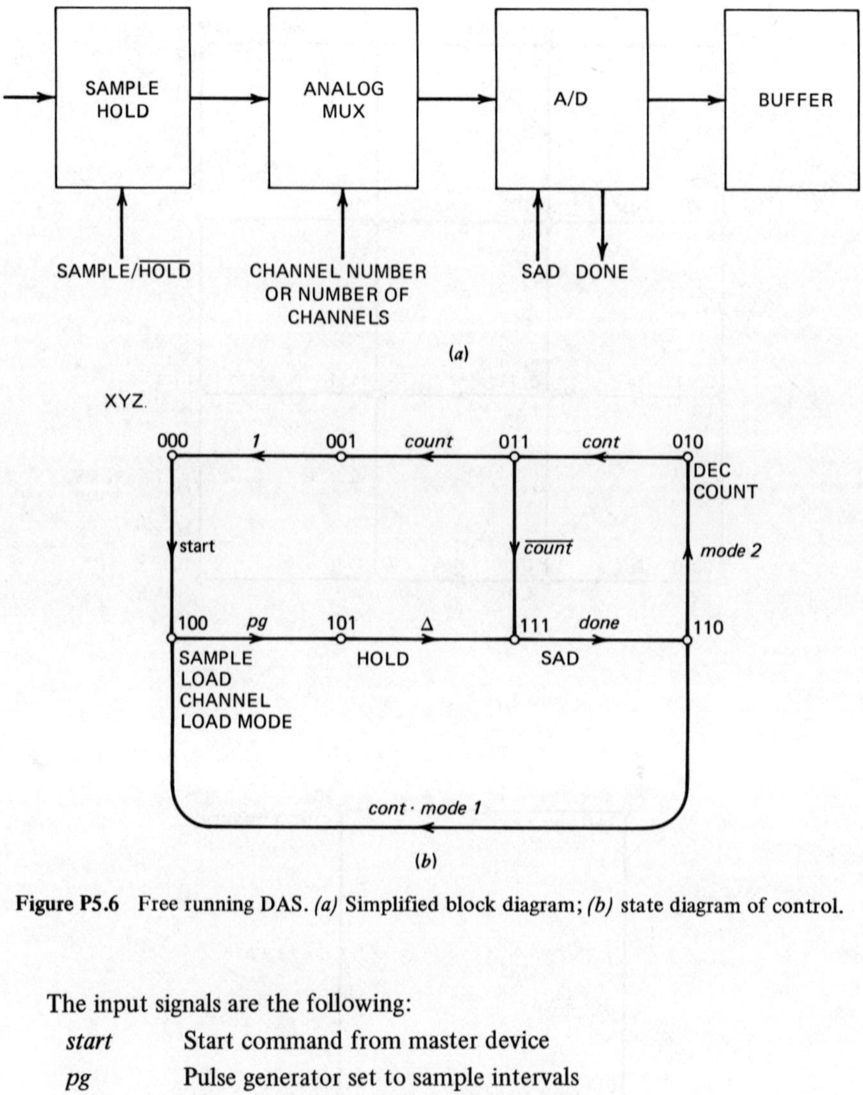

Figure P5.6 Free running DAS. *(a)* Simplified block diagram; *(b)* state diagram of control.

The input signals are the following:

start	Start command from master device
pg	Pulse generator set to sample intervals
Δ	Delay to allow sample/holds to settle
done	Analog-to-digital conversion complete
mode 1	Single channel test mode
mode 2	Multiple channel mode
cont	Continue command from master
count	Counter is a zero.

The output signals are the following:

SAMPLE	Sample/hold device to sample
LOAD CHANNEL	Channel number or number of channels to be sampled
LOAD MODE	Single or multiple channel input
HOLD	Sample/hold device to hold
SAD	Start A-to-D converter
DEC COUNT	Decrement counter containing number of channels to be sampled

Design the control circuit shown in part (*b*) using multiplexer techniques.

P5.7 Figure P5.7 shows the simplified state diagram for an interface circuit between two computers. The smaller computer (SC) is acting as a front-end device for the main-frame computer (MC). The MC uses this circuit to determine the status (busy) of one of the peripheral devices, say a printer, connected to the SC. When the MC sends the proper function number (*fn no*), the interface requests the attention (ATTEN) of the SC. The SC responds (*go*) causing the interface to send a signal (ACTIVE) to the MC which indicates that communication is established. There is an automatic transition to the next state. The ACTIVE signal thus becomes a pulse. The MC requests status with a *full* request. In state MNP = 110 the interface responds to the MC with a STATUS signal. Coincidentally a coded word is transmitted from the SC to the MC on the data bus. When the MC reads the data bus it transmits an *inactive* signal that terminates transmission.

Design this circuit using multiplexer techniques.

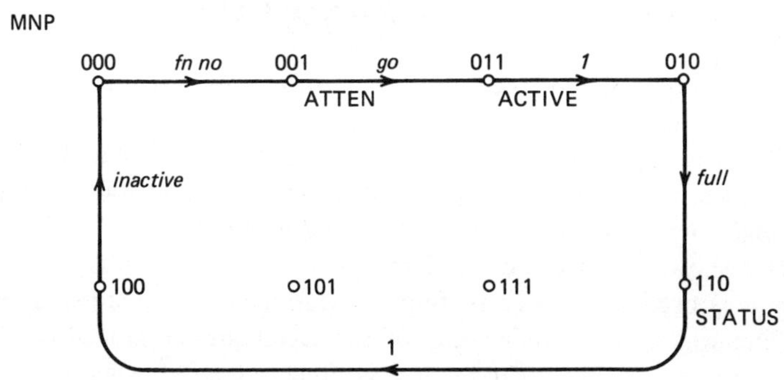

Figure P5.7 Status state diagram.

CHAPTER 6

Clock-Driven Sequential Circuits

The design techniques described in Chapters 4 and 5 give a strong technique for the development of sequential circuits. However, in general, such circuits have one important limitation. The limitation is that the value of only one state variable can change in a transition. This restriction leads to design difficulties. For example, consider a circuit that counts in a binary sequence. A binary counter is a particularly useful circuit. It can be used to count events, point to different events in an orderly sequence, and so on. The count from 01 to 10, from 011 to 100, and so on requires that the value of more than one variable be changed in a transition.

In general, transitions that require that the value of more than one variable change at a time are accomplished by the use of (JK) flip-flops. In this chapter the construction and operation of the JKFF is explained, and the JKFF is used to design sequential circuits.

6.1 ACTIVITY IN SYNCHRONOUS SEQUENTIAL CIRCUITS

For a better understanding of the objectives of this chapter, the special problem of counter design is examined. In this section an attempt is made to design a counter using RSFFs. The difficulties encountered are analyzed. From an understanding of these difficulties, a modified flip-flop suitable for multiple transitions is designed in the following section [3]. The design is given in problem P4.1 also.

The circuit shown in Figure 6.1 is intended to count the number of times the event (input x) occurs. After every fourth occurrence of x there is an output K from the counting circuit. Application of the techniques developed in Chapter 4 gives the result that variable B is to be set according to the following equation:

$$S_B = \bar{A}x + Ax$$

Similarly, its reset operation is given by the equation

$$R_B = \bar{A}x + Ax$$

Clearly, the technique breaks down because the rule regarding single-distance step changes is violated.

Examination of Figure 6.1 shows that what is needed in this circuit is to set B

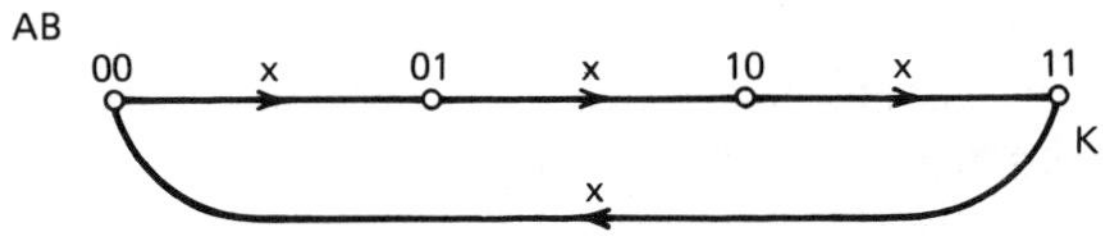

Figure 6.1 State diagram for a four-event counter.

when it is low and reset B when it is high, on each input event x. So for this diagram an equation for setting B is

$$S_B = \bar{B}x$$

and similarly the equation for resetting B is

$$R_B = Bx$$

Figure 6.2 shows the connection for the reset part of these equations only. The operation of the RSFF is such that it misbehaves in this application. In the circuit, the input signal x can be regarded as a series of pulses or equivalently as a clock signal. The circuit is shown in symbolic form in Figure 6.3. Its misbehavior can be viewed in two ways. If the set signal is on, that is, S = 1, and a pulse occurs on the x line, the output B goes high causing a reset signal, that is, R = 1. Thus the condition RS = 11 occurs. When the RSFF has both the set and reset inputs high, there is no way of predicting the final settling state of the flip-flop. The other way to look at the same problem is that there is a normal feedback from the output of gate 603 to the input of gate 604, and to the input of gate 602. This causes the circuit to misbehave because of the conflicting demands of the two feedback circuits. The conflicting demands can be seen by examining the circuit in detail. For B = 1 the output of gate 602 must be zero. Therefore the two inputs to gate 604 are 1, 0 so the output of gate 604 attempts to go high. For B = 0 the output of gate 602 must

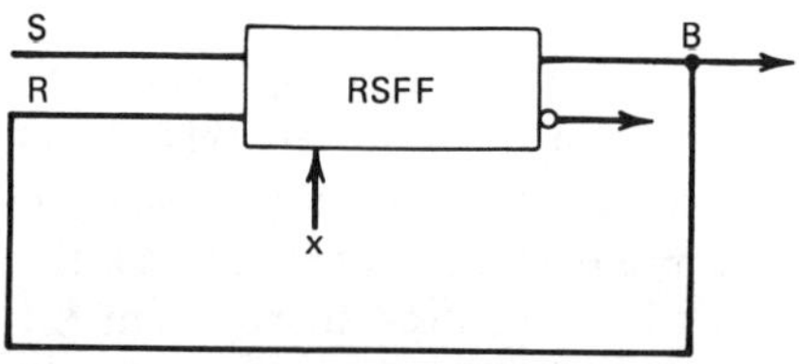

Figure 6.2 Feedback in an RSFF.

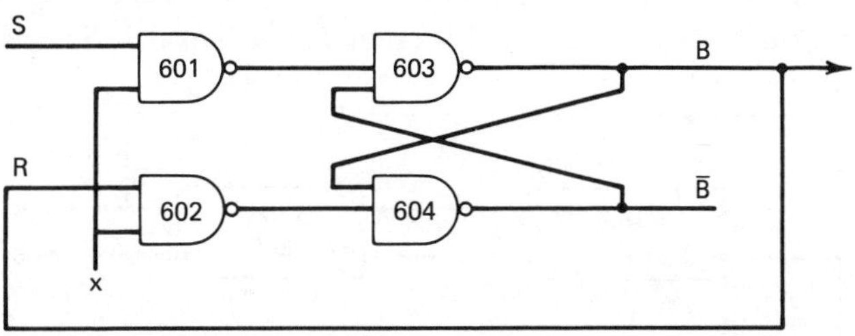

Figure 6.3 RSFF with double feedback.

be 1. Therefore the inputs to gate 604 are 0 and 1, and again 604 attempts to go high. A similar case exists for $\overline{B} = 0$ and 1 if a feedback loop from $\overline{B}$ to S is included in the diagram. Clearly, this is not the solution to the problem.

6.2 TWO-PHASE SYNCHRONOUS LOGIC

The solution to the problem is given in Figure 6.4, which shows two RSFFs in series. The input x is *and*ed with both R and S inputs to the first RSFF; and its complement $\overline{x}$ is *and*ed with the R and S inputs of the second RSFF. If feedback loops from B to R and $\overline{B}$ to S are included there are no false operations. On an input S to RSFF 605, its output Q goes high. If B is low, it remains low because $\overline{x} = 0$. When $\overline{x} = 1$, the value of Q is passed to RSFF 606 and B goes high. No misbehavior is caused by the feedback from B to R because $x = 0$. With this arrangement of RSFFs and feedback from B to R and $\overline{B}$ to S operation is hazard free for any transition. The result is that design is not restricted to the limitations of single variable (unit distant) transitions within the state diagram [3].

6.3 OPERATION OF THE JKFF

The circuit in Figure 6.4 is redrawn in Figure 6.5. The main inputs are designated J and K because of the need to identify the fact that this flip-flop operates quite differently from the RSFF. The x input becomes a clock. Feedback loops around the device are included since presumably this device is intended to operate in this mode. Inputs S and C are included to make the flip-flop operate in the normal RSFF sense. The reset terminology R is relabeled as the clear terminal C, which is the desired notation as specified by industrial standards book because S and C respond to low, 0, inputs rather than high, 1, inputs.

Figure 6.6 gives the JKFF action diagrams [3, 12]. Figure 6.6 (*a*) shows JK action in map form. The present and future states are q and Q respectively. Again, this is a slightly different version of the Karnaugh map in that it gives the values of J and K, which cause action rather than outputs. For $qQ = 00$, there is no transition in the map, the state remains at 0, and J must be guaranteed to be 0, but K is optional. This is shown in the state representation of action, Figure 6.6(*b*), with the loop from 0 to 0 designated $\overline{J}K + \overline{J}\overline{K}$. For a transition described by $qQ = 01$, the state change is from 0 to 1. For this event J must equal 1. Since there is feed-

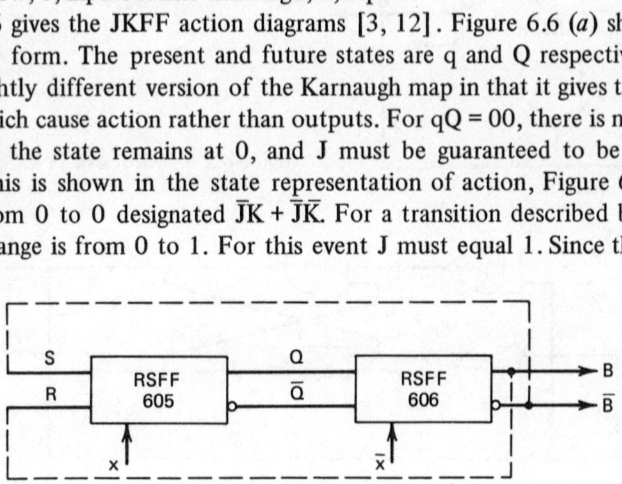

Figure 6.4 RSFFs in series.

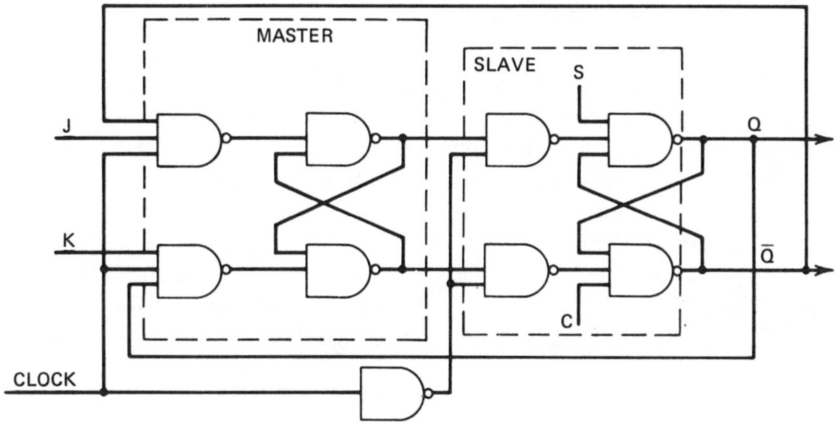

Figure 6.5 JKFF internal circuit diagram.

back from the output Q to the K terminal, the occurrence of K does not cause operation. Therefore K is optional. This is designated in the state representation by the transition with inputs $JK + J\bar{K}$. Similarly, for qQ = 10, there is a transition of state from 1 to 0. If the output value Q is to go low, the K signal must be present. Because the input J is *nand*ed with $\bar{Q}$ by the feedback loop and $\bar{Q} = 0$, J is optional. This is represented in the JK map as $JK = \phi 1$ and in the state representation form as $JK + \bar{J}K$. For the condition where qQ = 11 there is no action, since the output value Q is 1, $\bar{Q} = 0$, and since J is *nand*ed with $\bar{Q}$ a further input J cannot cause any action. Therefore J is optional. However, an occurrence of K guarantees reset of the device, and this is not permitted, so K must be 0. This is represented in the JK map as $JK = \phi 0$ and in the state representation as $J\bar{K} + \bar{J}\bar{K}$.

Notice that the value JK = 11 is defined. Repeated occurrence of JK = 11 causes the device to toggle from 0 to 1 and 1 to 0 and so on. This is the exact operation sought when the RSFF was modified to achieve the JKFF action.

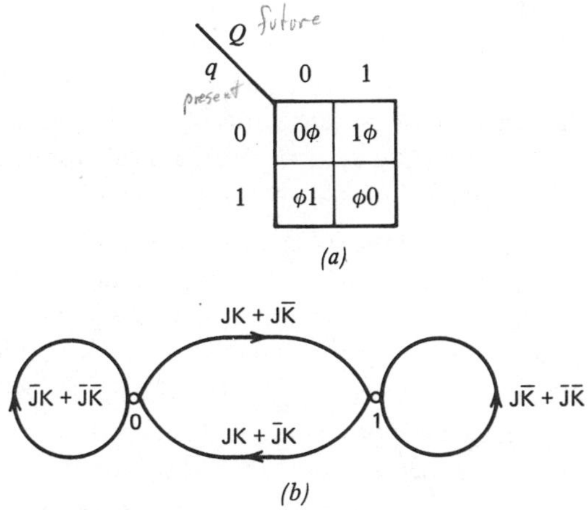

Figure 6.6 Action representation for a JKFF. *(a)* JK action map; *(b)* state representation.

6.4 DESIGN FLEXIBILITY OF JKFFS

The JKFF represents a further broadening in the possibilities of design. There are few restrictions on the input variables. Also the RSFF had the limitation that the two inputs must not be high at the same time. In the JKFF case this input is permitted and causes a toggle action in the flip-flop. Previously transitions between states in a state diagram were limited to unit distances. This restriction no longer exists. The JKFF is ideally suited to counting the occurrences of a single input. This is called a type I application. The JKFF also responds to multiple inputs that occur sequentially. This is called a type II application. The JKFF gives much better utilization of secondary state variables, but the cost of achieving this is the introduction of a clock.

6.5 STEPS IN THE DESIGN OF CLOCK-DRIVEN SEQUENTIAL CIRCUITS USING JKFFS

The synthesis of clock-driven sequential circuit design using JKFFs is achieved in the following steps:

1. The circuit is identified according to its inputs. If the circuit has one or few inputs to which the response must be varied, it is of type I, and the design synthesis follows steps 2, 3, and 4. If the circuit has a multitude of inputs to which there is generally a different response to different inputs, it is of type II, and the synthesis follows steps 5, 6, and 7.
2. A Karnaugh map for the JK conditions of each state variable for each input is drawn. This map has the form with the inputs designated across the top denoting columns in the map. The state variables are designated down the left-hand side of the map, and designate rows in the map. To complete this map the row designation is regarded as the present state q. The cells within the map are regarded as the future state Q in the transition response to the appropriate input. When completed this map should contain 1s and 0s only.
3. The map from step 2 is converted to an action map using the action map given in Figure 6.6(*a*). Starting with the present status or row designation and completing with the cell value representing Q, a JK action map is made for each state variable.
4. A symbolic diagram of the expression derived from step 3 is drawn.
5. A loop is drawn to enclose all the states where a value of a specific state variable equals 1. Since transition in and out of this loop requires excitation, input lines intersecting the loop designate the excitation. Arrows entering the loop designate the J excitation, and arrows exiting the loop designate the K excitation.
6. Write an S of P expression for the J excitation and another for the K excitation for a specific state variable. Each product must contain the input variable and the remaining state variables associated with the initial state in the transition. Inclusion of the remaining state variables guarantees that the circuit has progressed to the proper stage. Repeat steps 5 and 6 for each state variable.
7. Draw a symbolic diagram of the expressions derived from step 6.

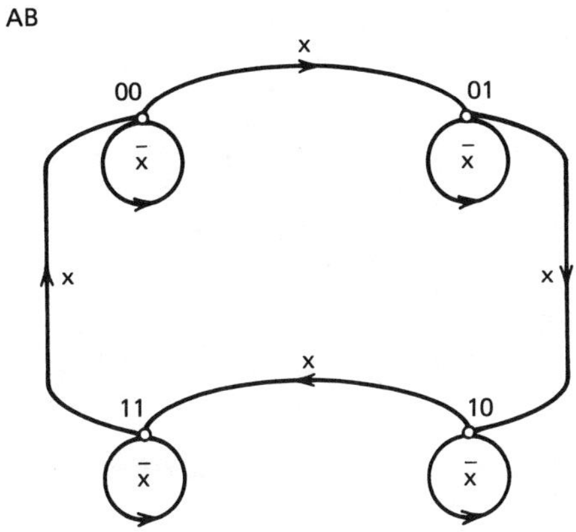

Figure 6.7 Design of a four-count counter.

6.6 EXAMPLE OF A COUNTER DESIGN USING JKFFS

The state diagram of a zero to three counter is given in Figure 6.7. In the first step of the synthesis, step 1, this is identified as a circuit with only one input. The next step in the synthesis, step 2, is to draw a Karnaugh map showing present and future states. This map is shown in Figure 6.8. The JK action map is shown in Figure 6.9. Its construction is accomplished by following step 3 of the synthesis. The action map for variable A and the column for the input x = 0 is completed as follows: Figure 6.8 shows that for ab = 00, A = 0. Thus the value of A is unchanging, and aA = 00. The inputs are JK = 0ϕ, since aA = 00 is the equivalent of qQ = 00 in Figure 6.6(*a*). Similarly, for ab = 01, A = 0. Again this is a transition for aA = 00, which can again be read from Figure 6.6(*a*) for the equivalent condition qQ = 00, where JK = 0ϕ. This entry is made in the action maps. For the case where ab = 11, A = 1. In this case, aA = 11 and is equivalent to qQ = 11 in Figure 6.6(*a*). The entry is JK = ϕ0. Similarly, for ab = 10, A = 1, or aA = 11, and the same entry is appropriate.

To complete the second column for the variable A of Figure 6.9(*a*), where x = 1, the following procedure is used: In the first row ab = 00, A = 0. In this case aA = 00 and JK = 0ϕ as read from the JK action map where qQ = 00. For the row with ab = 01, A = 1. Thus aA = 01, which is equivalent to qQ = 01 in the JK action map.

ab \ *x*	0	1
00	00	01
01	01	10
11	11	00
10	10	11

AB map

Figure 6.8 K map of present and future states.

$ab \backslash x$	0	1
00	0ϕ	0ϕ
01	0ϕ	1ϕ
11	$\phi 0$	$\phi 1$
10	$\phi 0$	$\phi 0$

(a)

$ab \backslash x$	0	1
00	0ϕ	1ϕ
01	$\phi 0$	$\phi 1$
11	$\phi 0$	$\phi 1$
10	0ϕ	1ϕ

(b)

Figure 6.9 Action map construction. *(a)* JK action map for variable A; *(b)* JK action map for variable B.

This entry is JK = 1ϕ. For x = 1 and ab = 11, A = 0. Thus aA = 10, and JK = $\phi 1$ as obtained from the JK action map for the equivalent condition qQ = 10. The last entry is for x = 1, ab = 10, and A = 1. The value of aA = 11, equivalent to qQ = 11, for which the JK action status is JK = $\phi 0$. This entry completes the action map for the variable A.

A similar procedure is followed to complete the action map for the variable B. The next step in the synthesis is to determine values for the input variables. Reading from Figure 6.9(*b*), the variable B control equations are

$$J_B = x$$

and similarly,

$$K_B = x$$

The equations for the control of variable A are

$$J_A = xB$$

and

$$K_A = xB$$

Step 4, the realization of this circuit, is shown in Figure 6.10.

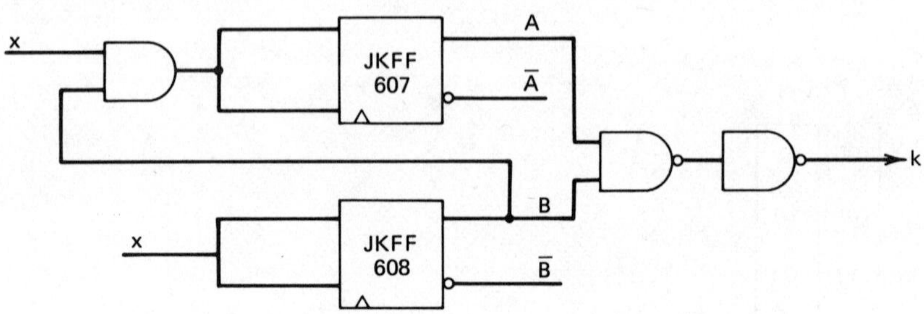

Figure 6.10 Realization of four-count synchronous counter.

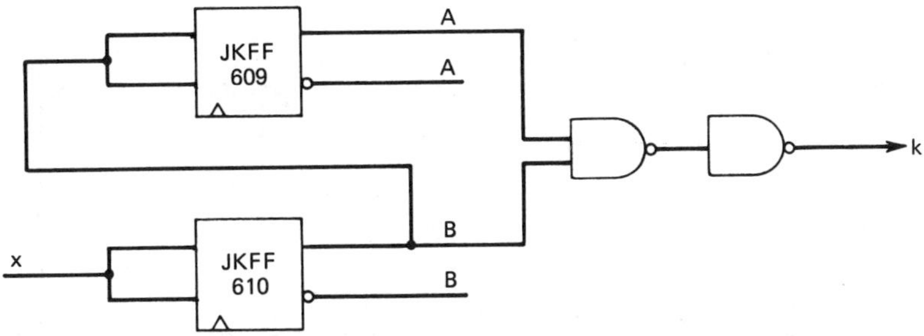

Figure 6.11 Four-counter ripple circuit or asynchronous-counter circuit.

6.7 RIPPLE COUNTERS IN REALIZATION

Examination of the foregoing equations leads to an interesting thought.

$$J_B = K_B = x$$

and

$$J_A = K_A = xB$$

It would seem reasonable to say that x is not required twice as an input to A. The control equations for variable B are a function of x, and by use of this argument the JK inputs for variable A reduce to

$$J_A = K_A = B$$

Examination of Figure 6.7 shows that indeed this is a valid supposition. Whenever B becomes high, there is a subsequent action of the variable A. Thus this counter can be realized using the version of the circuit shown in Figure 6.11.

Figures 6.10 and 6.11 represent two ways to build counters, and the student should be aware that both methods exist. The realization shown in Figure 6.10 is termed a synchronous counter, and that shown in Figure 6.11 is a ripple (or asynchronous) counter.

6.8 BCD SYNCHRONOUS-COUNTER DESIGN

As a further example to illustrate the technique, the BCD (10) counter is designed as shown in state variable form in Fig. 6.12 [16]. The first step in the synthesis is to identify the circuit according to its inputs. This is again a type I circuit. It is the same example as shown in Figure 6.7 except that the counter is from 0 to 9 rather than 0 to 3. As a result this example is a more detailed illustration of the use of the JKFF action map shown in Figure 6.6. Synthesis follows steps 2, 3, and 4 of Section 6.5. So the next step in the synthesis is to convert the state diagram into a Karnaugh map. Again, this is done by regarding the identification of each row as the present state, or q variable, and the cell in the Karnaugh map as the next state or Q variable.

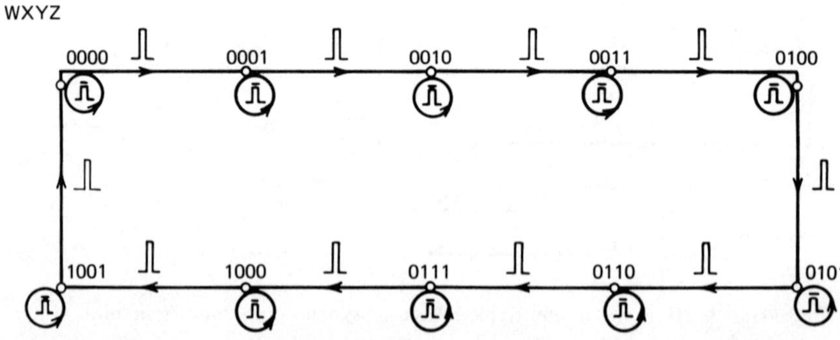

Figure 6.12 Counter using JKFFs.

The Karnaugh map for the variables W, X, Y, Z and input ⎍ is shown in Figure 6.13.

The action map for each of the variables W, X, Y, Z is shown in Figure 6.14. The entries are made from Figure 6.13 by determining the value of the present and future state values for each variable. For example, the values where wW = 00 can be found by examination of Figure 6.6(*a*) for the equivalent condition qQ = 00. A similar procedure is followed for all values of wW, all values of xX, all values of yY, and all values of zZ. The JK entries are placed into the action maps for each variable. The control equations for the variables W, X, Y, and Z are obtained by the

wxyz \ ⎍	0	1
0000	0000	0001
0001	0001	0010
0010	0010	0011
0011	0011	0100
0100	0100	0101
0101	0101	0110
0110	0110	0111
0111	0111	1000
1000	1000	1001
1001	1001	0000

WXYZ map

Figure 6.13 Variable map for a BCD counter.

wxyz	0	1
0000	0ϕ	0ϕ
0001	0ϕ	0ϕ
0010	0ϕ	0ϕ
0011	0ϕ	0ϕ
0100	0ϕ	0ϕ
0101	0ϕ	0ϕ
0110	0ϕ	0ϕ
0111	0ϕ	1ϕ
1000	$\phi 0$	$\phi 0$
1001	$\phi 0$	$\phi 1$

(a)

wxyz	0	1
0000	0ϕ	0ϕ
0001	0ϕ	0ϕ
0010	0ϕ	0ϕ
0011	0ϕ	1ϕ
0100	$\phi 0$	$\phi 0$
0101	$\phi 0$	$\phi 0$
0110	$\phi 0$	$\phi 0$
0111	$\phi 0$	$\phi 1$
1000	0ϕ	0ϕ
1001	0ϕ	0ϕ

(b)

wxyz	0	1
0000	0ϕ	0ϕ
0001	0ϕ	1ϕ
0010	$\phi 0$	$\phi 0$
0011	$\phi 0$	$\phi 1$
0100	0ϕ	0ϕ
0101	0ϕ	1ϕ
0110	$\phi 0$	$\phi 0$
0111	$\phi 0$	$\phi 1$
1000	0ϕ	0ϕ
1001	0ϕ	0ϕ

(c)

wxyz	0	1
0000	0ϕ	1ϕ
0001	$\phi 0$	$\phi 1$
0010	0ϕ	1ϕ
0011	$\phi 0$	$\phi 1$
0100	0ϕ	1ϕ
0101	$\phi 0$	$\phi 1$
0110	0ϕ	1ϕ
0111	$\phi 0$	$\phi 1$
1000	0ϕ	1ϕ
1001	$\phi 0$	$\phi 1$

(d)

Figure 6.14 JK action map for a BCD counter. *(a)* Action for variable W; *(b)* action for variable X; *(c)* action map for variable Y; *(d)* action map for variable Z.

action map of Figure 6.14. From Figure 6.14(*a*),

$$J_W = \text{⎍}XYZ$$
$$K_W = \text{⎍}Z$$

From Figure 6.14 (*b*),

$$J_X = \text{⎍}YZ$$
$$K_X = \text{⎍}YZ$$

From Figure 6.14(*c*), the control equations for variable Y are

$$J_Y = \text{⎍}\bar{W}Z$$
$$K_Y = \text{⎍}Z$$

However, in this case it is probably just as easy to make it

$$K_Y = \text{⎍}\bar{W}Z$$

in the implementation. From Figure 6.14(*d*), the equations for the variable Z are

$$J_Z = K_Z = \text{⎍}$$

Again, it is possible to eliminate the ⎍ from the variables W, X, and Y, and this counter will act like a ripple counter rather than the synchronous counter developed in the example.

This design technique gives an easy and effective way of designing circuits that move through a given sequence by the change of input of a single variable. Since such circuits are so easy to design and so useful, probably someone else has already worked the problem. Thus this type of circuit is generally available as an MSI chip at moderate cost. An understanding of the technique is advocated, but it seems pointless to develop it into a highly skilled procedure. The more useful design procedure is one involving a circuit not previously encountered. These are usually multi-input or type II circuits.

6.9 MULTIEVENT PULSED SEQUENTIAL CIRCUITS

The remainder of this chapter is devoted to type II circuit design and follows steps 5, 6, and 7 of the synthesis given in Section 6.5.

Figure 6.15 shows a block diagram of a distributed computer system. In this system a number of minicomputers are tied together to give the system its distributed aspect. Normally, a user logs onto the system by entering the appropriate command from one of the terminals (CRTs) shown either on node 1 or node 2. As an example, assume that input to the system is from one of the CRTs on node 1. Access to the plotter on node 2 and a program running on large computer number 1 is available. The technique is illustrated by design of the interface that connects large computer 1 to node 1 (minicomputer).

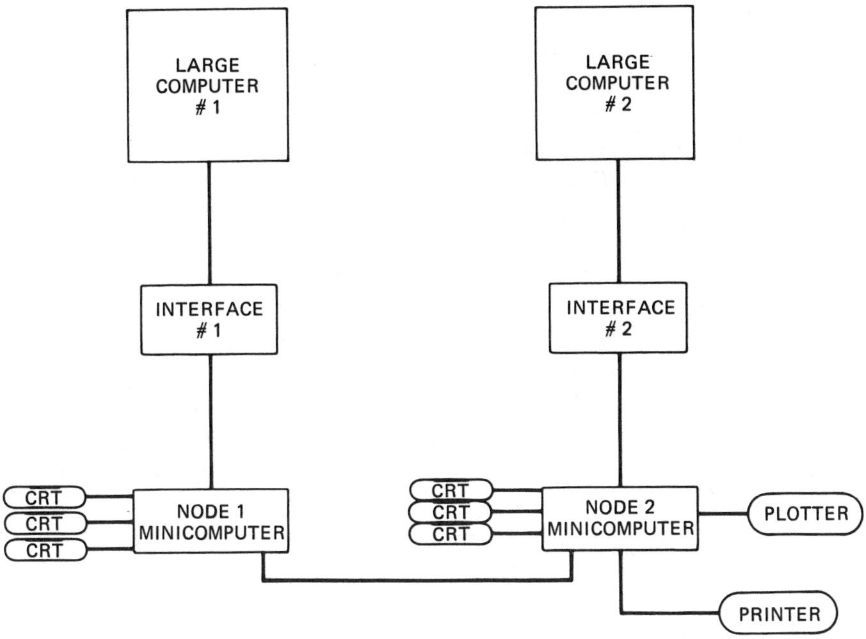

Figure 6.15 Distributed computer system.

A detailed explanation follows. Figure 6.16 shows a state diagram for the generalized circuit for input-output routines in interface number 1. Normally the interface is in the idle condition where ABC = 000. Upon receipt of an input signal function and its number (*Fn No*), which is in code from the large computer, there is a transition to state ABC = 001. Here an output signal ATTENTION is sent to the minicomputer signifying that service is required. When the minicomputer is available, it responds with a *Go* signal, which causes a transition in states ABC = 011. If, for any reason, the interface is inactive or the minicomputer is not operational, an input signal *Interrupt Not* ($\overline{Inter}$) is generated. This allows the main computer to proceed without interruption when part of the system is down.

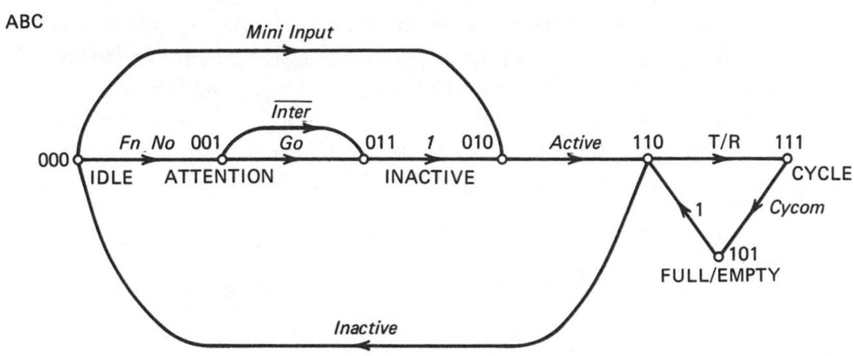

Figure 6.16 Generalized interface circuit state diagram.

In state ABC = 011 an output signal is sent from the interface to the large computer saying that it has accepted the coded function as valid, INACTIVE, and is waiting for the next function. Immediately upon transmission of this signal to the large computer, there is an automatic transition within the interface to the state ABC = 010. This automatic transition is required because the large computer does not accept repeated INACTIVE signals. There is a facility for the minicomputer to initiate data transfers between the two computers. An input signal from the minicomputer, *mini input*, causes a transition from state ABC = 000 to ABC = 010 to bypass the regular sequence. The response *Active* from the larger computer, indicating that it is ready to transmit data, is an input signal to the circuit and causes a transition from state ABC = 010 to 110. A signal is received from either the main computer or the minicomputer dependent upon the direction of data flow. This is the input signal *Transmit/Receive* designated as *T/R*. It causes a transition to state ABC = 111 where an output signal CYCLE is sent to the minicomputer, which is a request for data transmission or receipt of data. When the minicomputer has transmitted or received the appropriate data it responds with a *Cycle Complete* (*Cycom*) input signal, causing a transition in the interface state diagram to ABC = 101. In this state a FULL/EMPTY signal is sent to the main computer, as appropriate for the direction of data transfer. Immediately upon transmission of this signal there is an automatic transition back to the state 110. Again, the automatic transition is required because the main computer does not accept repeated functions from peripherals.

The loop ABC = 110, 111, 101 is repeated until all data is transferred between computers. When it is done the main computer initiates an input signal *Inactive* to the interface, informing it that the requested function is complete. The *Inactive* function is the main computer's way of saying goodbye, and returns the interface to ABC = 000 or idle state, ready for immediate response to any further requests for data transfer.

6.10 SYNTHESIS OF MULTIEVENT CLOCK-DRIVEN SEQUENTIAL CIRCUITS

The foregoing circuit is quite easily realized by use of the design synthesis explained in Section 6.4. Figure 6.17 shows the diagram designating action of the JKFFs. Each loop is superimposed on the diagram encircling the condition where one of the state variables is on (1). This is step 5 in the synthesis technique. The J excitation for flip-flop A is determined from the arrows entering the loop in Figure 6.17(*a*) in accordance with step 6. Thus

$$J_A = \textit{Active}\ B\overline{C}$$

The K excitation for this variable is given by arrows exiting the A loop. Thus

$$K_A = \textit{Inactive}\ B\overline{C}$$

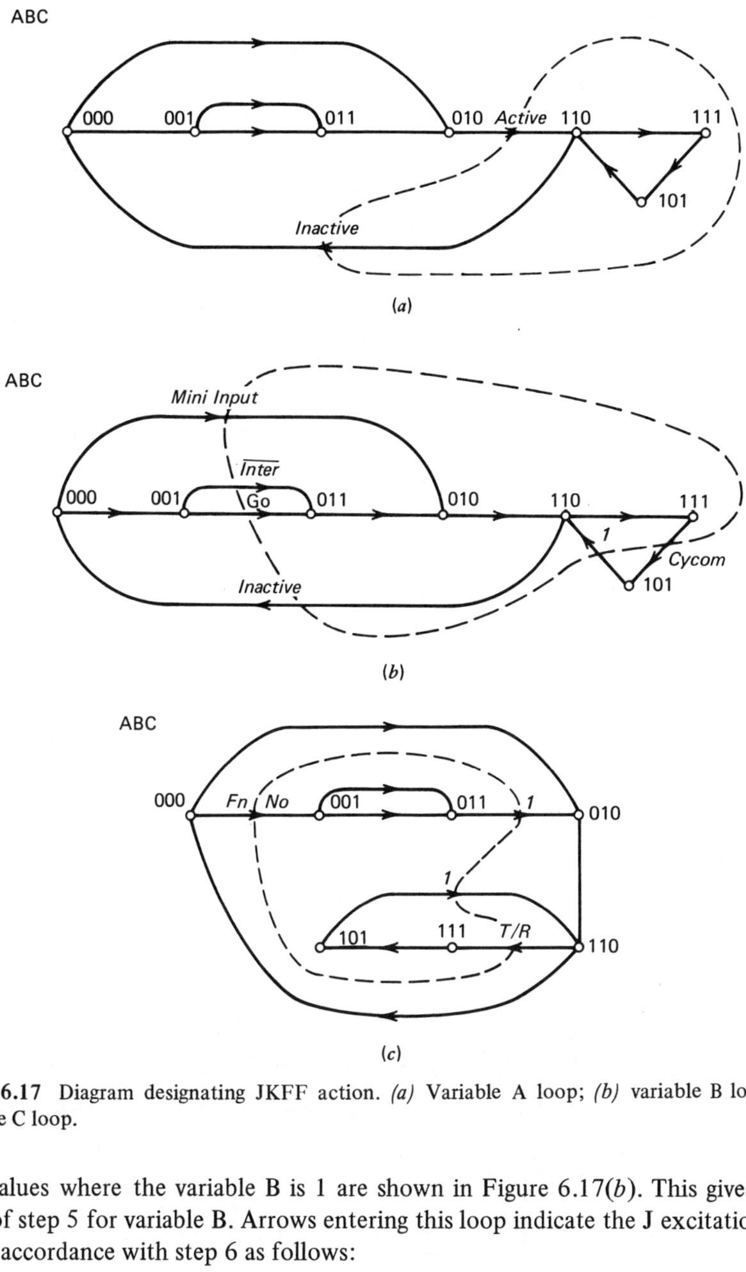

Figure 6.17 Diagram designating JKFF action. *(a)* Variable A loop; *(b)* variable B loop; *(c)* variable C loop.

The values where the variable B is 1 are shown in Figure 6.17(*b*). This gives a repeat of step 5 for variable B. Arrows entering this loop indicate the J excitation and are in accordance with step 6 as follows:

$$J_B = \textit{Mini Input}\ \bar{A}\bar{C} + (\textit{Go} + \overline{\textit{Inter}})\bar{A}C + AC$$

Arrows exiting this loop designate K excitation and are given by the equation

$$K_B = \textit{Cycom}\ AC + \textit{Inactive}\ A\bar{C}$$

The positive conditions for variable C are encircled by the loop in Figure 6.17(*c*) by following step 5. Arrows entering this loop give the J excitation for flip-flop C according to step 6.

$$J_C = Fn\ No\ \bar{A}\bar{B} + T/R\ AB$$

The K excitation for the variable C is given by arrows exiting the loop of Figure 6.17(*c*). The equation is as follows:

$$K_C = \bar{A}B + A\bar{B}$$

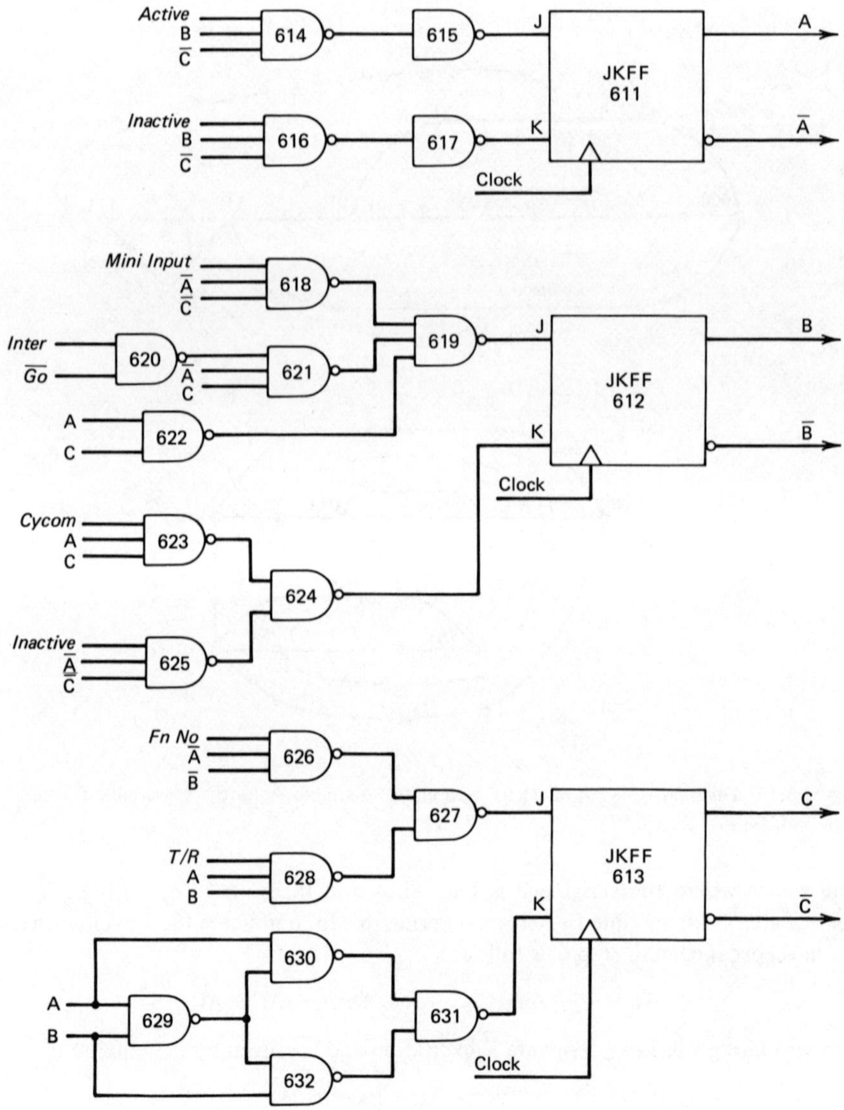

Figure 6.18 Realization of generalized interface circuit.

C \ AB	00	01	11	10
0	IDLE			
1	ATTENTION	INACTIVE	CYCLE	FULL/EMPTY

Figure 6.19 K map of output for generalized interface circuit.

Realization of this circuit is shown in Figure 6.18. The feedback connections for the variables A, B, and C are not shown, although, as always, such connections are implied. The output functions are as follows:

$$\text{ATTENTION} = \bar{A}\bar{B}C$$

$$\text{INACTIVE} = \bar{A}BC$$

$$\text{CYCLE} = ABC$$

$$\text{FULL/EMPTY} = A\bar{B}C$$

The Karnaugh map for the output functions is given in Figure 6.19. The idle condition is ignored since it is not a genuine output.

6.11 PROGRAMMABLE CIRCUITS

In Figure 6.16 an interesting design philosophy is introduced. The inputs *Transmit/Receive* and the outputs FULL/EMPTY are selectable. To take advantage of this feature it is possible to send a control word CW to an interface control register when the interface is in the IDLE state. The control word is decoded in a combinational circuit. This decoder determines which modes the selectable inputs and outputs are to assume.

The philosophy can be extended considerably. For example in prototype I of the above interface two separate circuits were used for high-speed and low-speed data transfers. In prototype II a single circuit and a control word were used for both speeds. Savings in design, construction time, power, and component consumption resulted.

Another extension of the philosophy is to have the control word select transitions in the state diagram. This concept is illustrated in the problem set for this chapter. Another variation of this approach is given in Chapter 9.

Figure 6.20 illustrates a simplified version of a popular method of constructing interfaces. In the IDLE state the control register is loaded with the control word b7-b0. On a start command there is a transition to state ABC = 100 and an inter-

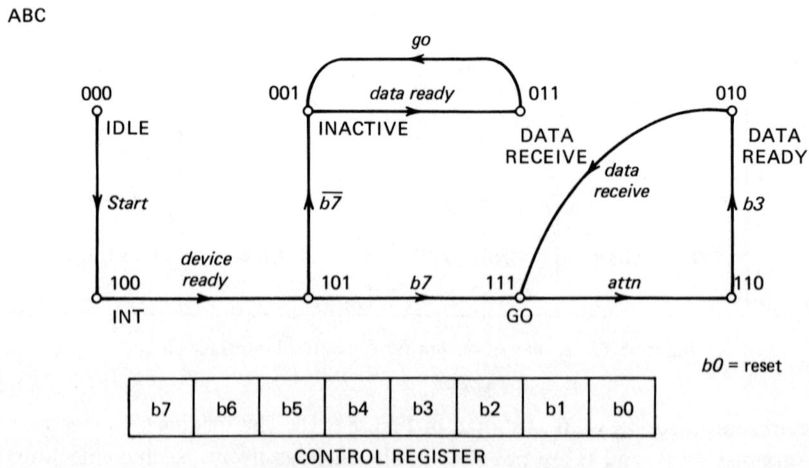

Figure 6.20 Programmable interface.

rupt is sent to the device. It responds with a *device ready* signal that causes a transition to state ABC = 101. Depending upon the value of bit b7 in the control register the interface responds with data transfer in different directions. Transition ABC = 001, 011 represents input, and transition ABC = 111, 110, 010 represents output. In the transition from ABC = 110 to 010 bit b3 can be used to delay the process. Such delays may be used to synchronize hardware and software.

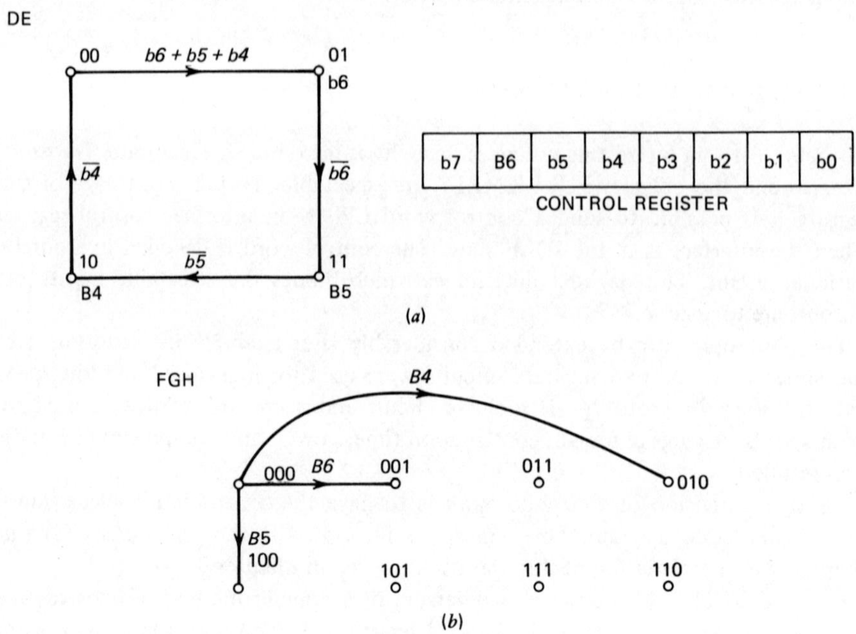

Figure 6.21 Nested circuits. *(a)* Executive circuit; *(b)* interface action circuit.

6.12 NESTED CIRCUITS

In the foregoing enhancement, which led to a programmable interface, the sequential operation of a circuit is determined by a combinational or decoding circuit. A natural follow-up is to have sequential operation of a circuit determined by a master or executive sequential circuit. This leads to a nesting of sequential circuits. The state diagram of Figure 6.21 illustrate this philosophy. Figure 6.21(*a*) shows the executive circuit. It is programmed by the same control register used above. The executive circuit is considered to be event driven in this case. In the executive circuit any one of the bits b6, b5, or b4 causes a transition to DE = 01. However if two bits are set, the executive selects the MSB. The interface action circuit, which is assumed to be clock driven, is forced to respond differently for each bit. Of course a simple extension that makes this concept more powerful is to have a separate action circuit for each output B6, B5, or B4.

6.13 PROBLEMS

P6.1 Using JKFFs and the type I design synthesis, design the up *U*, down *D*, counter shown in Figure P6.1.

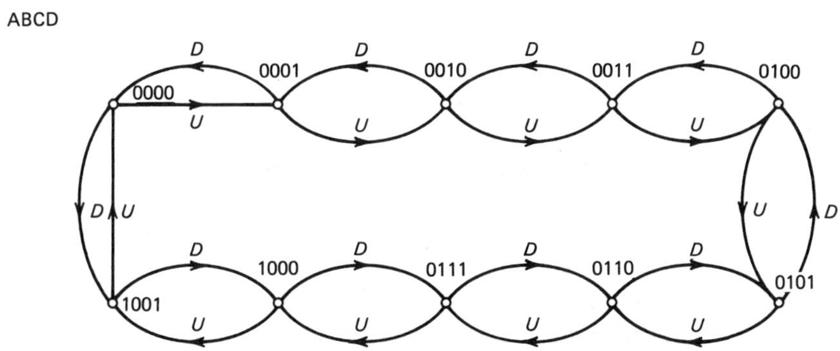

Figure P6.1 Up-down counter.

P6.2 Design the counter shown in Figure P6.2 using type I design synthesis.

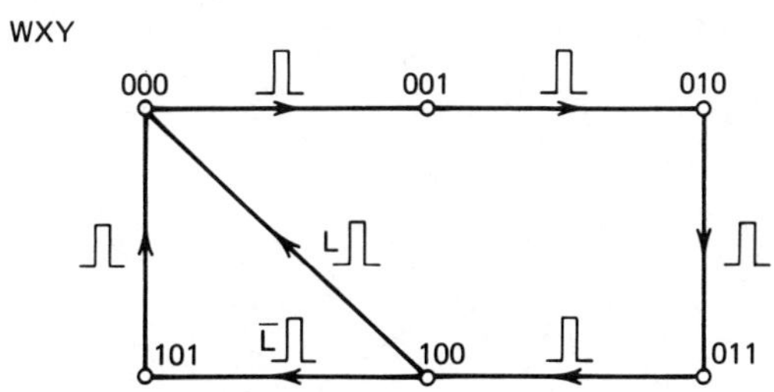

Figure P6.2 Count skipper.

P6.3 Design the counter shown in Figure P6.3 using type I design synthesis.

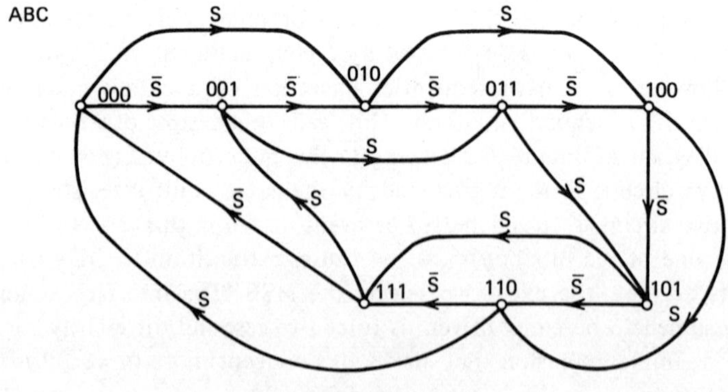

Figure P6.3 Skip counter.

P6.4 Design the counter shown in Figure P6.4 using type I design synthesis.

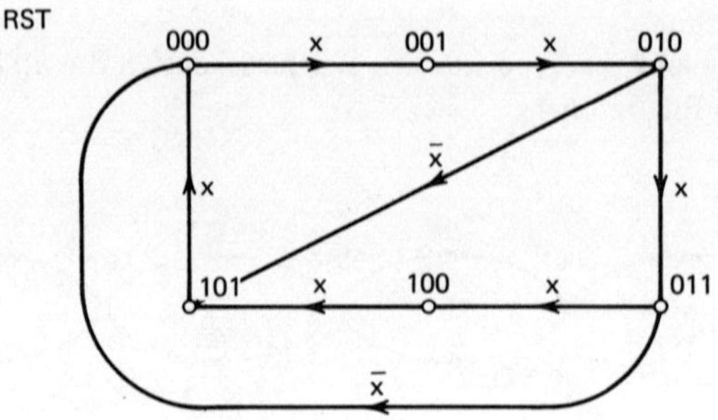

Figure P6.4 Jump counter.

P6.5 Design the counter shown in Figure P6.5.
Partition the problem by drawing circuit a with input E only and circuit b with inputs L and M. Use type I design synthesis to solve circuit a and type II design synthesis to solve circuit b. Combine the two partial solutions to obtain the total solution.

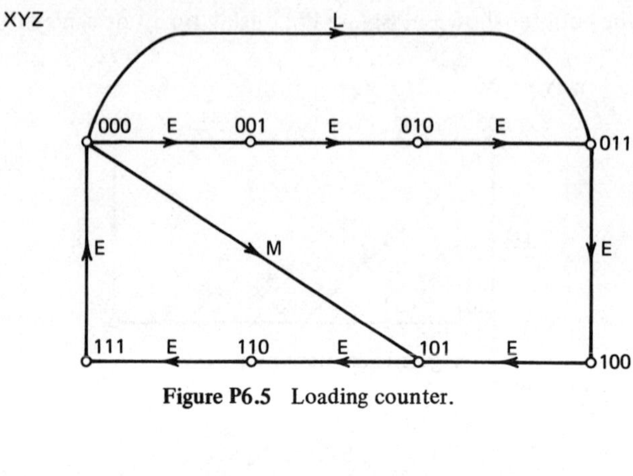

Figure P6.5 Loading counter.

P6.6 A certain interface is constructed such that one of its control signals (*ca1*) may be programmed to respond to a high or a low transition. Bit b1 of an 8-bit register is used to provide this feature. The circuit state diagram is shown in Figure P6.6 [30]. This state diagram is illustrative only. Some details are omitted and no claim to knowledge of the manufacturer's circuit is made. There is a transition from ABC = 000 to 001 when b0 is high. If bit b1 is high there is a transition to state ABC = 101. When *ca1* goes high a transition to state ABC = 111 occurs. An output (FLAG) signal is issued. A transition to state ABC = 101 from 111 occurs on a *read* data lines instruction. Similarly, b1 low causes a transition from state ABC = 001 to 011. When *ca1* goes low a transition to state ABC = 111 occurs. An output (FLAG) signal is issued, and an automatic restoration ABC = 011 occurs on a *read* data lines instruction.

Design this level and enable control circuit using type II synthesis.

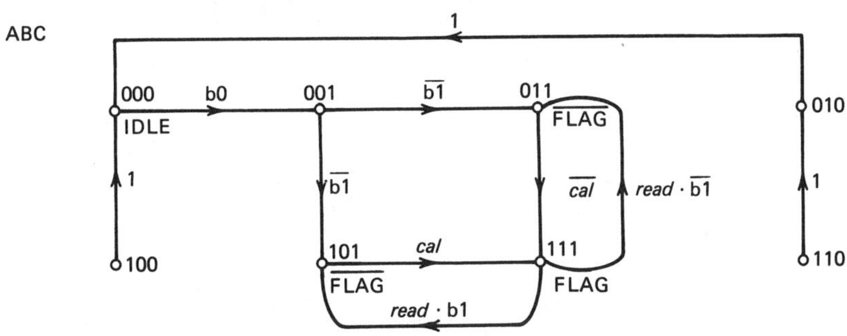

Figure P6.6 Level and enable control circuit.

P6.7 In the same interface as described in P6.6, a second control signal, *ca2*, is programmable for level and enable control and also for input or output. Bits b5, b4, and b3 are used to provide this feature.

The circuit state diagram is shown in Figure P6.7. Again this is an illustrated example that is incomplete and not a manufacturer's circuit. This circuit illustrates programmable nested circuits.

Bit b5 is used to select the input or output mode, as shown in Figure P6.7(*a*). If the input mode is selected the operation is as shown in Figure P6.7(*b*). Figure P6.7(*b*) resembles Figure P6.6 closely. There is a transition from state EFG = 000 to 100 as a result of executive decision ($\overline{\text{I}}$/O is low). The circuit is enabled when bit b3 is high. Bit b4 high causes the FLAG to go from low to high as a result of *ca2* and resets as a result of a *read* data instruction. Bit b4 low causes the FLAG to go from low to high as a result of $\overline{ca2}$ and resets as a result of a *read* data instruction.

If the output mode is selected using bit b5, the operation is as shown in Figure P6.7(*c*). There is a transition from state HJK = 000 to 001 caused by the executive circuit. The output CA2 is used with a *read* data instruction when bit b4 is low. Bit b4 low causes a transition to state HJK = 101 and output CA2 goes high. On a *read* data instruction CA2 goes low on the state transition to HJK = 100. CA2 is restored to high on *ca1* high if b3 is high or on chip *enable* if b3 is low.

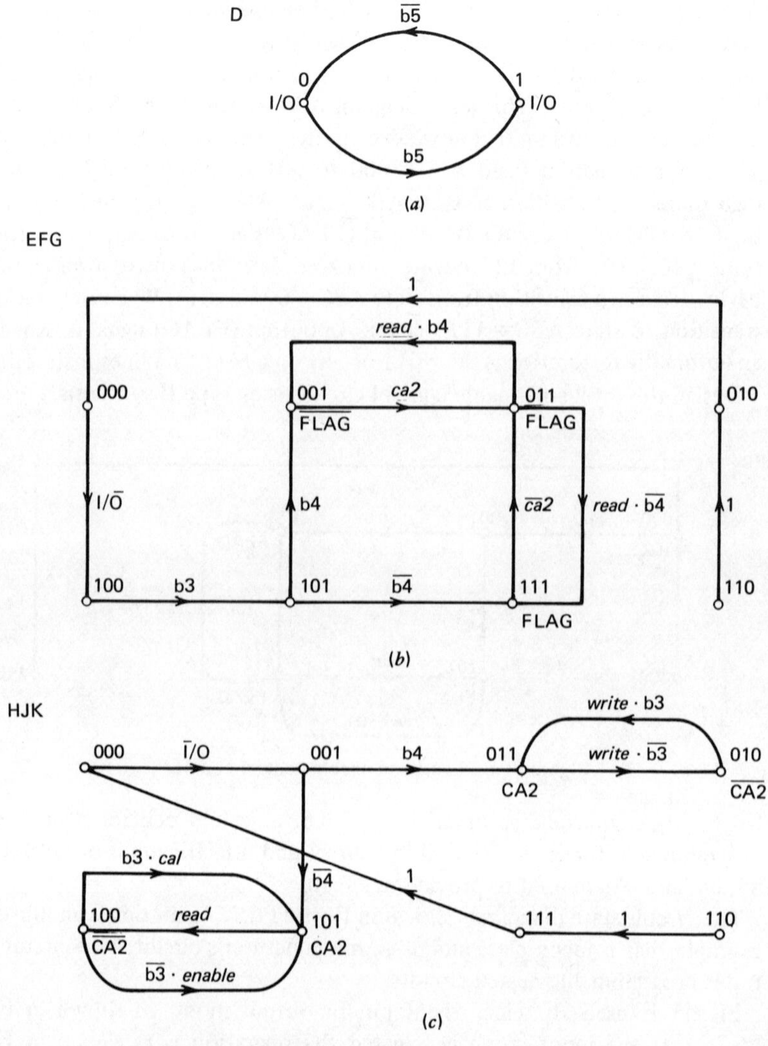

Figure P6.7 Control bit programming. *(a)* Executive circuit; *(b)* input, level, and enable control circuit; *(c)* output, *read/write*.

In a similar way the CA2 output is used with a *write* data instruction when bit b4 is high. Bit b4 high causes a transition from state HJK = 001 to 011. CA2 follows the value of bit b3 in the transition HJK = 011, 010.

Assume the executive circuit is an event-driven circuit and the control circuits are clock driven.

Design these circuits using type II synthesis.

P6.8 Another interface that can be programmed for different modes of operation can be represented by the state diagrams shown in Figure P6.8.

This interface is programmable for modes 0, 1, and 2 as well as input/

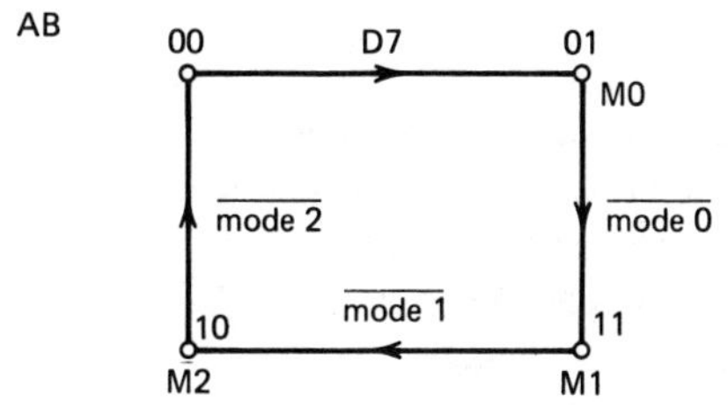

MODE	D6	D5
0	0	0
1	0	1
2	1	0

(a)

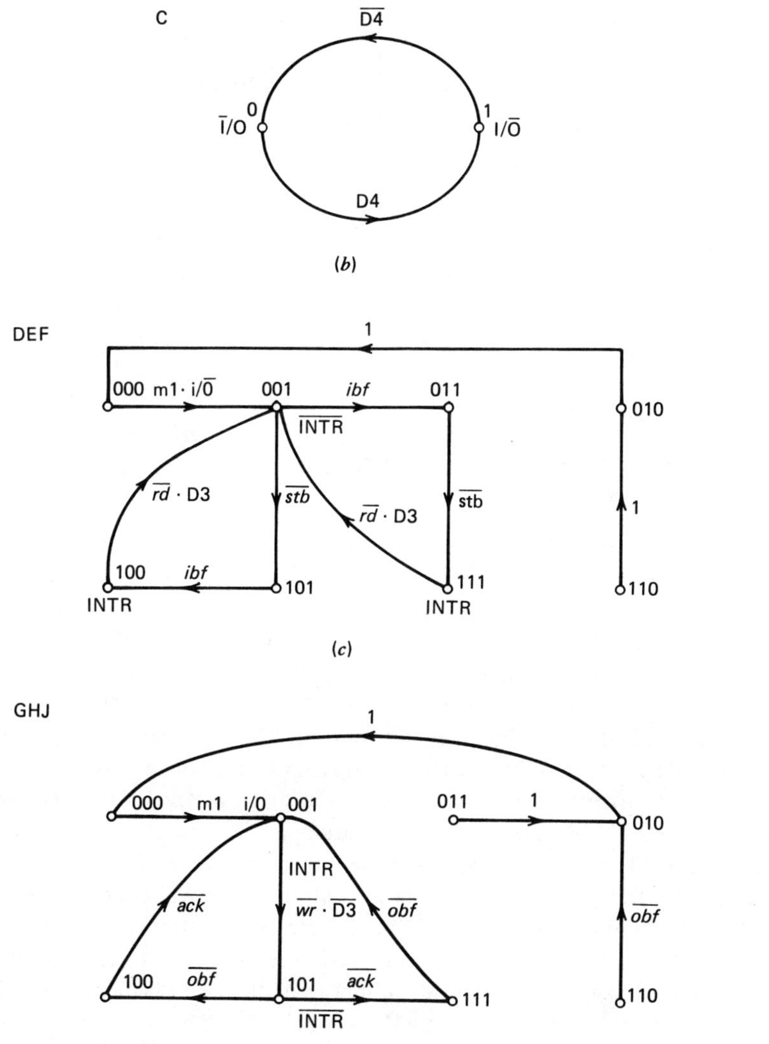

Figure P6.8 Mode 1 programming. *(a)* Mode executive circuit; *(b)* input/output executive circuit; *(c)* input mode 1 control circuit; *(d)* output mode 1 circuit control.

output. In this problem only mode 1 input and output is illustrated. Mode 1 provides a straightforward handshake with input or output selectable [31].

The mode selection executive circuit state diagram is shown in Figure P6.8(*a*). Bit D7 high indicates that mode selection is being done. Bits D6 and D5 are used to select the mode in accordance with the table.

The input/output executive circuit state diagram is shown in Figure P6.8(*b*). Bit D4 high selects data input; D4 low selects data output.

The input mode 1 control circuit state diagram is shown in Figure P6.8(*c*). The executive circuit causes a transition to state DEF = 001. Input signal $\overline{stb}$ ($\overline{strobe}$) loads input data. It is a request for service from the peripheral device. Input signal *ibf* (*input buffer full*) indicates the data has been loaded. It is an acknowledge signal to the peripheral device. Signals $\overline{stb}$ and *ibf* may occur in any order and cause a transition to state DEF = 100 or 111 where the interrupt INTR signal is issued. On a $\overline{rd}$ (*read*) instruction, and if bit D3 is high, the interface makes a transition to state DEF = 001, which drives the interrupt low $\overline{\text{INTR}}$.

The output mode 1 circuit control state diagram is shown in Figure 6.8(*d*). The executive circuit shows a transition to state GHJ = 001 where the interrupt INTR signal is issued. On a $\overline{wr}$ (*write*) instruction, and if bit D3 is low, the interface makes a transition to state GHJ = 101, which drives the interrupt low $\overline{\text{INTR}}$. Input signal $\overline{obf}$ (*output buffer full*) indicates data is available to peripheral device. It is a request for service from the peripheral device. Input signal *ack* (*acknowledge*) is a response from the peripheral device. Signals $\overline{obf}$ and *ack* cause a transition back to state GHJ = 001.

Design a clock-driven circuit for the state diagrams of Figure P6.8(*c*) and (*d*) using type II synthesis.

P6.9 The interface described in P6.8 can be programmed for mode 2 operation. In this mode the interface is capable of both input and output. The mode executive circuit of Figure P6.8(*a*) is used to place the interface in mode 2.

The input/output executive circuit state diagram of Figure P6.9(*a*) is used to ensure that the interface generates an interrupt for each *read* and *write*.

Starting with the mode executive in mode 2 and the I/O executive in state KL = 00 the interface is set for output. There is a transition to state QRS = 011 by way of 000 and 001. In this state an interrupt INTR signal is issued. A $\overline{wr}$ (*write*) instruction causes a transition to QRS = 111. The interrupt is cleared $\overline{\text{INTR}}$, *qrs* causes a transition in the I/O executive circuit to state KL = 11. Inputs $\overline{obf}$ and $\overline{ack}$ described in P6.8 cause transitions to either state QRS = 001 or 010 where the output circuit waits for the input circuit.

The input makes a transition to state MNP = 001 in a mode 2 (*m2*) input from the executive circuit. The interrupt is cleared $\overline{\text{INTR}}$. Input signals *ibf* and $\overline{stb}$ cause transition in the input circuit state diagram. If the I/O executive circuit is in state KL the transition is to state MNP = 110. The interrupt signal INTR is issued, and MNP causes a transition in the I/O executive circuit to state JK = 10. On a $\overline{rd}$ (*read*) instruction there is a transition to state

MNP = 001. The interrupt is cleared $\overline{\text{INTR}}$. It also causes a transition in the executive circuit to JK = 00.

Design this circuit.

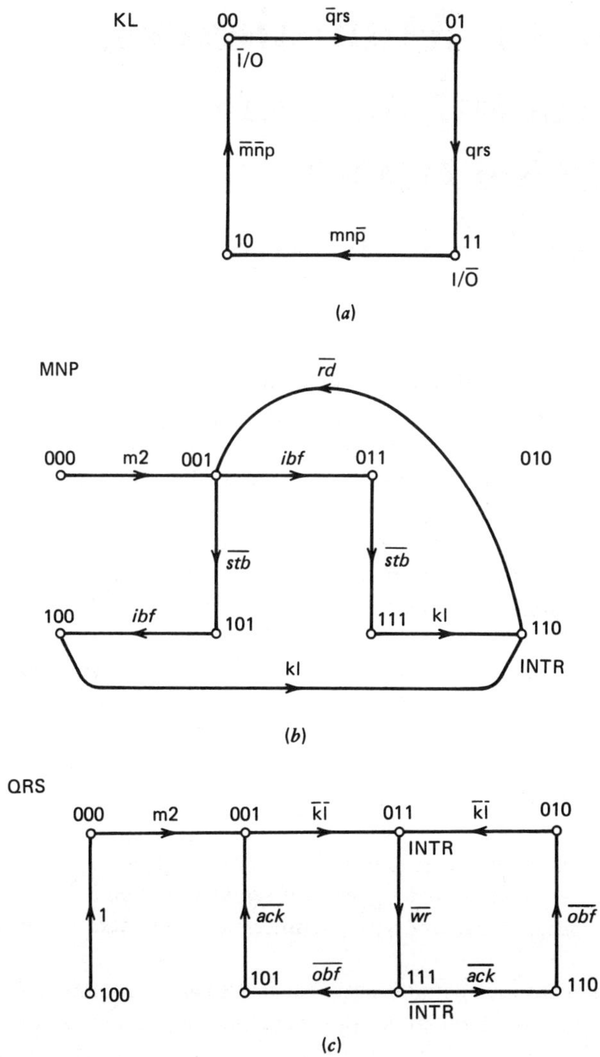

Figure P6.9 Input/output mode 2 control. *(a)* Input/output executive circuit; *(b)* input circuit; *(c)* output circuit.

CHAPTER 7

Counters, Multiplexers, and Demultiplexers in Clock-Driven Sequential Circuits

In Chapter 6 basic methods for developing the implementation of pulsed sequential circuits using JKFFs are developed. In that chapter two types of circuits are described: the type I basic circuit design is intended for applications that utilize one input variable, and the type II basic circuit design technique is applicable to multi-input variable cases. The type I, or single-input variable, circuits are very well known, and at present a great number of MSI chips using these principles are on the market. It then seems appropriate to develop techniques using type I circuits to replace the components normally used in type II circuit design. Indeed, such a utilization of chips leads to many basic design savings. In this chapter it is proposed to outline a method of designing multiple-input clock-driven sequential circuits (type II) using single-input devices (type I) [9, 10, 17].

7.1 COUNTERS AS MEMORY DEVICES

The general requirement in sequential circuits is to represent states by secondary variables. Since a counter is designed as a sequential circuit, it provides a natural way of representing states and gives a functional sequential circuit design that is compact.

A typical synchronous counter available as a single MSI chip is shown in Figure 7.1. The outputs are denoted A, B, C, D. The basic input to the circuit is the enable line. When it is high, pulses coming in on the clock line are counted in the normal fashion. Generally, a counter has an extra output called "carry," which goes high when ABCD = 1111, say. In order to increase the versatility of this circuit, there is usually a load line with its associated inputs. The load facility is used as follows: Counters are usually built to count from 0 to 9 (BCD) or 0 to 15 (binary). However, either counter can be modified in a straightforward manner. For example, if a BCD counter is to be used to achieve an 8 count, the counter can be modified by simply connecting the "load" line to the "carry" line. The associated

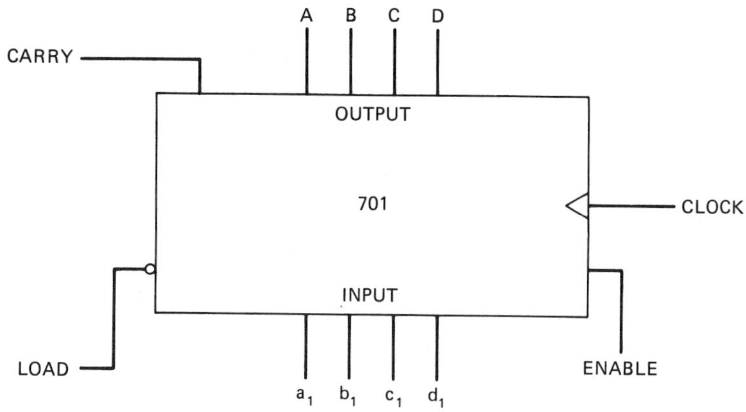

Figure 7.1 Typical MSI counter chip.

load inputs are connected such that $a_1b_1c_1d_1 = 0010$. When the count reaches 9 (BCD) the carry line energizes the load line, and the value $2(a_1b_1c_1d_1 = 0010)$ is loaded to the counter. Subsequent clocks count from 2 to 9, rather than 0 to 9, thus giving an 8 count rather than a 10 count. In this manner, any counter can be modified to count any number of clock pulses less than the maximum value of the counter.

This idea can be used to design sequential circuits. On an enable signal, a counter operates in a straightforward manner. On the load signal any value can be loaded onto the input terminals, and the counter takes the new value and passes it to the counter output. Thus the counter can be made to move from state to state in a normal binary sequence or loop from any desired state to any other desired state.

7.2 STANDARD MSI CIRCUIT FOR CLOCK-DRIVEN SEQUENTIAL CIRCUIT DESIGN

Figure 7.2 shows a multi-input configuration for designing sequential circuits using MSI components. The counter (702) is a basic synchronous unit. (It is important that ripple counters not be used in this application.) The counter progresses in the normal binary sequence when a clock pulse and an enable signal from the forward *mux* (703) are present. The forward multiplexer combines the proper state status (select lines) and the corresponding inputs to move in a normal progression. The loop *mux* (704) signifies the states and input variables for which the counter is to progress in a nonbinary sequential mode. Counters with a synchronous load terminal are recommended. Input terminals a_1, b_1, c_1, d_1 are connected to a combinational circuit to input the correct state destination. The *demux* (705) takes states from the counter onto its select lines and thus routes a signal to the proper outputs. The output of the counter must be connected to the select lines on the multiplexers as well. These connections, although not shown, are always implied.

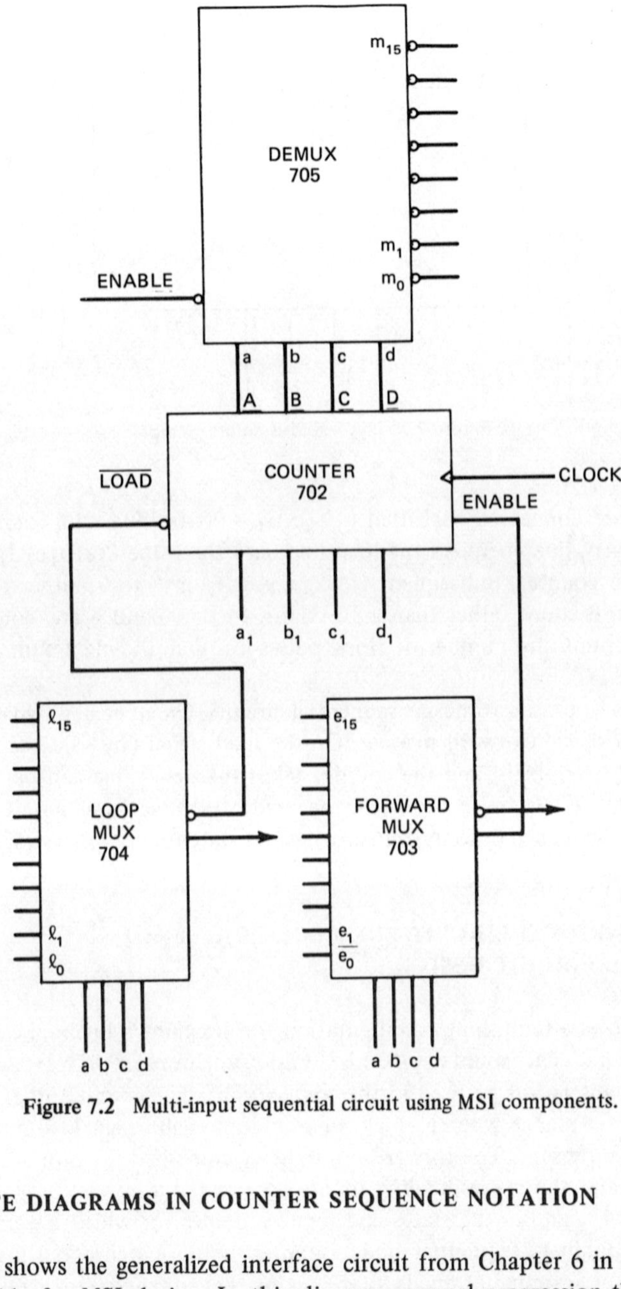

Figure 7.2 Multi-input sequential circuit using MSI components.

7.3 STATE DIAGRAMS IN COUNTER SEQUENCE NOTATION

Figure 7.3 shows the generalized interface circuit from Chapter 6 in state diagram form suitable for MSI design. In this diagram, normal progression through states follows a binary sequence. Inputs that cause a sequential binary transition in the counting sequence are called forward inputs. Inputs that cause state-to-state transitions that are not in binary sequence are called loop inputs. Thus a transition from state ABC = 000 to 001 is a binary transition, caused by an input function of the

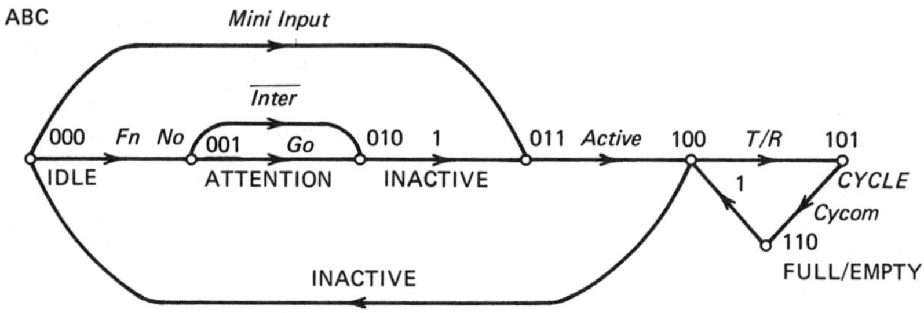

Figure 7.3 Counter application to the generalized interface state diagram.

forward *mux*. Similarly, the transition from ABC = 001 to 010 is a binary transition. Thus all inputs in the sequence ABC = 000, 001, 010, 011, 100, 101, 110, come from the forward *mux*. However, a transition from ABC = 000 to 011 is not a binary sequence; it is therefore a loop input and must originate from the loop *mux*. Similarly, the transitions from ABC = 110 to 100, and from ABC = 100 to 000 are not binary and are therefore assigned to the loop *mux*.

The forward multiplexer is shown in Figure 7.4. Its design is obtained from the state diagram by solving the forward *mux* multiplication equation as follows:

$$[E]\begin{bmatrix} a \\ b \\ c \end{bmatrix} = [FM]$$

The input matrix [E] is completed as follows: For column 0, ABC = 000, the input line that causes a transition from this state is labeled *Function Number* (*Fn No*). This is column 0, or input e_0. For column 1, ABC = 001, the input that causes a

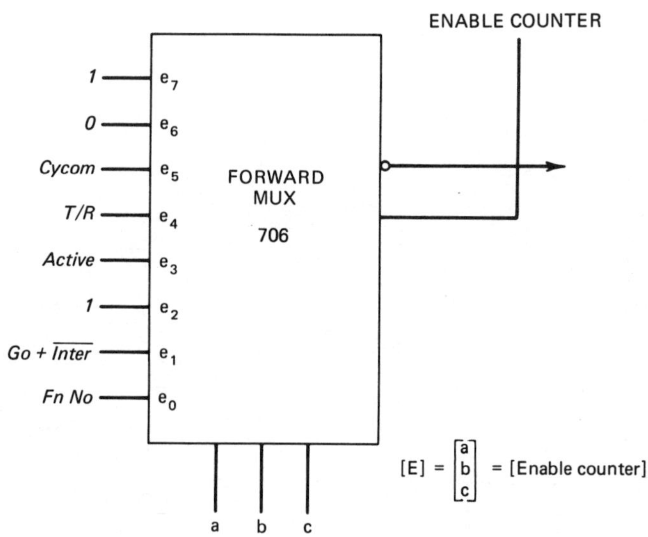

Figure 7.4 Forward multiplexer design.

transition from this state is labeled $Go + \overline{Inter}$. Thus $Go + \overline{Inter}$ forms column 1, or input e_1. Similarly, for ABC = 010 the input causing the transition is *1*. Thus *1* forms column 2, or input e_2. For ABC = 011 the input causing transition is *Active.* Thus *Active* forms column 3, or input e_3. For ABC = 100 the transition is caused by *T/R* so that *T/R* forms column 4, e_4, in the input matrix [E].

Similarly, for ABC = 101 the transition is caused by the input *Cycom.* Thus *Cycom* becomes column 5, e_5, in the input matrix. For ABC = 110 there must not be a transition to ABC = 111. Therefore there is no forward count from ABC = 110, and the input in column 6, e_6, of the input matrix [E] is 0. ABC = 111 is not defined in the state diagram. However, it is desirable to have a natural transition from state 7, ABC = 111, into the idle state ABC = 000. Thus a 1 is placed in column 7, e_7, of the input matrix.

The loop multiplexer is shown in Figure 7.5. The loop *mux* multiplication equation is as follows:

$$[L]\begin{bmatrix} a \\ b \\ c \end{bmatrix} = [LM]$$

Examination of the state diagram shows that the input matrix [L] can be completed as follows: There is a desired loop from state ABC = 000 to 011. This loop input is called *Mini Input.* It represents a transition from state ABC = 000, so *Mini Input* is connected to l_0. For ABC = 001 there is no loop, so $l_1 = 0$. For ABC = 010 no loop is shown in the state diagram; thus the input to $l_2 = 0$, and similarly $l_3 = 0$. However, when ABC = 100, there is a transition to ABC = 000 when the input *inactive* occurs. Thus the input to loop mux l_4 is *inactive.* For ABC = 101 no loops are present. Thus the input to $l_5 = 0$. For ABC = 110 there is a loop to 100 on a natural, 1, transition. Thus the input to l_6 is 1.

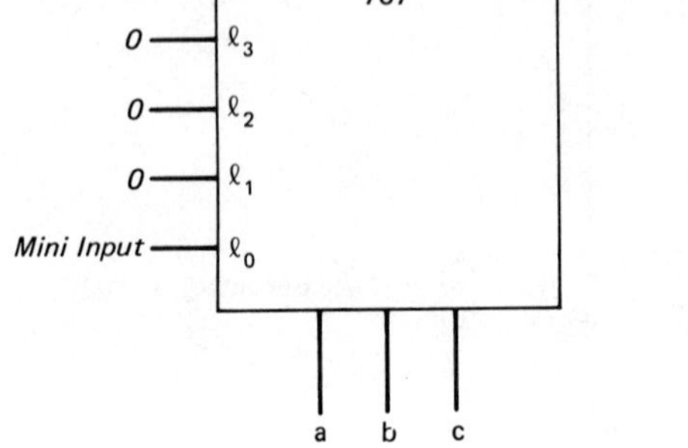

Figure 7.5 Loop multiplexer design.

C \ AB	00	01	11	10
0	IDLE	INACTIVE	FULL/EMPTY	
1	ATTENTION			CYCLE

Figure 7.6 K map demultiplexer design.

The state ABC = 111 is not shown in the diagram. However, there is a design transition from this state to 000 as given in the forward multiplexer loop design. It is not desirable to make this type of transition in the loop multiplexer, so the input to $l_7 = 0$.

The design for the demultiplexer is shown in Figure 7.6 Outputs are entered in the appropriate cell of this Karnaugh map. The demultiplexer is shown in Fig. 7.7.

The preliminary arrangement for the combinational circuit design to load lines a_1, b_1, c_1 on the counter is shown in Figure 7.8. From the loop *mux* design (Figure 7.5), it is evident that the counter must be loaded with state destinations appropriate for inputs l_0, l_4, l_6. The destination state is entered in the cell equivalent to each of these inputs. Destinations are determined from each loop in the state diagram. For l_0 the loop destination is ABC = 011. Thus in cell ABC = 000 the destination, 011, is entered. Similarly, for l_4, ABC = 100. The destination is 000. This value, 000, is entered in cell ABC = 100. For cell ABC = 110 in the Karnaugh map, the loop destination is 100. Thus cell ABC = 110 has the value 100. This completes

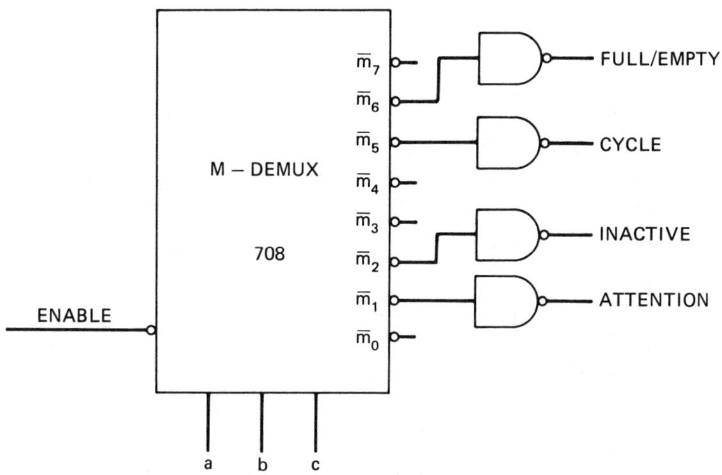

Figure 7.7 Demultiplexer circuit for general interface.

C \ AB	00	01	11	10
0	011	ϕ	100	000
1	ϕ	ϕ	ϕ	ϕ

Figure 7.8 Load circuit notation map $a_1 b_1 c_1$; $a_1 = B$, $b_1 = \bar{A}$, $c_1 = \bar{A}$.

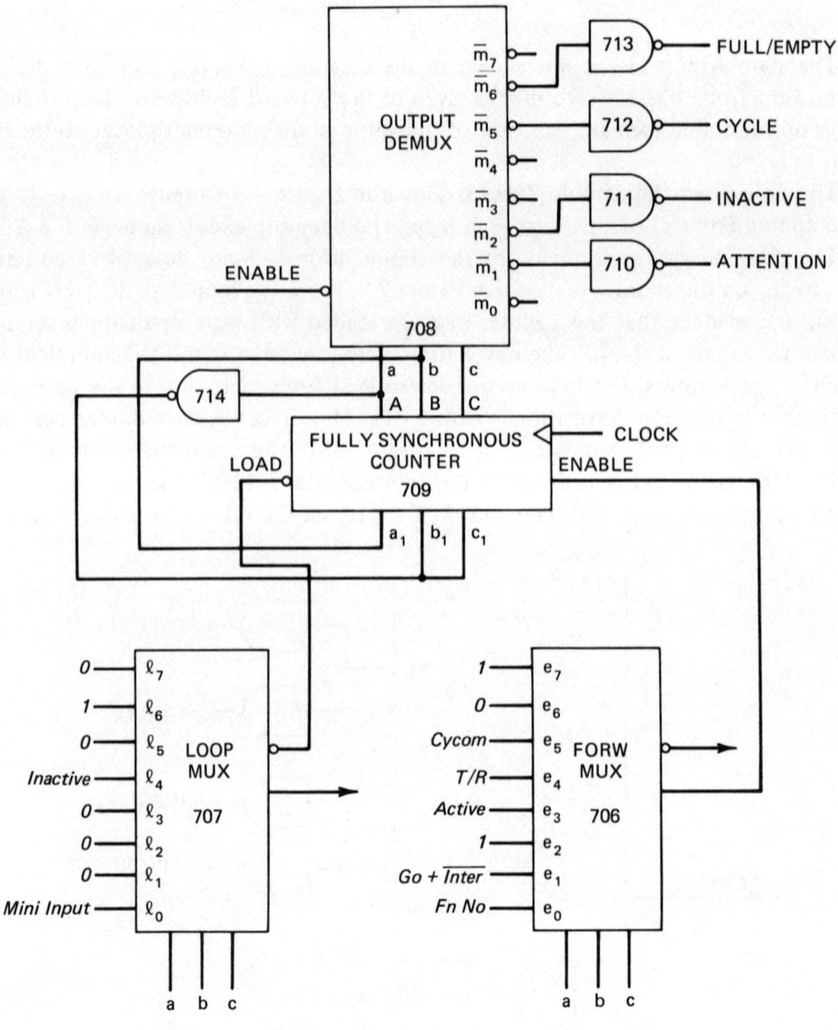

Figure 7.9 Generalized interface circuit using MSI components.

all the jump notation from the loop *mux* in Figure 7.5. All remaining cells in the Karnaugh map of Figure 7.8 receive don't care designation. This then becomes the map of the input values a_1, b_1, c_1. The logic function for the load lines as determined from the Karnaugh map are

$$a_1 = B, b_1 = \bar{A}, \text{and } c_1 = \bar{A}$$

These values are entered into the load lines of the counter. Figure 7.9 shows the completed generalized input interface circuit design using MSI components. The back connections from the output of the counter ABC to the select lines of the forward and loop *muxes* are not shown; but again these connections are implied and must be made when realizing the circuit.

7.4 SINGLE-OUTPUT CIRCUITS

In Chapter 6 the point is made that sequential circuits may be of two types: single and multiple inputs. In fact this distinction also applies to the output. The general approach in which each state is associated with a distinct output signal is described. However, there is a large application of sequential circuits in which each state simply designates the appropriate level of a single output. One important example is data transmission, which for reasons of distance and/or coding standards must be serial. Figure 7.10 shows a configuration for parallel to serial data conversion using MSI components.

As an illustrative problem the letter K is to be transmitted in ASCII code over an asynchronous communication line. The most common asynchronous communication line format is shown in Figure 7.11

Since this is an asynchronous format the start bit and two stop bits are used to frame the transmitted character. They are so named because frequently the start bit is used to start the receiver's clock, and similarly the stop bits are used to stop the receiver's clock.

In digital communications terminology, the start bit at logic zero level is called a space. The stop bits at logic one level are called marks. Since the operator of teletype or CRT is certain to pause, transmission between characteristics is filled with marks.

The hexadecimal equivalent of K (4B) in seven-character code is found in Appendix B Table B.3. Framing 4B with start and stop bits, the complete chararceristic cycle required is 010010111. A state diagram for the generation of K in this format using a BCD counter is given in Figure 7.12.

The design of the output *mux* is shown in Figure 7.10. Its design is obtained from the state diagram by solving the output multiplication equation as follows:

$$[S]\begin{bmatrix} a \\ b \\ c \\ d \end{bmatrix} = [SM]$$

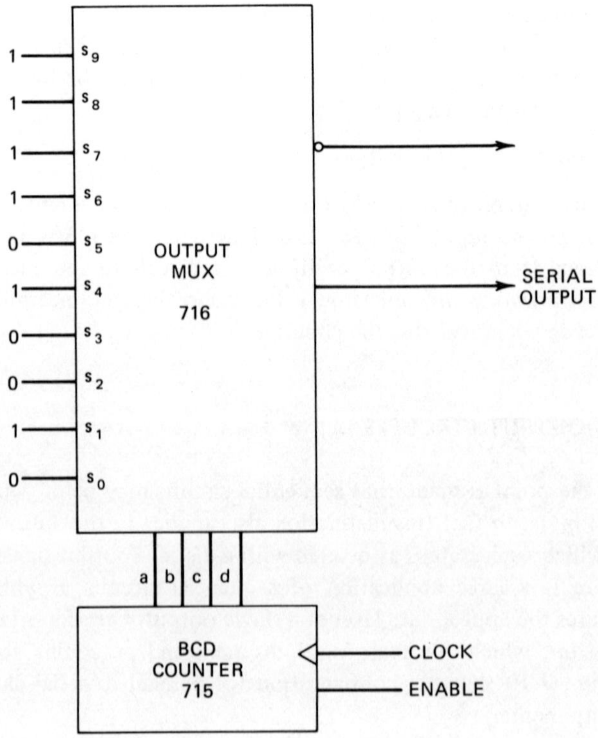

Figure 7.10 Single output sequential circuit using MSI components.

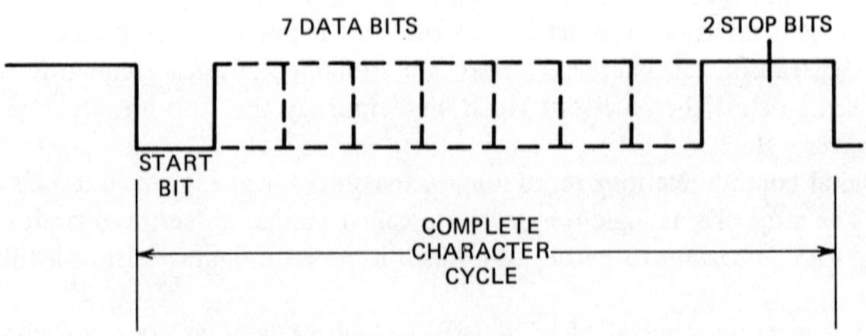

Figure 7.11 Asynchronous data character format.

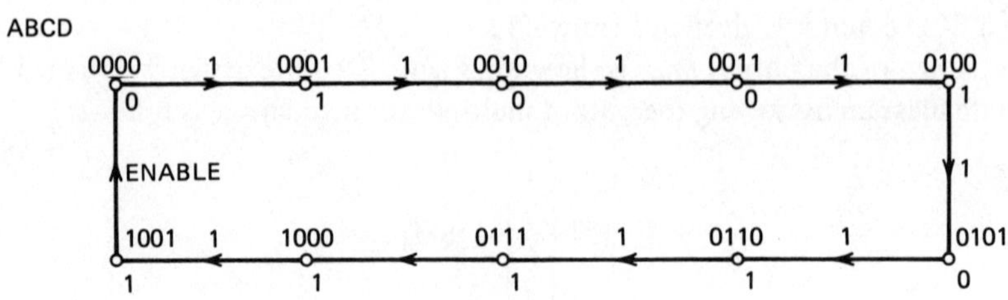

Figure 7.12 State diagram for data character.

The input matrix [S] is completed as follows: For column 0, ABCD = 0000, the output desired is 0, so $s_0 = 0$. For column 1, ABCD = 0001, the output desired is 1, so $s_1 = 1$. For column 2, ABCD = 0010, the desired output is 0, so $s_2 = 0$. Similarly

$$s_3 = 0$$

$$s_4 = 1$$

$$s_5 = 0$$

$$s_6 = 1$$

$$s_7 = 1$$

$$s_8 = 1$$

$$s_9 = 1$$

In this application the values of inputs s_0, s_8, and s_9 are fixed. Other characters are transmitted by changing the inputs to s_1 through s_7.

As described in Section 7.1, counters can be modified in a straightforward manner. Thus the cycle length is easily changed.

7.5 DESIGN FOR FLEXIBILITY IN CLOCK-DRIVEN SEQUENTIAL CIRCUITS

The method of design proposed in this chapter shows that a standard configuration of components can be used to design any circuit. In Chapter 6, the concept of programmability to improve flexibility is introduced. From a practical point of view it seems that an extremely efficient sequential circuit could be designed so that it would fit on one printed circuit (PC) board.

This approach is often used in one form or another. However, a further problem occurs as a result of this efficiency. The problem cannot really be solved, but it can be circumvented: the circuit designed generates or needs so many different data and information signals that the number of connections would increase the size of the board; the idea of time multiplexing data as given in Section 7.4 reduces connections.

The following principles are commonly used in microcomputer interfaces.

Figure 7.13 shows a state diagram for time multiplexing information to (2 or more control words) into the interface storage register.

The circuit is normally in the idle ABC = 000 position. On an input *rd* there is a transition to ABC = 001. In this state the value of a register is READ from the interface. On the next clock the circuit returns to the IDLE state. On an input *wr* there is a transition to ABC = 010. In this state the contents of WRITE are transferred to the interface register CW4. The interface sequential circuit tests bit *D5* of CW4. If bit *D5* = 0 there is a transition to ABC = 000 on the next clock pulse. However if bit *D5* = 1, there is a transition to ABC = 110 if s = 0 (s = D1D0), and

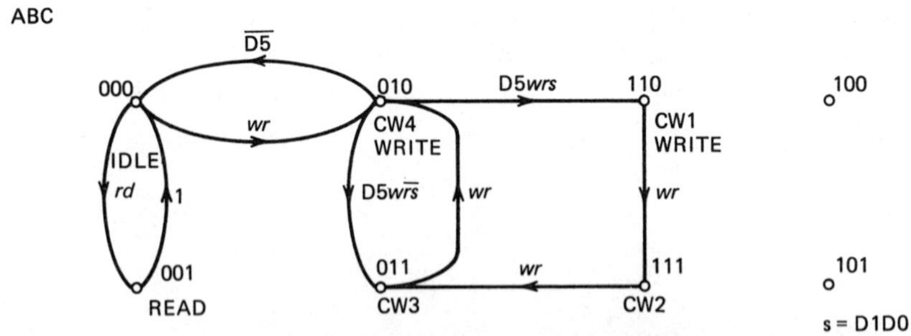

Figure 7.13 Time multiplexing information.

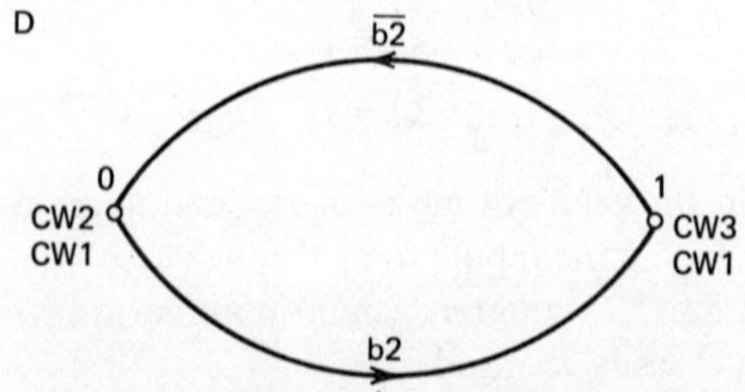

Figure 7.14 Time multiplexing 2 out of 3.

to ABC = 011 if s = 1 on input *wr*. In this state the contents of WRITE are transferred to the interface register CW1.

Interface registers CW2 (ABC = 111) and CW3 (ABC = 011) are each loaded on consecutive inputs of *wr*.

Notice the feature of this circuit. CW4 can be changed at any time. If CW4 register bits s = 1, access to registers CW1, CW2, CW3 is obtained. If s = 0 only two control words CW3 and CW4 are loaded into the register. Bit *D5* is called an internal reset.

The above technique used one set of lines to load two or more registers. There is another technique that uses two sets of lines to load three registers. The state diagram for this two out of three time multiplexing is given in Figure 7.14.

If two sets of lines are available the required state diagram is much simpler. However it is a more difficult interface to understand from the programming point of view.

In state D = 0 information can be sent to registers CW1 and CW2 on the two sets of lines. If CW1 bit *b2* = 1 there is a transition to state D = 1. In this state information can be sent to registers CW1 and CW3 on the two sets of lines.

7.6 PROBLEMS

P7.1 The state diagram shown in Figure P7.1 is used to generate the following serial information:

- Start bit. A high to low transition is used to cause the output to generate a starting signal.

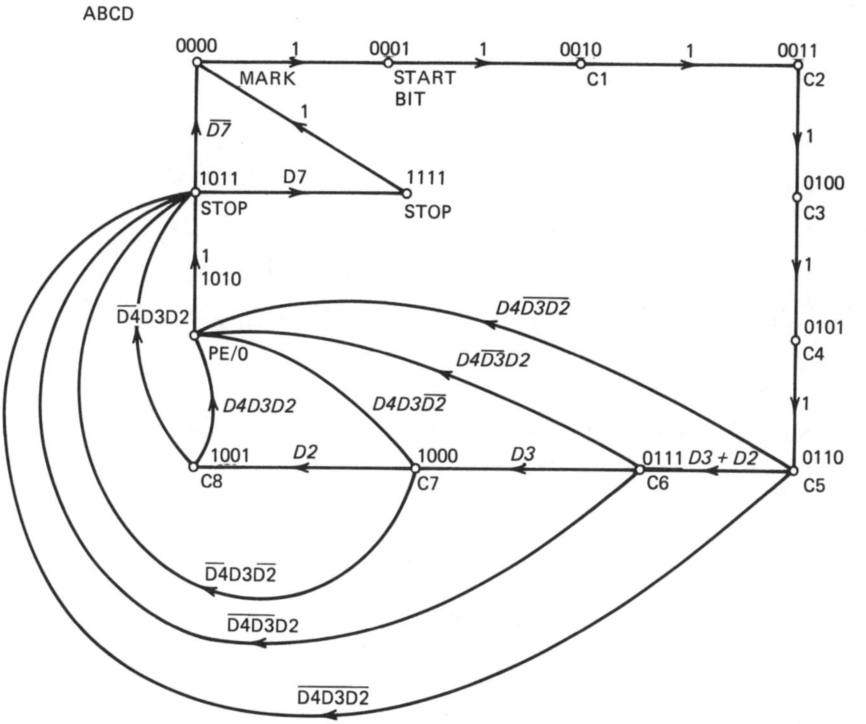

Figure P7.1 Character generator.

- Character bits. The circuit is selectable in the number of characters generated. $\overline{D3}\,\overline{D2}$ causes the circuit to generate five characters, $\overline{D3}D2$ causes six, $D3\overline{D2}$ causes seven, and $D3D2$ causes eight characters to be generated.
- Parity is generated on $D4$ and not generated on $\overline{D4}$. Another feature of the circuit, ignored for the purposes of this exercise, is that $\overline{D5}$ generates odd parity.
- The presence of $D7$ causes two stop bits to be generated. On $\overline{D7}$ only one stop bit occurs.

Design this circuit.

P7.2 The handshake routine for a serial interface might be described in hardware terms using the state diagram in Figure P7.2.

Starting from the IDLE state EF = 00 the interface receives the signal *data set ready* ($\overline{dsr}$) from a peripheral device. There is a transition to state EF = 01 where an output of response signal, data terminal ready (DTR), is issued. The control register of the interface is checked. If $D0 = 1$, enable transmission, a transition to state EF = 10 occurs. An output signal request to send ($\overline{\text{RTS}}$) is sent to the peripheral device. It responds with the input signal *clear to send* ($\overline{cts}$), which causes a transition to state EF = 11. The output signal RUN is issued until the input signal *disable* ($\overline{DO}$) is received. Design this circuit.

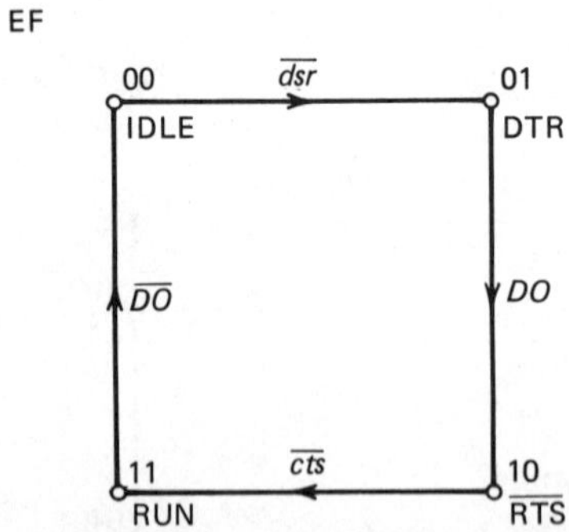

Figure P7.2 Serial interface start-up.

P7.3 Control of interface handshake custom for serial transmission of data is often accomplished by programming. As an exercise the state diagram for interrupting a microcomputer is given in Figure P7.3.

Starting in the IDLE state GHJ = 000 a transition to state GHJ = 001 occurs when the RUN signal from P7.2 is received. In this state a FLAG is sent to the microcomputer. The microcomputer tests bits $D0$ and $D2$ of the interface. If bit $D0 = 1$ it is interpreted as a transmit enable, T X EN, and the microprocessor responds with data and a write, $\overline{wr}$, command. A transition to state GHJ = 010 occurs. The flag is removed. On transmission of data by the input signal *out* a transition to state GHJ = 011 occurs. The flag is again issued, and the microprocessor tests bit $D0 = 1$ again. The interface cycles through states GHJ = 010 and 011 to transmit data.

If the microcomputer finds that bit $D2 = 1$, it is interpreted as a receive enable R X EN, and an input $\overline{rd}$ signal is issued. This causes a transition to state GHJ = 110. The flag is cleared. When data is received on an input *in* a transition to state GHJ = 111 occurs. Bit $D2$ is tested, and when $D2 = 1$ the microcomputer generates the input signal $\overline{rd}$. Data is received as the interface cycles through GHJ = 110 and 111. The interface is reset when register bit $D6 = 1$.

Design this circuit.

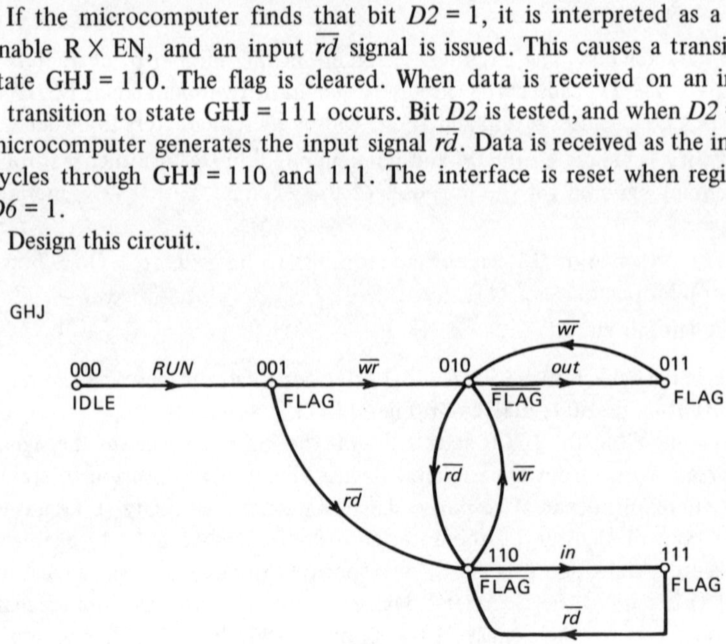

Figure P7.3 Transmission state diagram.

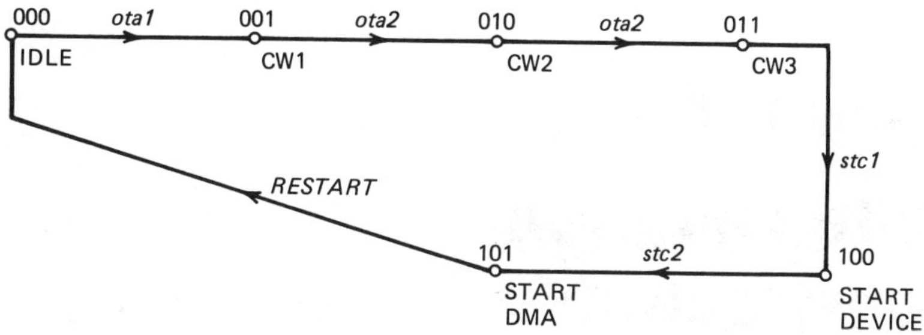

Figure P7.4 DMA transfer.

P7.4 A certain minicomputer manufacturer used three 16-bit words to initialize a direct memory access interface. The state diagram for register access is given in Figure P7.4. CW1 contains the peripheral device address. CW2 contains the first location in memory. CW3 contains the number of words required.

The input signal *ota1* causes a transisiton to XYZ = 001 where CW1 is sent to the interface. Similarly *ota2* causes transition through XYZ = 010 and 011, and CW2 and CW3 are loaded into the interface control register. Input start command *stc1* causes a transition to state XYZ = 100 where the output START DEVICE is transferred to the peripheral device. Input *stc2* causes a transition to state XYZ = 101 where the DMA transfer begins.

Design this circuit.

CHAPTER 8

State Diagram Construction

So far in this book the reader has been asked to design circuits from the state diagram. The steps followed in this procedure are extremely important. However, one must realize that the state diagram is the foundation of the complete design procedure. Generation of a state diagram is accomplished by way of transfer of a word statement of a problem into a diagrammatical statement of the same problem. Many things are assumed here. The first is that a word statement of the problem exists. In reality it is seldom so. The designer is usually called upon not only to generate the state diagram but also to generate a word statement of the problem. In practice the problem is envisaged from a discussion, and the designer goes away to put the problem on paper. The first attempt at definition is followed by further discussion. The best comment that a designer can really hope for is as follows: "No, I don't know how it should work; however, it shouldn't work that way."

In this chapter for practical purposes it is assumed that a word statement of the problem exists. The techniques described concentrate on converting this word statement into a state diagram. The steps in converting from a word statement of a problem to a state variable diagram are given in the following sections.

8.1 FORMAT

As mentioned, state diagrams are used to convey all the information about the sequential circuit to be designed. In order to accomplish this information must be transmitted in very subtle ways.

Inputs are represented as lines joining states in the diagram. Arrows indicate the direction of the sequence of events. Inputs come from external sources. So that they can be clearly identified, arrows are placed at the center of the line, and the corresponding input name that causes the transition is placed immediately over that arrow. If the line is vertical the name is placed to the immediate right.

States are represented by nodes in the form of small circles. The names of the state variables are placed in the upper left-hand corner of the diagram. The value of each state variable is placed above or above and to the right of the circle. Outputs are placed below the line and to the right of the circles. This position indicates the general intent that outputs are to be associated with a given state.

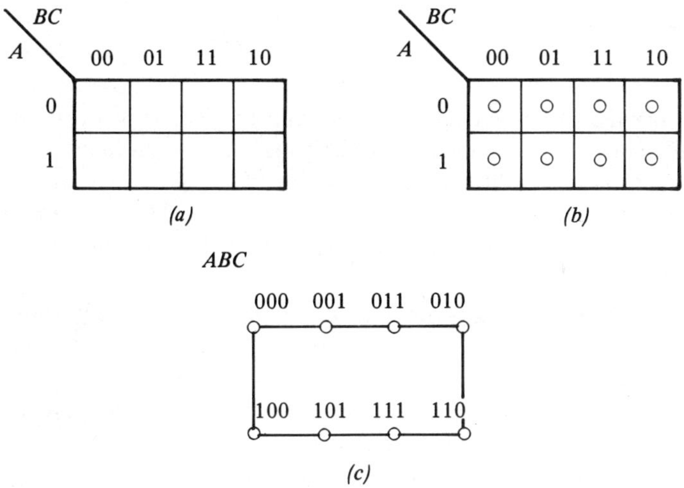

Figure 8.1 Cell-state relationship for unit distance to the columns. *(a)* K map; *(b)* conversion; *(c)* state equivalence.

8.2 A CELL-TO-STATE DUALITY

Figure 8.1 illustrates a method of converting a Karnaugh map to a state diagram with adjacent states one unit distance apart. In this process a four-cell map generates a four-state diagram, eight cells generate eight states, and so on. In Figure 8.1(*a*) the eight cells for a three-variable Karnaugh map are shown. The first variable(s) is assigned to the rows, with the remaining variables assigned to the columns. In Figure 8.1(*b*) a state (small circle) is placed within each cell. Each state assumes the values of the variables assigned to the cell within which it is located. Figure 8.1(*c*) shows the generated state diagram from Figure 8.1(*b*) with the Karnaugh map removed. The variable names (ABC) are placed in the upper left-hand corner of this diagram. The value of each variable is placed directly above the state with which it is associated.

Each state in this diagram is one unit distance from its adjacent state.

This conversion is useful in the design of state diagrams for all event-driven circuits and for pulse sequential circuits with multiple inputs to be realized using JKFFs. It is used specifically in Sections 8.4 and 8.5.

8.3 CONSTRUCTION OF THE PRELIMINARY STATE DIAGRAM

Construction of a preliminary state diagram can be accomplished by using the following steps:

1. Determine the inputs and outputs of the logic system to be designed and draw a block diagram placing inputs on the left and outputs on the right. The difficulty to

be faced here is that inputs to one circuit are usually outputs to another and vice versa. So the problem must be broken down to identified pertinent input output (I/O).

2. Determine the cause and effect relationship between inputs and outputs. In a sequential circuit the basic premise for design is that a given input produces a corresponding output. In reality, a number of inputs in a particular sequence may be required to produce an output, or after an input or a number of inputs a number of outputs may result. It is the cause (input) and effect (output) that is the dominant information.

3. Determine the state from which a sequence of events normally starts. One way to determine this state is to find the state in which the circuit would normally wait or in which the designer would like to see the circuit return on a reset command. This state is placed on the upper left of the diagram, and all state variables are usually assigned the value zero for the state.

4. Starting with the inputs, arrange them in a sequential progression of events.

5. When a state occurs from which the next input is optional depending upon an external decision, two or more output lines normally exit this state.

6. One should always check to see that inputs that should lead to the same state destination do so. An understanding of the redundancy checks given in Chapter 3 helps in the determination of destination states.

7. The procedure should be continued until all states are part of a loop or lead into a loop. Although it is possible for a state to be terminal it is not desirable. Thus a terminal state must be regarded as a probable error in construction of the state diagram.

8.4 COMPLETION OF THE STATE DIAGRAM FOR EVENT-DRIVEN SEQUENTIAL CIRCUITS

The state diagram for event-driven sequential circuits can be completed using the following steps:

1. From the preliminary state diagram, determine the number of states required. Lay out 2^n states where $2^n \geqslant$ the number of states required. These states should be laid out such that four states form a two-by-two figure; eight states form a four-by-two figure; or sixteen states form a four-by-four figure.

2. Determine the state variable names (NP, ABC, WXYZ, etc.) such that there are n variables and assign each state a value for each variable. Use the duality method of Section 8.2. Be sure that adjacent states change in the value of one variable only.

3. Place preliminary sequential states on the diagram. If there is an even number of states in a loop (2, 4, 6, 8, 10, etc.), the reader should be able to find a sequential progression from state to state to complete the loop exactly. If there is an odd number of states, an extra natural (input = 1) transition must be generated so that the loop can be completed exactly.

4. In many problems parts of a sequence may be repeated one or more times. This leads to minor loops. Minor loops also must have an even number of states. This may require the addition of natural transitions. However, another solution, which if acceptable is very attractive, is to return to the initial state in the minor loop by reversing the direction of sequencing through states. In the case where the return requires reversing through a number of states, it may be necessary to utilize the same input for many transitions.

5. All minor or additional loops in the state diagram must be placed on the diagram.

6. All states not included in the main sequential loop or secondary loops should be directed to the idle state using natural transitions. It is assumed that all circuits are provided with a reset to the idle state capability.

8.5 COMPLETION OF STATE DIAGRAMS FOR CLOCK-DRIVEN CIRCUITS USING JKFFS

From the word statement of the problem, the sequential progression through states in a pulse-driven circuit may well be defined. Indeed, for a great number of circuits, the sequence through the states requires a binary progression or a reverse binary progression. Under such a condition the assignment of variables and the value of variables in each state is done in strict accordance with design requirements.

For the other cases, where the states are not defined, a procedure similar to that outlined for event-driven circuits can be used and is as follows:

1. From the preliminary state diagram, determine the number of states required. Lay out 2^n states where 2^n is greater than or equal to the number of states required. These states should be laid out in the normal 4, 8, 16 type of matrix.

2. Assign state variable names such that there are n state variables. In this case it is convenient in the design of the state diagram to change the value of one variable only in a transition. However, since JKFFs are capable of double transitions, this is not a requirement; it is simply a matter of convenience that generally leads to a simpler implementation.

3. Place the major loop from the preliminary states on the diagram. However, here there is no requirement for an even number of transitions.

4. All additional or minor loops in the state diagram must be placed on the diagram.

5. All states that are not included in any of the defined sequences should be directed to the idle state by use of natural transitions.

8.6 COMPLETION OF THE STATE DIAGRAM FOR PULSE SEQUENTIAL CIRCUITS USING MSI COMPONENTS

When the values of the variables in the state diagram are not defined, the following procedure should be used to design state diagrams for MSI implementation:

1. From the preliminary state diagram determine the number of states required. Lay out 2^n states where 2^n is equal to or greater than the number of states required.

2. Determine the state variable names such that there are n variables.

3. Identify the main loop in the system; starting from 0, number this loop progressively in a binary sequence. For example, for a three-variable state diagram, ABC, the main loop should be numbered as follows: ABC = 000, 001, 010, 011, 100, and so on. The next largest minor loop should then be numbered in a similar binary sequence following until all states are given a number.

4. The diagram should be examined to guarantee that where branching occurs at least one of the transitions follows a binary sequence. If this is not the case, an extra state should be added with a natural transition exiting from this state to achieve the binary sequence. This condition is not mandatory but is extremely useful in simplifying the design.

5. All states that are not included in any of the loops should be directed into the loops using natural transitions.

8.7 EXAMPLE OF AN EVENT-DRIVEN SEQUENTIAL CIRCUIT STATE DIAGRAM CONSTRUCTION

The following paragraphs give a problem statement for an emergency power start-up circuit.

A diesel and ac generator set is used for emergency power in a chemical plant. When the voltage on the main bus falls below 40% rated, the diesel fuel line is opened and the starter motor is energized from the battery for 3 s. After a further 3-s delay, the speed of the set is checked, and if it is below 10% full speed, the start sequence from energization of the starter motor is to be repeated. After the third unsuccessful attempt at starting the set is to be locked out with fuel lines closed. A manual reset is required to return the system to the start of the sequence.

If the speed is equal to or above 10% full the starting circuit is to hold until 95% full speed is reached.

The field voltage is to be applied to the ac generator, and when the speed reaches 98% of rated the emergency supply breaker is to be closed. Shutdown is initiated by a manual command. Control is tranferred to another circuit for this operation.

Design the state diagram for this control circuit.

Solution. The first step of the design is given in Section 8.3. The steps are executed as follows:

1. The inputs and outputs must be determined from the word statement of the problem. The inputs are the following:

(a) V_{40} Voltage below 40% rated. This is clearly an input signal and is the initiating signal required to start the whole process.

(b) Δ_3 Time delay of 3 s. There are two 3-s time delays required to control this circuit. The most logical approach seems to be to design a 3-s time delay unit to interact with this circuit.

(c) $\bar{S}_{10}$ Speed below 10% full rated. This input signal from the generating unit is required to check that starting occurred.

(d) *R* Repeat starting sequence. Here the designer has some choice. The starting sequence may be repeated three times. It seems more efficient to assume (design) an external circuit to count the number of starts attempted.

(e) *US* Unsuccessful start. This signal signifies that three unsuccessful attempts at starting have occurred.

(f) *Reset* Return to normal starting sequence.

(g) S_{95} Rated speed of 95% has been reached.

(h) S_{98} Rated speed of 98% has been reached.

(i) *SD* Initiate the shut-down sequence. On the shut-down command, the circuit could return to the idle-on position (a natural transition) and this would meet the design specifications according to the word statement of the problem. However, the author suggests that here the designer has an opportunity to use his judgment. It may be in order to ask the designer of the shut-down circuit to provide a shut-down complete signal *SDC*, since that designer is likely to be well known. In this way, one could guarantee that the start-up and shut-down circuits never lose synchronism.

The outputs are the following:

(a) OFL Open fuel line.

(b) EN Energize the starter motor from the battery. This is the operation that cranks the diesel.

(c) CFL Close fuel line. On an unsuccessful start-up attempt, the fuel lines must be closed.

(d) FV Field voltage is applied to the ac generator when the speed reaches 95%.

(e) CB Close breaker. The emergency supply breaker is closed when the unit reaches 98% full speed.

(f) SSD A start shut-down sequence command is to be issued to the shut-down circuit.

(g) X An output must be generated from this circuit to count the number of start-up attempts. This output is added by the designer for inter-logic circuit communication.

The input-output block diagram for this event-driven circuit is shown in Figure 8.2.

2. The next step in the design is to determine a cause and effect relationship. This relationship is given in Table 8.1.

Table 8.1 **Cause and Effect**

Cause (input)	Effect (output)
V_{40}	OFL, EN
Δ_3	X
$US \cdot \bar{S}_{10}$	CFL
R	EN
S_{95}	FV
S_{98}	CB
SD	SSD
Reset	

3. The state from which the sequence normally starts is immediately before input V_{40}. This is the situation where the main system is operating normally with the voltage well above 40%.
4. The first event to occur is loss of system voltage. The preliminary state diagram is shown in Figure 8.3. Starting from the idle state the first input V_{40} causes a transition to the next state. This causes the output open fuel line OFL and energizes the starter motor EN. Next a sequence of two 3-s delays occurs. The first 3-s delay determines the length of time that the motor is cranked; the second 3-s delay allows time for the diesel to accelerate.

The diesel may or may not start. Thus the sequential circuit must have branching capabilities explained under step 5. If the speed has not reached 10%, $\bar{S}_{10}$, and if there have not been three start-up attempts $\overline{US}$, a return R for another attempt at starting is made using a minor loop. If after the third attempt start-up US is unsuccessful, $\bar{S}_{10}$, a transition to a new state is provided, the command close fuel line CFL is issued, and this state is maintained until a *Reset* signal returns the circuit to the starting point.

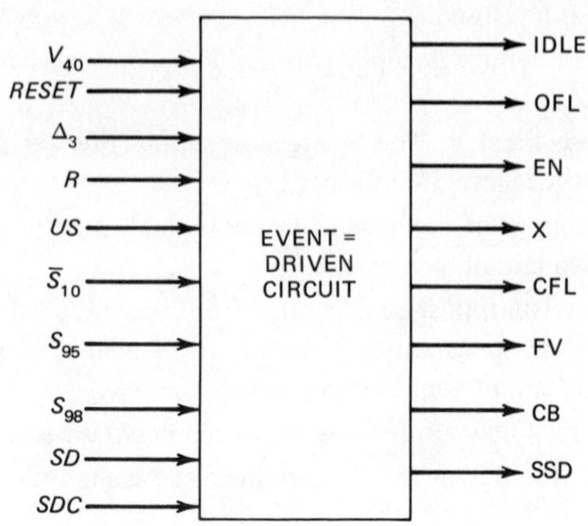

Figure 8.2 Block diagram of problem formulation.

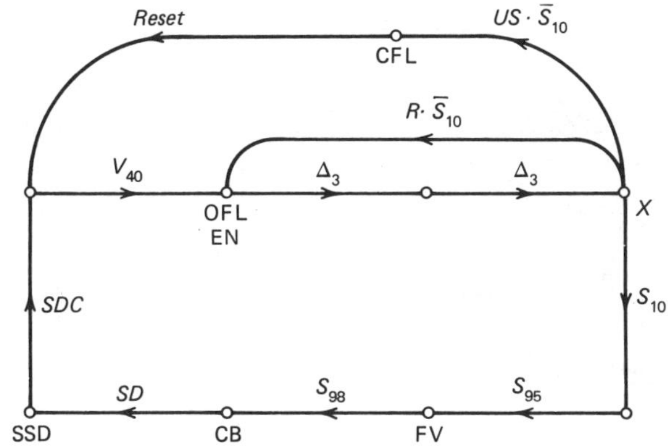

Figure 8.3 Preliminary state diagram for diesel starting.

If the start-up is successful S_{10}, there is a transition to another state. There is a further transition when 95% full speed is achieved. In this state field voltage FV is applied to the ac generator. At 98% full speed there is a transition to a state where the output close circuit breaker CB is issued. The diesel unit is now on line, and emergency power is being supplied.

With a shut-down request *SD* there is a transition to the next state, and an output, start shutdown SSD, is given to the shut-down circuit. A shut-down complete signal *SDC* returns the device to the idle state. Note here that the problem statement does not specifically say that the circuit is to be returned to the idle state, but this is always an implication under step 7 of the state diagram construction. All states in this diagram are unique so step 6 is satisfied.

The preliminary state diagram is now complete. The major loop and all minor loops are closed.

Completion of the state diagram construction as described in Section 8.4 is achieved as follows:

1. Inspection of Figure 8.3 shows that at least nine states are required. Since 2^n must be greater than or equal to 9, n is chosen to be equal to 4, so a map with 16 states and 4 variables is required. Such a map is shown in Figure 8.4.
2. The variables in this map are labeled ABCD.
3. The IDLE state is placed in the upper left-hand corner. The transition caused by V_{40} is laid out. However, at the designer's choice, the outputs OFL and EN are assigned to separate states by addition of a natural transition.

The major loop has nine transitions because of the addition of the natural transition in the larger minor loop. The major loop transition must be increased by the addition of another natural transition. The major loop is placed on Figure 8.4.

4. The addition of the first natural transition also increases the length of the largest minor loop from five to six transitions. This is an even number of transitions which permits the larger minor loop to be placed on the diagram.

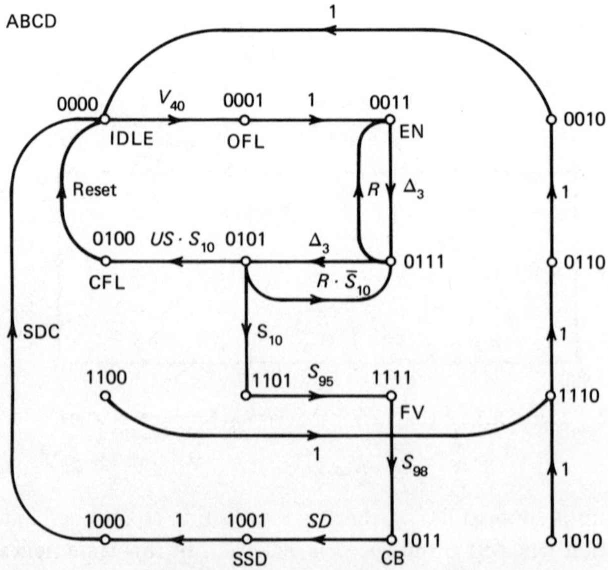

Figure 8.4 State diagram of emergency diesel start-up.

5. The smaller minor loop contains three transitions. This is an odd number. Thus there are two ways to overcome this problem.
   (a) Increase the loop by adding a natural transition.
   (b) Return to the start of the minor loop by reversing the sequence. It is the author's practice to attempt to complete small minor loops with an odd number of transitions by reversing the direction of transitions in the major loop. So in this case the direction of transition in the major loop is reversed.
6. There are five remaining unused states. They are routed into the sequence such that the entry point is the idle condition.

One task remains to be done—construction of the restart state diagram. The signals restart R and unsuccessful start US must be generated. The problem definition for this circuit can be given as follows.

A circuit must be designed to have an output US when the transition ABCD = 0011, 0111, 0101 in the emergency power start-up circuit is made three times in succession. It must have an output R when the sequence ABCD = 0011, 0111, 0101 is completed, but not when the sequence is partially completed, that is, ABCD = 0011, 0111. To accomplish this design, the output X introduced by the designer to state ABCD = 0101 of the start up is used.

The first step in the construction of this state diagram is to identify the inputs. Inputs to the restart circuit are outputs from the emergency starting circuit. The outputs EN and X are chosen since the sequence of events in which these two outputs occur represents the sequence in which the states ABCD = 0011, 0111, 0101 are transgressed. So the restart circuit can be returned to its initial state, the IDLE

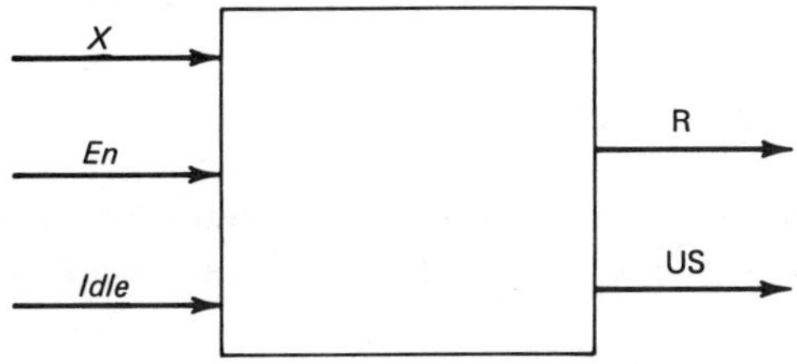

Figure 8.5 Block diagram of restart circuit.

output form the emergency start-up circuit is selected as another input. The outputs are R and US.

The input output block diagram is shown in Figure 8.5. The preliminary state diagram is shown in Figure 8.6. This diagram is constructed such that when the input X has occurred three times, the output unsuccessful start US is generated. This output is transmitted to the emergency diesel start-up circuit where it is anded with $\bar{S}_{10}$ to determine if the third start-up attempt is also unsuccessful. On the first and second occurrence of the input X, the transitions are to states with an output R. With input *En* the transition is to a state with output $\bar{\text{R}}$. This is to guarantee that the output R occurs only after the sequence *En*, X and that $\bar{\text{R}}$ occurs only after the sequence X, *En*.

The preliminary diagram has six states. Thus $2^n \geqslant 6$, $n = 3$, and there are eight states. Figure 8.7 shows the restart state diagram. Variable names KLM are assigned.

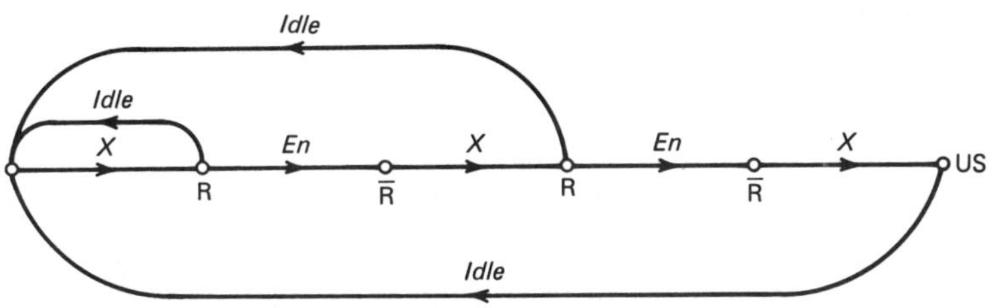

Figure 8.6 Preliminary restart state diagram.

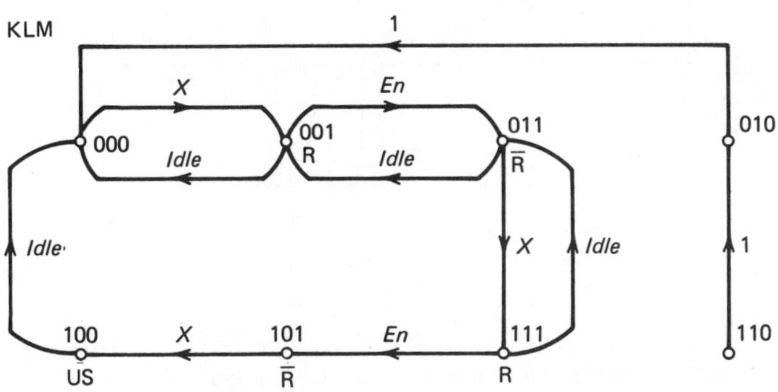

Figure 8.7 Completed restart state diagram.

It follows the straightforward progression of the preliminary state diagram. Note that in this diagram the major loop involves six states and there are two minor loops. Since all the transitions in these loops are even, no modifications in the transitions are required to guarantee that the loop can be completed exactly. However the largest minor loop is completed using reverse transitions. This overcomes the problem created when six-state and four-state loops occur together.

8.8 EXAMPLE OF A PULSE SEQUENTIAL CIRCUIT STATE DIAGRAM CONSTRUCTION

The following is an example of a pulse-driven sequential circuit. It is a direct memory access interface between a microprocessor and a data bank (peripheral device).

Figure 8.8 gives a block diagram of the basic structure for a microprocessor (μP). The central microprocessor unit CPU is controlled by a program normally resident in the read only memory ROM. The conventional method of loading data into the read write memory (random access memory, RAM) is from the input-output port (I/O).

Under these circumstances time is wasted. For example, the large amount of data to be acquired would normally pass from the I/O to the RAM by way of the CPU. If loaded directly from the I/O to the RAM as indicated by the dotted lines in Figure 8.8, half the time could be saved. This method of acquiring data is called direct memory access (DMA).

A problem statement for the design of a control circuit to handle DMA is as follows:

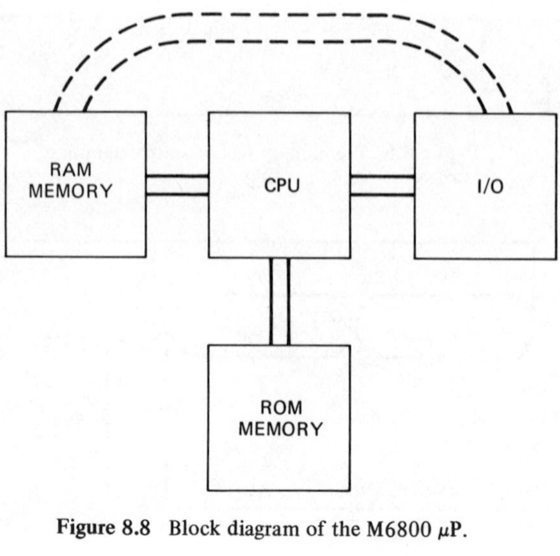

Figure 8.8 Block diagram of the M6800 μP.

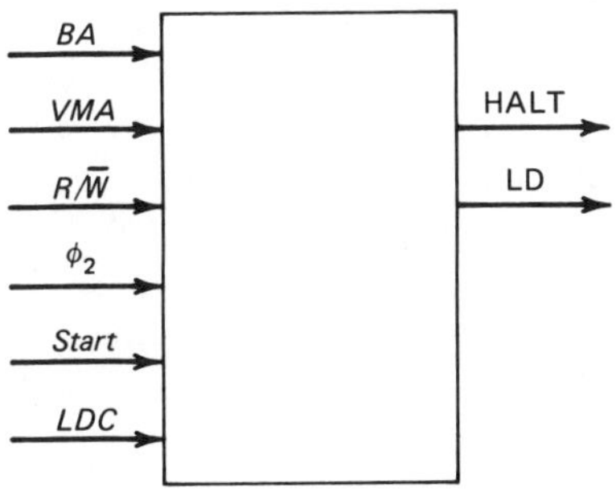

Figure 8.9 Input-output block diagram for M6800 DMA.

The HALT line serves as a DMA request signal. The bus available (BA) line serves as the DMA acknowledge signal. The ϕ_2 signal is used as a clock. The DMA subsystem must provide a valid memory address (VMA) signal to enable the RAMs for transfer and an $R/\overline{W}$ signal to establish direction of flow.

Can you draw the input output block diagram? Rereading the above paragraph, we find that HALT serves a request to DMA so it must be an output from the interface. *BA* is an acknowledge so it is input to the interface.

Since the DMA circuit provides the signals *VMA* and $R/\overline{W}$ they are inputs to the interface. But to be of any use the interface must be provided with a start command, and the peripheral device must be informed when data is available or required. Thus an assumption is needed. Assume a register is connected to the I/O. It must be loaded (LD) before the *VMA* signal on input $R/\overline{W}$. The load should be after the *VMA* on output $\overline{R}/W$. For proper handshaking the peripheral device connected to the register should respond with a load complete (*LDC*).

The input-output block diagram is given in Figure 8.9. The sequential state diagram is shown in Figure 8.10. The sequence follows the events as described above. Some variations are possible; for example, the order of $R/\overline{W}$, $(\overline{R}/W)$, and *BA* could be reversed. Note that the loop is completed by the inclusion of $\overline{Start}$.

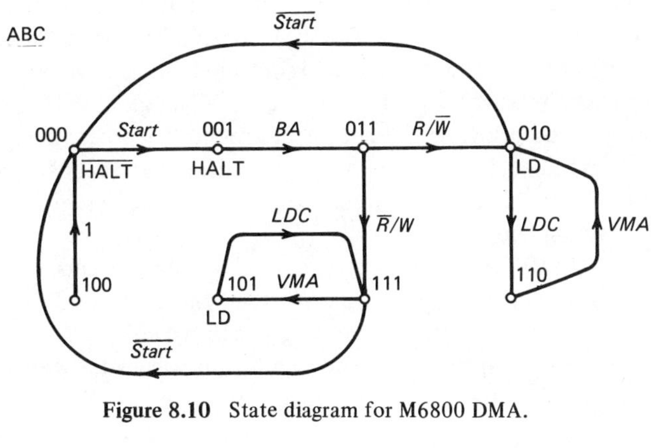

Figure 8.10 State diagram for M6800 DMA.

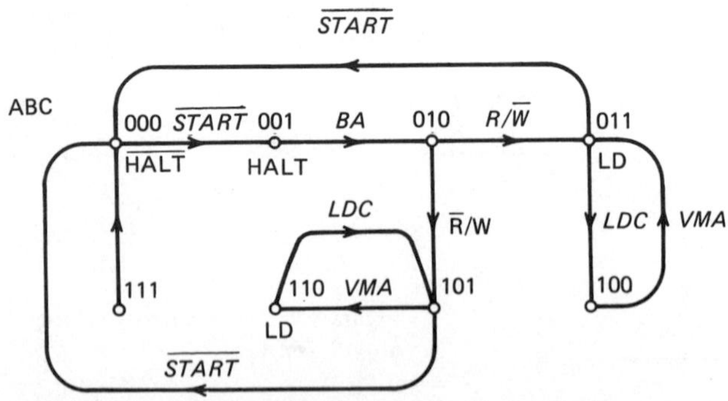

Figure 8.11 DMA state diagram for MSI implementation.

8.9 CONSTRUCTION OF STATE DIAGRAMS FOR PULSE-DRIVEN MSI CIRCUITS

The preliminary state diagram for the DMA microprocessor given in Figure 8.8 is renumbered in Figure 8.11 for MSI sequential circuit construction.

8.10 A GENERALIZED APPROACH TO LOGIC CIRCUIT DESIGN

As circuit complexity increases, the number of states in a sequential circuit can increase rapidly. Design techniques do not place any restriction on the number of states in a sequential circuit; however, from the viewpoint of layout, wiring, construction, and debugging, circuits with a large number of states present difficulties. From the practical limitations eight is a convenient maximum number of states.

When more than eight states are required it is good practice to generate a secondary circuit. This secondary circuit in turn should consist of no more than eight states. In this way each state in a primary circuit can generate a secondary circuit with eight states. Thus 64 states can be generated from an eight-state primary diagram. This is sufficient to tackle very sophisticated problems. Cascading in this manner, the number of states possible is infinitely large, while the individual circuits designed are relatively easy to construct and debug.

Using this cascading principle the author has found that where possible it is good practice to use primary circuits of two to four states and secondary circuits of six to eight states. When this division is possible, the resulting design is most efficient in the use of space and components.

In a large number of clock-driven sequential circuits, events occur at a slow rate compared to the clock. In this type of circuit there is virtually no limit on the total number of states or state transitions in a sequential circuit.

When the speed of occurrence of some events approaches the speed of the clock in a clock-driven sequential circuit, increasing the number of states increases the total time required to traverse the complete loop. Sometimes this increase in time causes problems. One way to overcome this problem is to fix the number of states within a clock-driven sequential circuit to correspond to the maximum time permitted to make the transitions through the complete loop. The remaining events are relegated to secondary circuits of the event-driven type.

8.11 THE DESIGN AND CONTROL OF COMPLETELY GENERALIZED STATE DIAGRAMS

There is a great similarity between state diagrams when the number of states is limited. Thus, a generalized state diagram can be used to realize a number of circuit functions by fitting an additional register. With, say, half a dozen such circuits it is possible to perform up to a 100 different functions.

Functions can be performed in any given order and repeated any number of times by simply loading the instruction register. This simply requires that the proper instruction in the proper sequence be directed to the instruction register. This task is easily done by storing the instructions sequentially in a memory bank and providing a pointing register to facilitate fetching the instructions in the proper sequence.

This generalized arrangement can be defined as a machine. The sequential circuits that fix the operation of this machine are defined as the control. Each time an instruction is executed the machine is said to have executed an instruction cycle, which may contain a number of machine cycles.

The power of this sequential or control circuit can be increased by including a register that can operate as a combinational logic device or a controlled element. This register may be considered to be part of the total process under control. Since the register can be manipulated, it gives this part of the process a degree of intelligence.

Chapter 9 outlines the development of registers. It ends with a register suited to arithmetic and logic operations intended to increase the power of sequential circuits as universal process controllers.

8.12 PROBLEMS

P8.1 Design the state diagram for the following toy.

A farmer (player) is taking a fox (F), a goose (G), and a sack of wheat (S) to market. He must cross a river in a boat and can take only one item in the boat (fox, goose, or wheat) at one time or travel empty (E).

He cannot leave the goose and wheat together on either bank because the goose will eat the wheat. Similarly he cannot leave the fox and goose together on either bank because the fox will eat the goose. Design the toy so that there are three outputs.

C: continue and make next move.

L: game lost because goose or wheat has been eaten.

W: game won, player has fox, goose, and wheat on far bank.

P8.2 Figure P8.1 shows a space/bar coding for a right-hand number 9 using the grocery industry's Universal Product Code (UPC) format [30]. You have undoubtedly seen the UPC on the groceries you use. There are 95 space/bars in the symbol itself bracketed by a guard bar and nine spaces on each side. The 95 space/bars consist of the following:

(a) A starting sequence 101.
(b) Six 7-bit characters as shown in Figure P8.1
(c) A center pattern character 01010
(d) Six 7-bit characters as shown in Figure P8.1
(e) A stopping sequence of 101

Table P8.1 UPC Characters

Decimal Value	Left Characters (Odd Parity–0)	Right Characters (Even Parity–E)
0	0001101	1110010
1	0011001	1100110
2	0010011	1101100
3	0111101	1000010
4	0100011	1011100
5	0110001	1001110
6	0101111	1010000
7	0111011	1000100
8	0110111	1001000
9	0001011	1110100

The numbers are coded differently depending on whether they are on the left or right-hand side of center. Table P8.1 shows this code.

Design a state diagram that generates a pulse at the start of each 7-bit character.

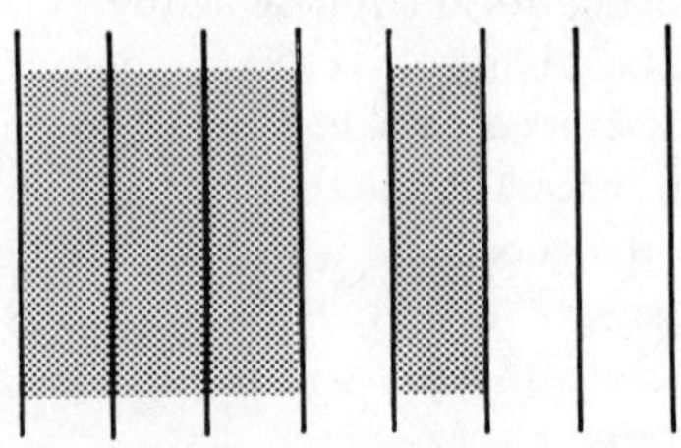

Figure P8.1 Space/bar code for a right-hand 9, which is 1110100.

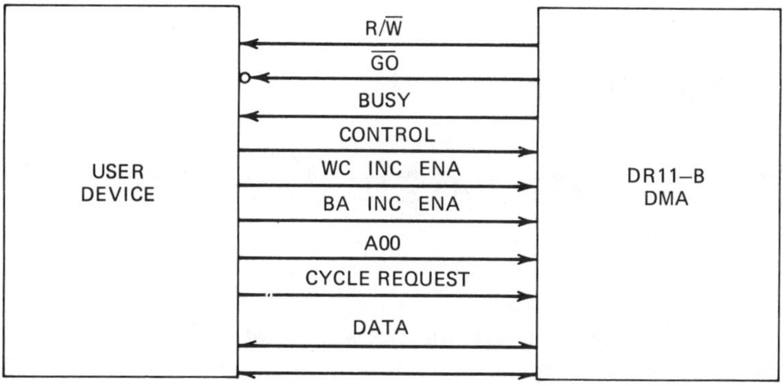

Figure P8.2 General purpose DMA.

P8.3 Figure P8.2 is an approximate diagram of the direct memory access general purpose interface for a popular minicomputer.

The signal $\overline{GO}$ is sent to the user device to start and indicates the value of R/$\overline{W}$ is valid. The user device must respond with Control = 0, direction of data transfer from unibus to user device, or Control = 1, data transfers from user device to unibus.

WC INC ENA	Word count enable
BA INC ENA	Bus address increment enable
A00 = 0	Bus address bit indicates sequential word addressing
A00 = 1	Bus address bit indicates sequential byte addressing

This set of signals must be established before the CYCLE REQUEST signal is sent. CYCLE REQUEST causes BUSY = 1. At the completion of a single (word or byte) transfer BUSY = 0.

Draw a state diagram for the USER device. Assume the user device has a countdown facility that generates a transfer complete (TC) signal.

CHAPTER 9

Registers in Logic Design

Shift registers can be used in logic circuits in a number of ways. First, they are convenient for conversion of parallel data into serial data and vice versa. Such a conversion allows the designer to make a price and performance trade-off between the logic hardware and the transmission line capabilities of an interface [9]. Another use for shift registers is to generate different codes and sequences. Such codes and sequences have use in guaranteeing the integrity of transmitted and received data. In a number of engineering applications sequences are, of course, an extremely convenient way to generate periodic signals.

A further use for registers is to modify data. Basic operations in addition to load and shift, which are conveniently done in a register, are clear, set, and complement. In addition the logic operators and, or and exclusive or (xor) are readily implemented using registers.

9.1 DATA STORAGE USING REGISTERS

In many cases it is convenient to store results or transmit data to and from logic circuits. A register is ideal for this application. So for the purpose of this chapter a register is defined as a device that is composed of a number of flip-flops whose main purpose is to hold data temporarily. Figure 9.1 shows a simple register. As with all the examples in this chapter, the data is limited to a nibble simply for the convenience in drawing figures. In fact, registers are longer than 4 bits.

The device shown in Figure 9.1 has the ability to load 1 or 0 into an RSFF. An RSFF and inverter in this configuration is called a DFF (data or delay flip-flop). A register using these components is inexpensive. It has the disadvantage that it cannot retain old data once new data is available. Also the only way the registers can be cleared is to make all inputs equal to zero.

Another configuration for a register is given in Figure 9.2 This device gives better operation, but it should be noted that RS = 11 is not permitted. In this device, once data has been read into the register, it remains there until the clear line is energized. This may not be convenient if a number of read enable signals can be generated between line signals because this register continuously accumulates all incoming data until all flip-flops become equal to 1. As a result the device can only be used when a read enable is always followed by a clear line. This gives a requirement for two clocks. A simple technique commonly employed to overcome the clock problem is to and the set and reset inputs to the RS flip-flop with a clock

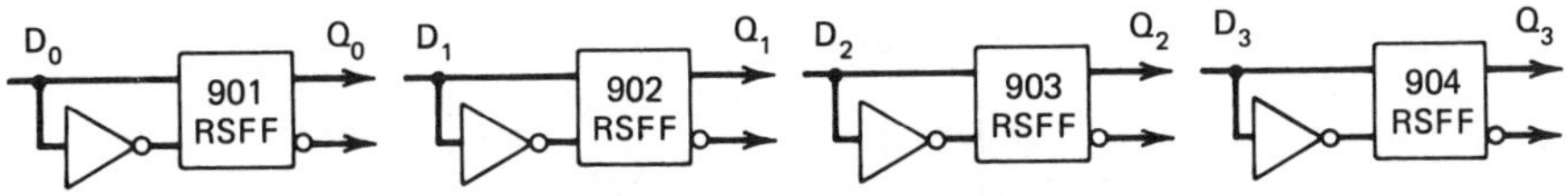

Figure 9.1 Storage register.

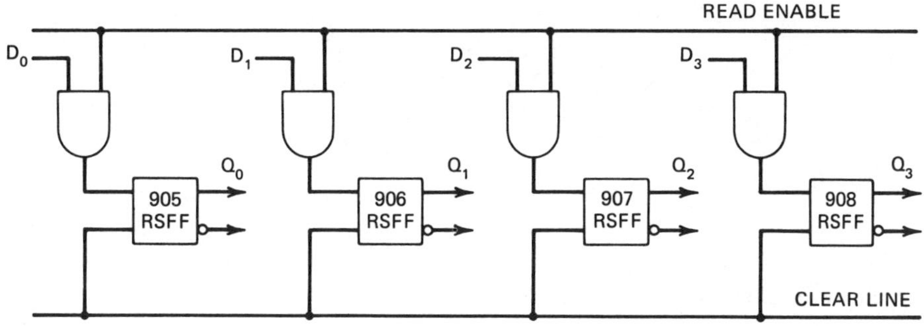

Figure 9.2 RSFF register to hold data.

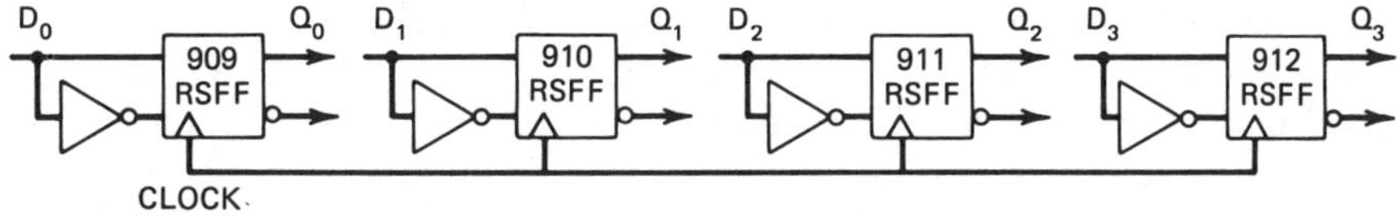

Figure 9.3 Clocked RSFF register.

signal. Such an arrangement is shown in Figure 9.3. When the clock = 0 data is held in the register. New data is accepted when clock = 1.

However, the basic problem of indeterminate output of the RSFF is described in Chapter 6. To guarantee stable data a master-slave system or JKFF is required.

9.2 SHIFT REGISTERS

A shift register is a device in which data can be shifted serially from flip-flop to flip-flop. Figure 9.4 shows the way JKFFs are used to perform the basic shift operation. Input data is read into the left-most flip-flop. With each clock, data is shifted one flip-flop to the right.

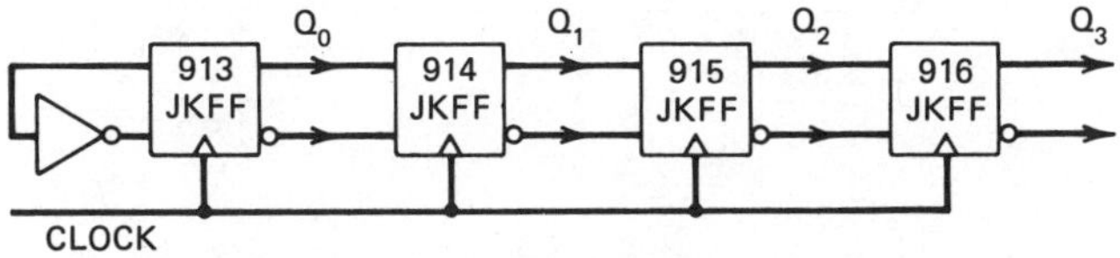

Figure 9.4 Basic shift register using JKFFs.

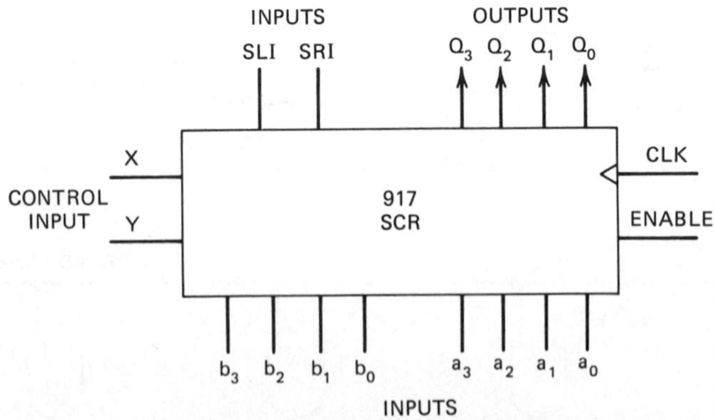

Figure 9.5 Shift register logic symbol.

Figure 9.5 shows a convenient logic symbol for a shift control register (SCR) as available in package form. Figure 9.6 gives a JKFF and multiplexer realization of this register.

The register can be loaded with one of two inputs, $A = a_3a_2a_1a_0$ and $B = b_3b_2b_1b_0$. There are four modes of control determined by the control inputs XY. When XY = 01, the register operates serially. New input is loaded into the most significant bit. All other bits are shifted to the next less significant bit (right). XY = 10 also causes serial action. New data can be loaded to the least significant bit of the register. All other bits are shifted to the next more significant bit (left). The inputs for serial operation are SRI for a right shift and SLI for a left shift. A ring shifter can be created by connecting SRI to Q_0 or SLI to Q_3. For XY = 00, the A inputs are loaded into the register. For XY = 11, the B inputs are loaded into the register. Both these loadings are referred to as parallel operation. The operation of the SCR is summarized in Table 9.1.

Table 9.1 SCR Function Description

Control Inputs		
X	Y	Function
0	0	Load Register with A inputs
0	1	Shift Right
1	0	Shift Left
1	1	Load Register with B inputs

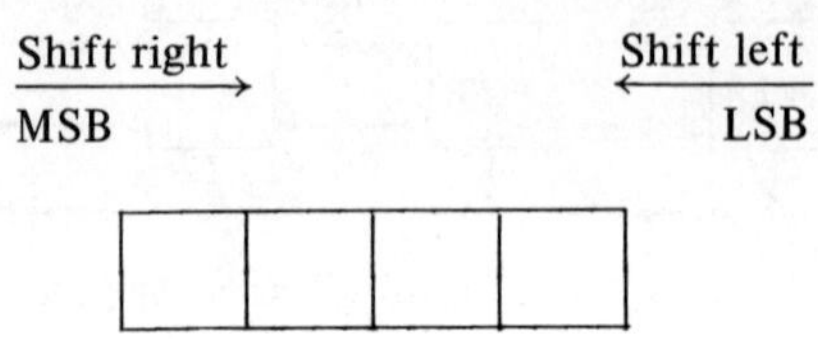

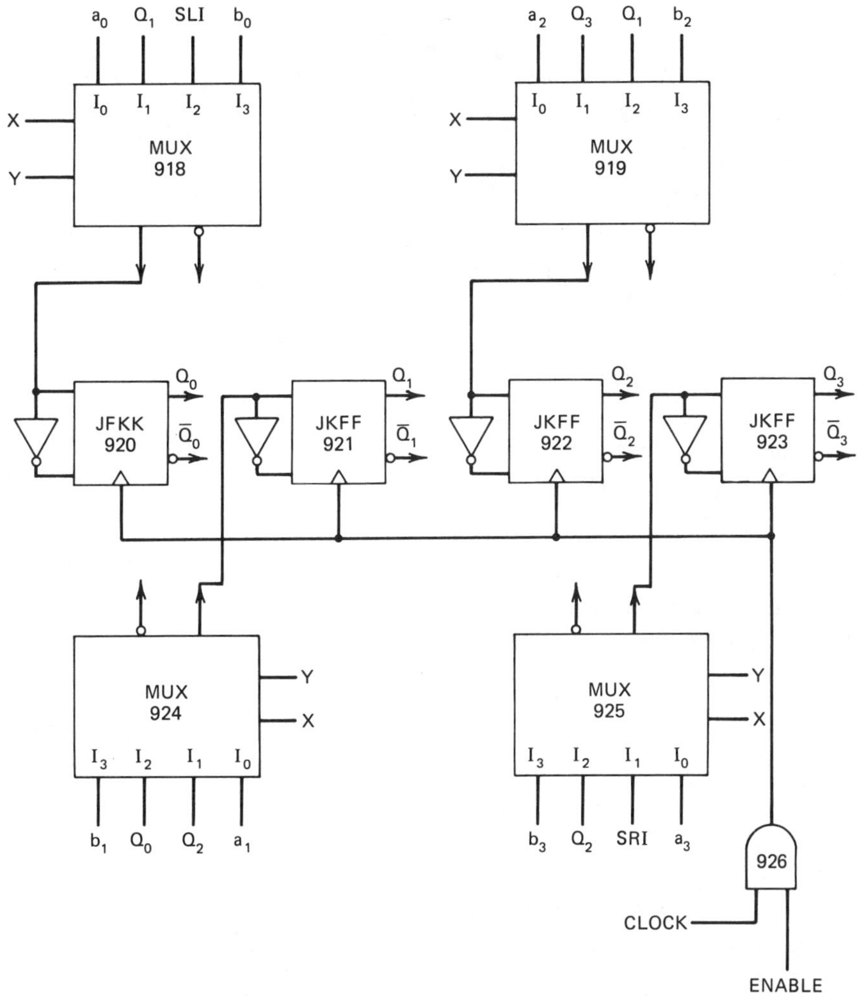

Figure 9.6 JKFF and multiplexer realization of basic SCR.

9.3 GENERATION OF CODES USING SHIFT REGISTERS

There are many applications where SCRs are extremely useful. An example is in interface design for the transmission of data between computers or a computer and peripheral device. Parallel data can be converted to sequential data for transmission, and this saves in transmission lines, cabling, and so on.

In addition an SCR can be connected to generate additional bits, which can be used to detect errors. A common illustration of this is the generation of a parity bit. Figure 9.7 shows a shift register connected for this application. The operation of this device is as follows. First let XY = 00 for one clock pulse. Data on inputs a_3, a_2, a_1, a_0 are loaded into the register. Then the control line is changed so that XY = 01 and the enable line is held high for five clock pulses. On the first four

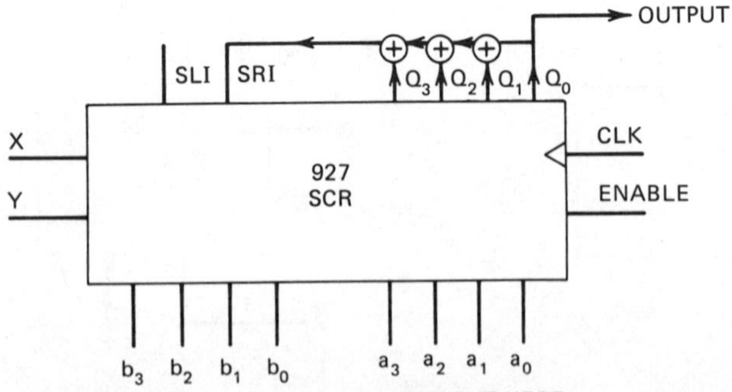

Figure 9.7 Generation of parity using a shift register.

clock pulses the output is equal to a_0, a_1, a_2, a_3, respectively, and on the fifth clock pulse the output equals $a_0 \oplus a_1 \oplus a_2 \oplus a_3$. The last bit, of course, is a parity bit.

The error detection and correction code developed by Hamming and given in problem P2.6 is repeated in Table 9.2 [2]. It can be generated using an SCR.

Errors can be detected and corrected in the following manner. We define the condition that the error is given by the error function E,

$$E = e_2 e_1 e_0$$

where

$$e_2 = W \oplus X \oplus Y \oplus 1$$

$$e_1 = W \oplus X \oplus Z \oplus m$$

$$e_0 = W \oplus Y \oplus Z \oplus n$$

Then there is a direct relationship between the value of E and the bit in error as given by Table 9.3.

Table 9.2 Hamming Correction Code

BCD				Hamming						
A	B	C	D	W	X	Y	1	Z	m	n
0	0	0	0	0	0	0	0	0	0	0
0	0	0	1	0	0	0	0	1	1	1
0	0	1	0	0	0	1	1	0	0	1
0	0	1	1	0	0	1	1	1	1	0
0	1	0	0	0	1	0	1	0	1	0
0	1	0	1	0	1	0	1	1	0	1
0	1	1	0	0	1	1	0	0	1	1
0	1	1	1	0	1	1	0	1	0	0
1	0	0	0	1	0	0	1	0	1	1
1	0	0	1	1	0	0	1	1	0	0

Table 9.3 Error Relationship

E	Variable in Error
001	n
010	m
011	Z
100	l
101	Y
110	X
111	W

As an example, let us attempt to transmit the number 4. The values of the variables are

ABCD	W	X	Y	l	Z	m	n
0100	0	1	0	1	0	1	0

Assume an error is made in transmission such that Z = 1, and the code transmitted is

$$0\ 1\ 0\ 1\ 1\ 1\ 0$$

Then

$$e_2 = 0$$

$$e_1 = 1$$

$$e_0 = 1$$

which makes

$$E = 3$$

Table 9.3 indicates that bit Z is in error.

Suppose the error is made in transmission such that l = 0, and the code transmitted is

$$0\ 1\ 0\ 0\ 0\ 1\ 0$$

Then

$$e_2 = 1$$

$$e_1 = 0$$

$$e_0 = 0$$

so that

$$E = 4$$

Table 9.3 indicates that bit l is in error.

The reader is invited to try this for any value of ABCD in the code and a single error in the transmission.

The initial problem is to generate the Hamming code W X Y 1 Z m n from the inputs ABCD. The reader is reminded that the equations for the variables in the Hamming code as determined in problem P2.5 are the following:

$$W = A$$

$$X = B$$

$$Y = C$$

$$1 = A \oplus B \oplus C$$

$$Z = D$$

$$m = A \oplus B \oplus D$$

$$n = A \oplus C \oplus D$$

Here are some further relationships that the reader may wish to verify.

$$m = n \oplus B \oplus C$$

$$1 = m \oplus C \oplus D$$

Notice that the Hamming code lists the letters in the sequence determined by the error function E. It is not necessary to transmit in this sequence. However, it is necessary to be aware of relationships given in Table 9.3.

Figure 9.8 gives the generation of the Hamming Code using an SCR with the variables D, C, B, A connected to inputs a_3, a_2, a_1, a_0, respectively. The operation is as follows. First, let XY = 00 for one clock pulse. This transfers the inputs a_3, a_2,

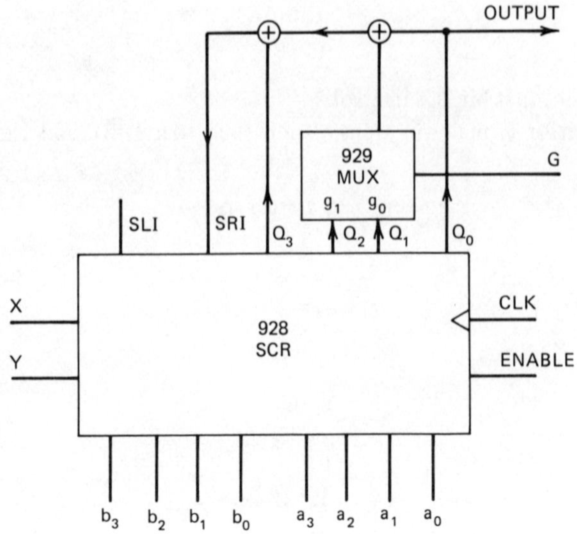

Figure 9.8 Generation of error detection/correction code.

a_1, a_0 into the registers such that

$$Q_3, Q_2, Q_1, Q_0 = a_3, a_2, a_1, a_0 = D, C, B, A$$

respectively. The control line is changed such that XY = 10, and the select line on the multiplexer is changed such that G = 1 for the second clock pulse. Since

$$Q_0 = A = W$$

the output then is equal to W. Multiplexer 929 selects Q_2. Thus

$$SRI = Q_3 \oplus Q_2 \oplus Q_0 = D \oplus C \oplus A = n$$

For clock pulses 3 through 8, the enable line is held high, and the value of G is made equal to zero; thus multiplexer 929 selects Q_1. Thus

$$SRI = Q_3 \oplus Q_1 \oplus Q_0$$

On clock pulse 3

$$Q_3 = n$$

$$Q_1 = C$$

and

$$Q_0 = B$$

and this causes

$$SRI = n \oplus C \oplus B = m$$

Since

$$Q_0 = B = X$$

the output is equal to X. On clock pulse 4

$$Q_3 = m$$

$$Q_1 = D$$

and

$$Q_0 = C$$

and this causes

$$SRI = m \oplus D \oplus C = 1$$

Since

$$Q_0 = C = Y$$

the output is equal to Y. On clock pulse 5

$$Q_3 = 1$$

$$Q_2 = m$$

$$Q_1 = n$$

$$Q_0 = D$$

Since

$$Q_0 = D = Z$$

The output is equal to Z. On the next three pulses the variables n, m, and l are placed on the output.

Thus the code is transmitted in the following sequence:

W X Y Z n m l

9.4 GENERATION OF SEQUENCES USING SHIFT REGISTERS

Another application for shift registers is in the generation of a periodic sequence [9, 18]. The author has used a periodic sequence to generate test signals for the input to large generators. The advantages of using such a sequence are the following:

1. The sequences are periodic and can easily be analyzed.
2. They are easily generated using a shift register.
3. They are easily realized on particular apparatus since the signal is of the binary type.
4. Such a signal is readily superimposed on a main signal and at such a level that it is indiscernible to consumers but readily available for testing.

To generate a periodic sequence, one must first select the sequence to be generated and use the steps given below. As an illustrative example, let us take the sequence 001101. The sequence can be generated using the following steps:

1. Place the sequence in a row form in a table. This step is illustrated in the first row of Table 9.4.

2. Repeat the row, left shifting one bit at a time until a square matrix is obtained as shown in Table 9.4.

3. The first n rows form the input. n is the least number required to generate a number $k = 2^n$ where $k \geqslant$ number of rows in the square matrix. A Karnaugh map is generated, taking the first n rows of the table as the input. If two columns in the input are the same, another row from Table 9.4 must be added to form the K map input. This process must be continued until no two input columns are alike. The

Table 9.4 Sequence Generation Table

0	0	1	1	0	1	
0	1	1	0	1	0	Input
1	1	0	1	0	0	
1	0	1	0	0	1	Output
0	1	0	0	1	1	
1	0	0	1	1	0	

Q_3 \ $Q_1 Q_2$	00	01	11	10
0	ϕ	0	1	1
1	1	0	ϕ	0

Figure 9.9 Karnaugh map for sequence generation.

next row in Table 9.4 forms the output. The function f, as determined from the Karnaugh map, gives the input to the shift register.

The Karnaugh map developed from Table 9.4 is shown in Figure 9.9. Realization of this sequential circuit is shown in Figure 9.10.

This circuit operates in the following way. First let XY = 00 for one clock pulse. This transfers the inputs a_3, a_2, a_1, a_0 into the register. However a_0 is not used. Thus

$$Q_3, Q_2, Q_1 = a_3, a_2, a_1$$

The values of a_3, a_2, a_1 are any three digits in the sequence. That is any number except 000 and 111. As an example select

$$a_3, a_2, a_1 = 1\ 0\ 0$$

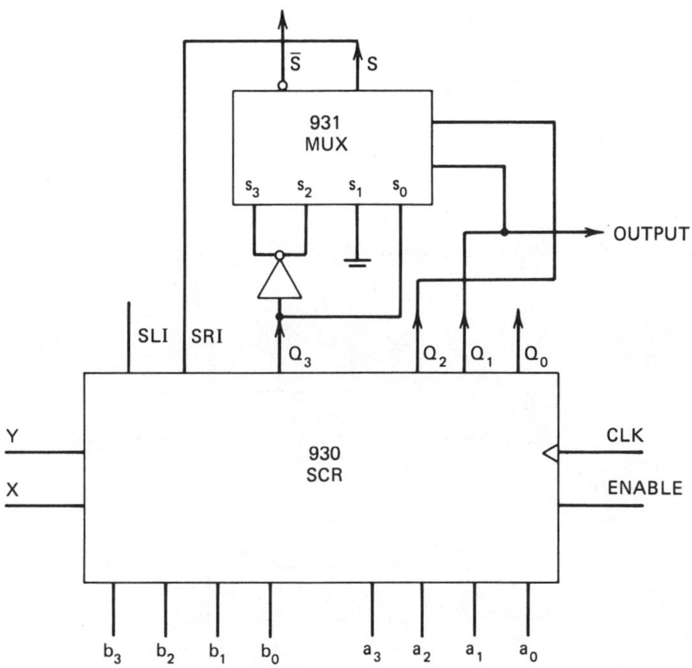

Figure 9.10 Sequence generator using SCR.

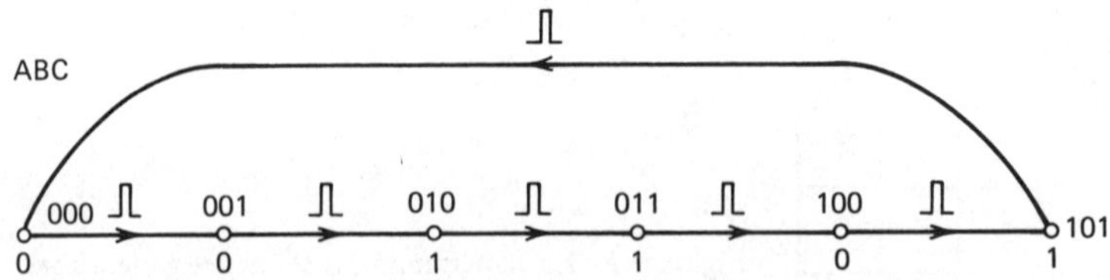

Figure 9.11 State diagram for sequence generation.

The control lines are changed to XY = 01 (shift right) and with each clock values of the variable are as follows:

Clock Number	SRI	Q_3	Q_2	Q_1	OUTPUT
1	1	1	0	0	0
2	0	1	1	0	0
3	1	0	1	1	1
4	0	1	0	1	1
5	0	0	1	0	0
6	1	0	0	1	1

The sequence continues repeating as long as the clock continues.

Generation of sequences can be done more directly using a counter and decoding circuits. To generate the sequence 001101, set up the required state diagram. Such a state diagram is shown in Figure 9.11. Using the technique described in Chapter 7, one can generate the required output sequence.

9.5 MULTIFUNCTION REGISTERS

The basic register shown in Figure 9.5 is capable of handling and manipulating data to great advantage. However, registers are designed to be much more flexible. For example, a register can be designed to clear, set, or complement. Figure 9.12 shows a convenient logic symbol for a basic multicontrol register (MCR). Figure 9.13 gives

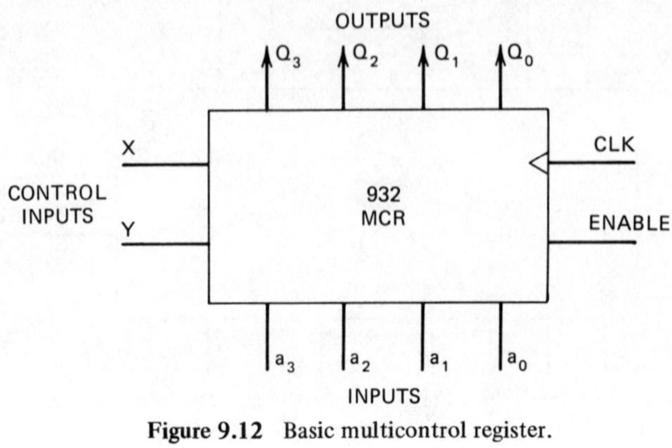

Figure 9.12 Basic multicontrol register.

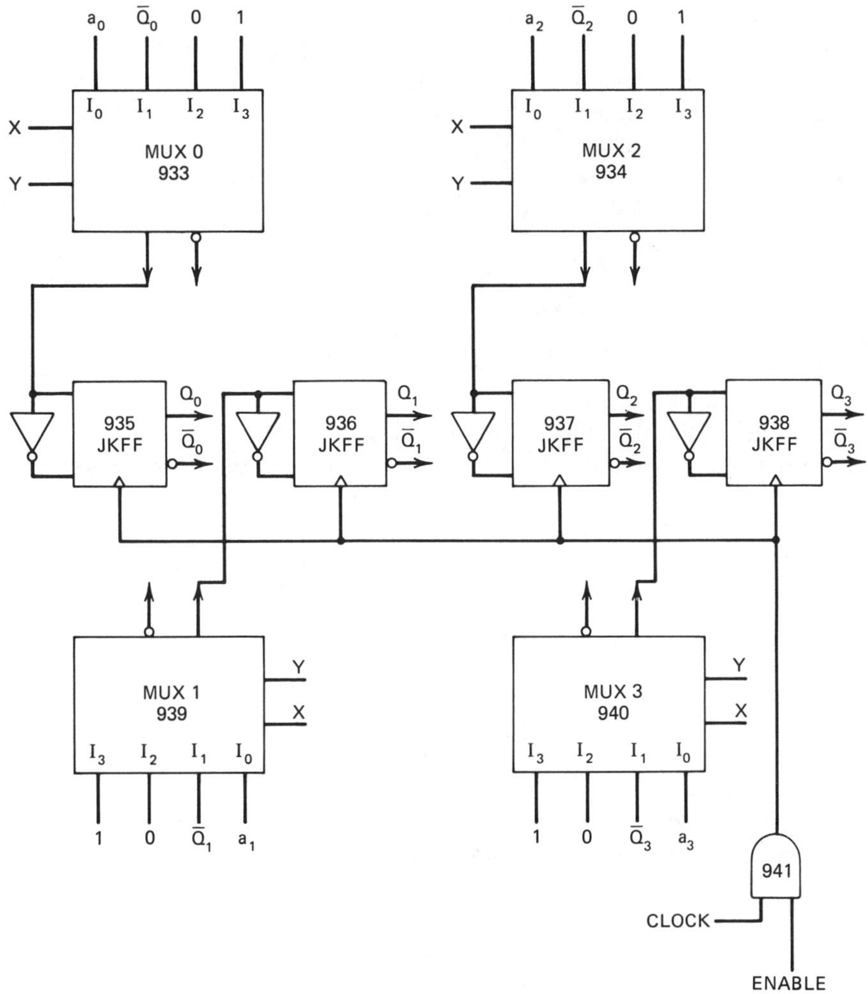

Figure 9.13 Realization of basic MCR.

Table 9.5 MCR Function Description

Control Inputs		
X	Y	Function
0	0	Load register with data on input lines
0	1	Complement register
1	0	Clear register
1	1	Set register

a JKFF and multiplexer realization of this register that has four modes of control as determined by control inputs XY. The load operation is accomplished by loading a's onto all I_0 multiplexer inputs. Normal complementation is accomplished by loading the negated form of each variable on the multiplexer inputs I_1. The clear operation requires the loading of zeros onto all I_2 multiplexer inputs. The set operation requires the loading of ones onto all I_3 multiplexer inputs. Table 9.5 shows the multifunction description of the device.

9.6 BINARY COUNTING REGISTERS

A logic symbol for a binary counting register is shown in Figure 9.14. This unit can accept external data, count up and down, or be cleared to zero. The realization is shown in JKFF and multiplexer form in Figure 9.15. The clear operation requires all zeros on the I_0 multiplexer inputs. Values placed onto the I_1 multiplexer inputs for the count-up operation are determined by the methods described in Chapter 6. This implementation requires a steering circuit between all J and K inputs, that makes $K = J$ when the context is $X \oplus Y$, rather than the simple not gate shown.

The equations for count-up are:

$$J_0 = K_0 = 1$$

$$J_1 = K_1 = Q_0$$

$$J_2 = K_2 = Q_0 Q_1$$

$$J_3 = K_3 = Q_0 Q_1 Q_2$$

In a similar way, the multiplexer inputs I_2 for the countdown operation are determined. The equations for countdown are:

$$J_0 = K_0 = 1$$

$$J_1 = K_1 = \bar{Q}_0$$

$$J_2 = K_2 = \bar{Q}_0 \bar{Q}_1$$

$$J_3 = K_3 = \bar{Q}_0 \bar{Q}_1 \bar{Q}_2$$

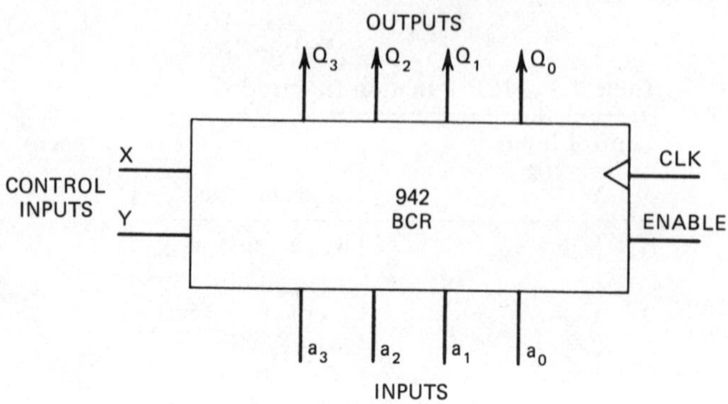

Figure 9.14 Basic binary counting register.

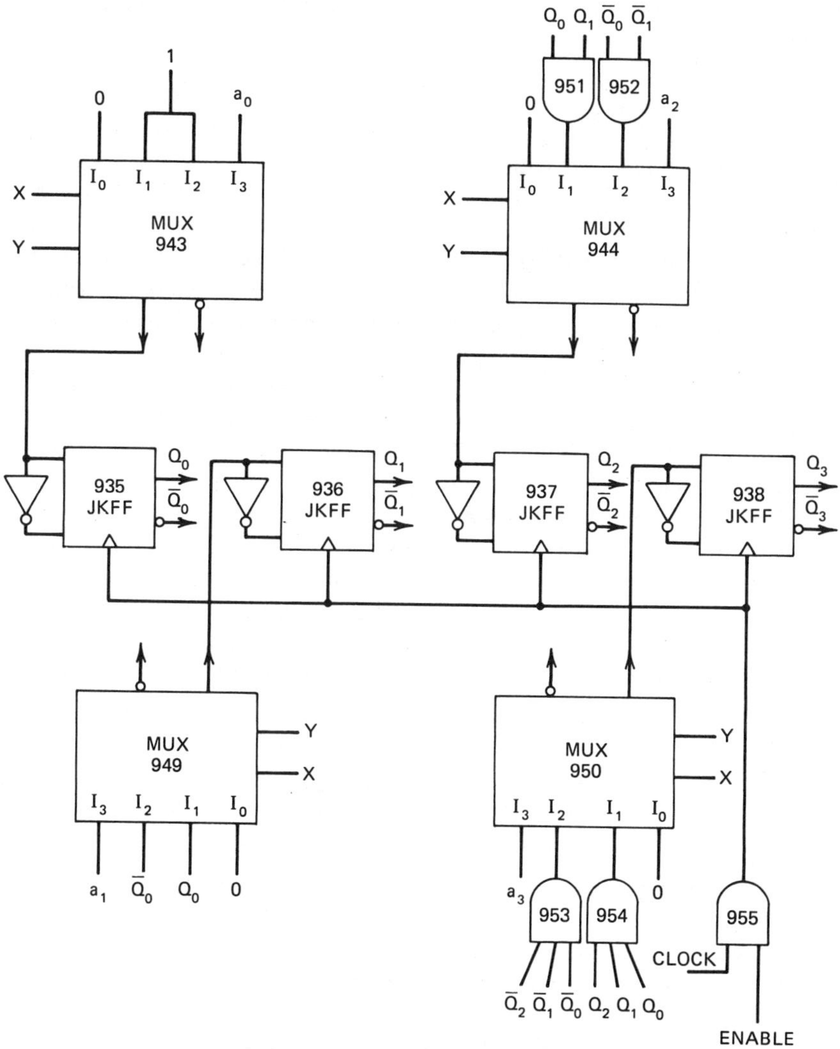

Figure 9.15 Realization of a basic BCR.

Table 9.6 BCR Function Description

Control Inputs		
X	Y	Function
0	0	Clear
0	1	Count-up
1	0	Countdown
1	1	Load

The load operation requires that appropriate values of a's be placed onto the I_3 multiplexer inputs. Table 9.6 gives the binary counting register functional description.

9.7 COMPLEMENTING REGISTER

Another useful register performs conventional and two's complement complementation. Figure 9.16 gives the logic symbol for a complementing register. The functional description device is given in Table 9.7. Figure 9.17 shows the realization of this complementing register using JKFFs and multiplexers. The clear operation is accomplished by loading zeros onto all I_0 inputs. Normal complementation is accomplished by loading the negated form of each variable on the multiplexer inputs I_1. The register is loaded on multiplexer inputs I_3. Two's complement values are loaded on multiplexer inputs I_2. The I_2 inputs are selected by a multiplexer to be complementing or noncomplementing. This selection is determined using a complementing detection circuit (CDC) designed using multiplexers 966 to 969.

If *complement in* is equal to 1 all these multiplexer outputs are equal to 1. All the multiplexer inputs to I_2 are complemented. However, if *complement in* is equal to 0 the CDC detects the first occurrence of a 1 starting from the LSB. It makes the output of the CDC equal to 1 for the first occurrence bit and all subsequent bits. Since COM 0 controls the Q_1 bit, COM 1 the Q_2 bit, and so on, all subsequent Q's after the first occurrence of a Q equal to 1 are complemented.

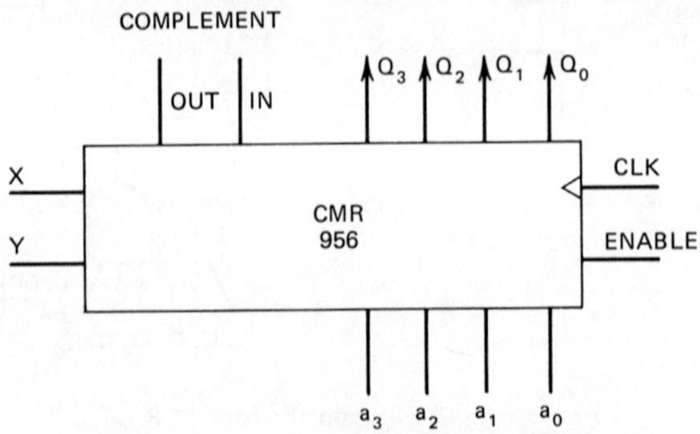

Figure 9.16 Logic symbol for complementing register.

Table 9.7 CCR Function Description

Control		
X	Y	Function
0	0	Clear
0	1	One's complement
1	0	Two's complement
1	1	Load

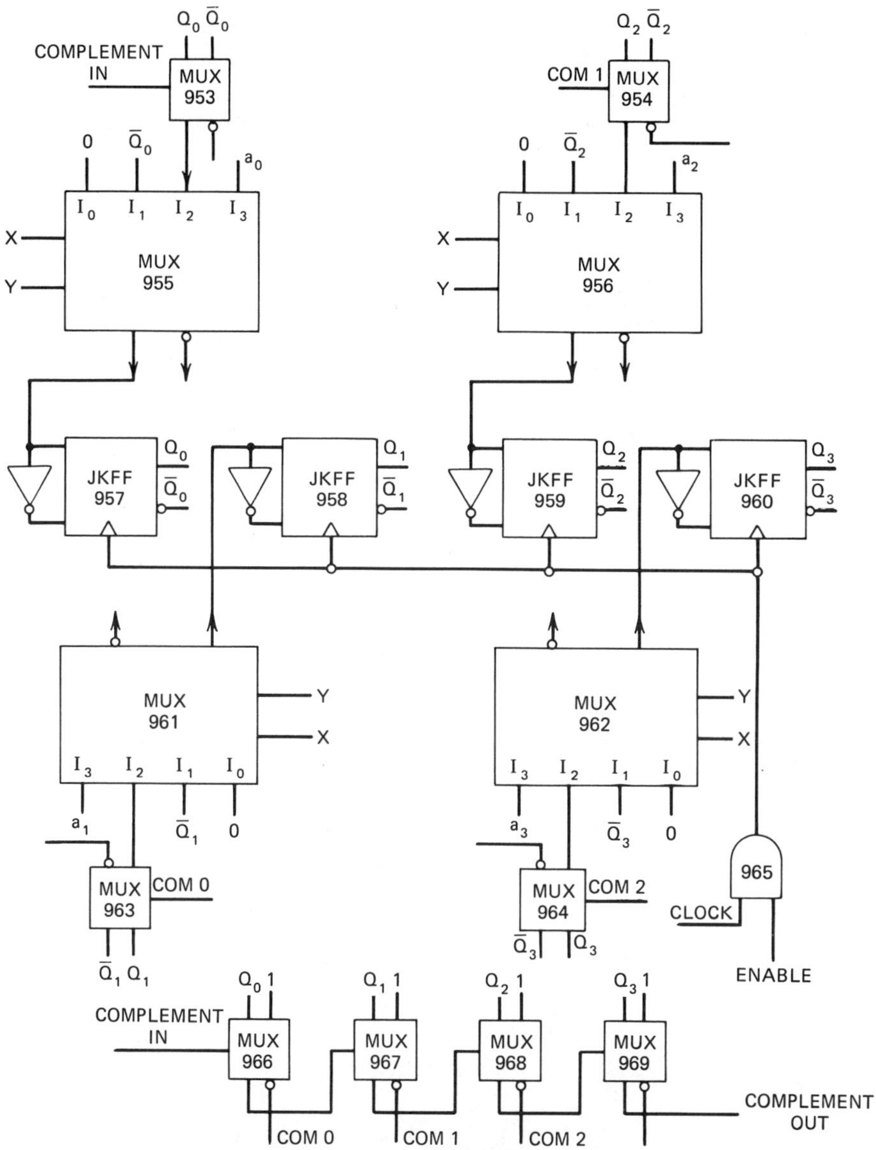

Figure 9.17 Complementing register.

9.8 SUPERPOSITION OF REGISTERS

It has been shown in the previous section how a register can be cleared, set, complemented, or loaded. It seems quite clear now that we can build a further register to perform the logic functions or, and, exculusive or (xor). A list of desirable functions that such a superposition register would give is given in Table 9.8. Indeed, the operations described in Table 9.8 describe the operations of an arithmetic logic

Table 9.8 **Arithmetic Logic Unit Operations**

Control Inputs			
X	Y	Z	Operation
0	0	0	No operation
0	0	1	Clear
0	1	0	Shift
0	1	1	Complement
1	0	0	Load
1	0	1	Or
1	1	0	And
1	1	1	Add

unit (ALU) which is a fundamental component of a computer. All logic and arithmetic operations performed in a computer use this fundamental set. Obviously the set is easily extended.

9.9 PROBLEMS

P9.1 In Section 9.3 a method of generating the Hamming code was given. The Hamming code is difficult to generate. In fact any similar code detects and corrects one error. A code that is simpler to generate is as follows:

$$e_2 = W \oplus Y \oplus Z \oplus l$$

$$e_1 = W \oplus X \oplus Y \oplus m$$

$$e_0 = X \oplus Y \oplus Z \oplus n$$

It is simpler to generate because it is cyclic.

$$l = W \oplus Y \oplus Z$$

$$m = X \oplus Z \oplus l$$

$$n = Y \oplus l \oplus m$$

Table P9.1 Error Relationship of Cyclic Code

E	Variable in Error
001	n
010	m
011	X
100	l
101	Z
110	W
111	Y

Table P9.1 shows the error relationship for this code. Design an SCR to generate this code.

P9.2 Design an SCR to generate the maximum length sequence 100010011010111.

P9.3 Design an SCR to generate the maximum length sequence 1001011.

CHAPTER 10
A Digital System

The basic purpose of a digital machine, be it microcomputer, minicomputer, or computer, is to provide a processing capability. In the early days of computing the process was usually a bookkeeping function. The next stage in the development was the processing of mathematical equations. As a natural development digital machines have been applied to all input/output situations where an intermediary process is required.

This chapter is intended to give an explanation of the components of a basic digital machine.

The descriptions in this chapter are most accurately applied to microcomputers; however they are meant to be general so that, with a small amount of further study, descriptions can be extended to include minicomputers. For purposes of definition a minicomputer is considered to have its main program stored in normal read-write memory, whereas a microcomputer has its main program stored in read-only memory [19–28].

In general a minicomputer is designed using a technology based on high-speed performance from a multitude of components. A microcomputer is based on a technology that allows many elements to be placed on one chip with low-energy consumption. This leads to the microcomputer having a lower cost with slower speed and/or shorter word length than a minicomputer. With technology advancing rapidly and the high competition between micros and minis, there is a cross-pollination. Thus some minis have micro components, and some micros approach minicomputer speeds. Presumably in the future there will be fewer differences between microcomputers and minicomputers.

The main processing unit in a microcomputer is a single large-scale integration (LSI) chip. It is sometimes called a central processing unit (CPU) and sometimes called a microprocessor (μP). For this reason a microcomputer is frequently called a microprocessor. The fact is, if the microcomputer is small it is almost always called a microprocessor.

The typical microcomputer system is shown in Figure 10.1. It consists of the following [30, 31]:

1. A central processing unit (microprocessor)
2. Random access memory (RAM)
3. Read-only memory (ROM)
4. Input/output (I/O) ports

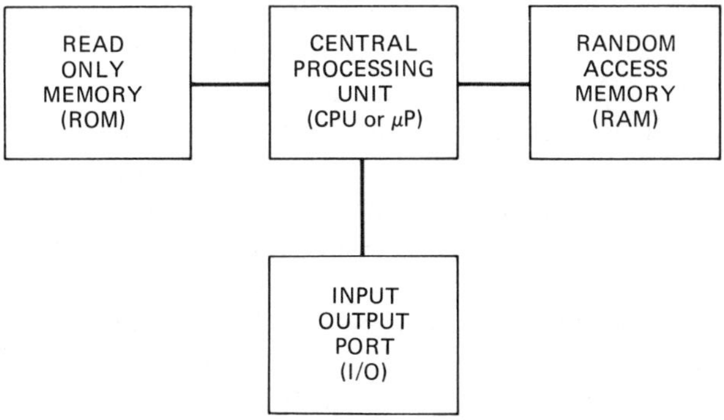

Figure 10.1 A typical microcomputer system.

10.1 ROM AND RAM

The ROM serves as a storehouse for instructions to control the operation of the CPU. The ROM is used for this purpose because:

- It is more compact than RAM assuming the same number of bits. This makes the normal ROM cheaper and faster.
- It is more secure than RAM.

Instructions stored sequentially in the ROM are referred to as a program. ROMs are also used for storing tables and constants.

Strictly speaking, a ROM is a very specific type of memory. However, in common usage ROM also stands for the family of ROM, which includes PROMs and EPROMs. The alternative forms of ROMs are:

- ROMs. ROMs are mass programmed by the manufacturer at the time the chip is made.
- PROMs or programmable ROMs. The contents of PROMs are programmed by the user with a special PROM programmer.
- EPROMs or erasable PROMs. The contents of an EPROM are programmed by the user but can be erased by exposure to ultraviolet light. This eases the programmer's chore. To prevent accidental erasure, the ultraviolet window can be covered.

ROMs are very cheap when programmed in huge quantities. In smaller quantities it is more practical to use PROMs. A PROM chip is easily programmed with the proper equipment. This device duplicates any program. A program that is going to be entered in PROM must be correct, since once entered it cannot be changed. A chip that is easier to use is an EPROM. This term at first might seem contradictory. However, an erasable version of PROM simplifies the equipment needed to debug a program. EPROM programs can be debugged on system hardware rather than development hardware. This is a mixed blessing.

ROMs in bulk are very inexpensive, PROMs are moderately inexpensive, and EPROMs are the most expensive.

The RAM serves as a data bank for the storage of temporary or working variables and parameters. It has read/write capability. RAM memory is of two types. The first, static memory, has the advantage that it is nonvolatile but the disadvantage that it is expensive. The second is dynamic memory. It has the disadvantage that it is extremely volatile and must be continually refreshed, but the advantage of being relatively inexpensive. Memory chips are provided with one or more chip select lines and two or more address lines. The chip select line is connected to the high, most significant bit lines through a decoding circuit. The address lines are connected to the lower-order address lines on the bus.

10.2 THE CPU

The CPU consists of control and timing circuits in the form of sequential circuits, an instruction decoding facility, an arithmetic logic unit, and a number of registers for the internal manipulation of data. The CPU configuration is shown in Figure 10.2. The CPU is connected to other devices in the system by way of a data bus, an address bus, and a control bus. The CPU unifies the system and controls the function of the system.

Operation of the basic components of a CPU is outlined in the following sections.

10.2.1 ALU Module

The ALU is used to do the basic arithmetic and logic functions in a CPU. A complete set of operations is given in Chapter 11 and consists of such fundamental

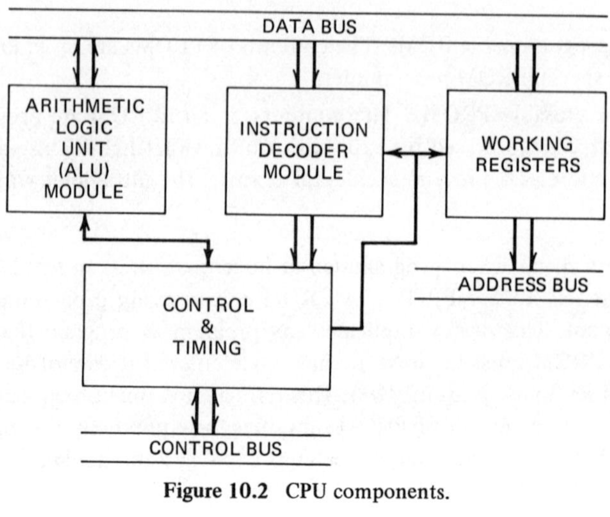

Figure 10.2 CPU components.

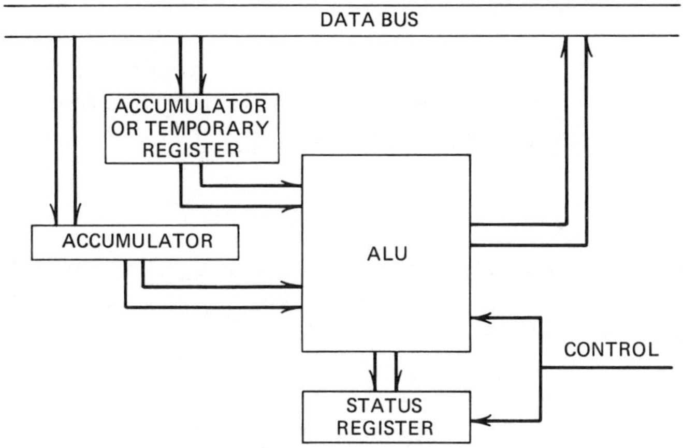

Figure 10.3 ALU module.

operations as:

- clear
- shift
- complement
- load
- or
- and
- exclusive or
- add

However, for an ALU module to function it is usually equipped with registers on its inputs. An ALU usually has two input registers, called accumulators, or a single accumulator and a temporary register, as shown in Figure 10.3. In addition an ALU has a status register. The status register typically contains the following flags to indicate the results of an ALU operation:

- negative
- zero
- overflow
- carry
- half-carry (carry between nibbles)

In addition, other flags that may occur are:

- parity
- interrupt mask–this bit is used to enable interrupts from I/O devices

10.2.2 Instruction Decoder

An instruction decoder is a combinational circuit. As with the ALU an additional register (instruction register) is required to hold data received from the data bus. The decoding arrangement is shown in Figure 10.4

10.2.3 Working Registers

Accessing registers is at least four times as fast as accessing memory. For this reason it is convenient to have a number of working registers for temporary storage. Two registers that work in association with the address bus are the program counter (PC) and the stack pointer (SP). The PC is used to point to locations in ROM. The SP is used to point to locations in RAM. These two registers usually have an increment/decrement circuit to progress sequentially through memory.

In general the PC points to the location of the next instruction to be executed, and the program is executed by incrementing the PC. However, as with the method described in Chapter 7, the program counter can be loaded from a register to permit looping in the ROM.

Program instructions that cause the program counter to loop are called jump, call, or return instructions. The call and return instructions are a special kind of jump instruction used to access subroutines. When a processor receives a call instruction it increments the PC and stores the PC contents in a stack. Thus the stack saves the address of the instruction that would normally be executed next. It makes this

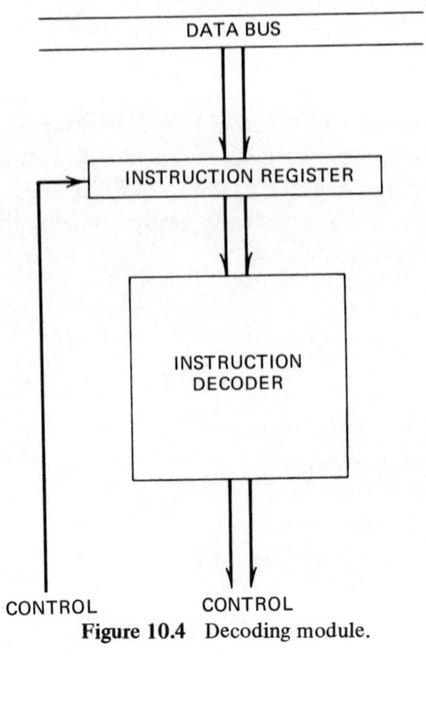

Figure 10.4 Decoding module.

instruction available when the subroutine is completed, which is when the return instruction is encountered. Thus the last instruction in any subroutine is a return, which causes the processor to resume execution of the normal program immediately following a subroutine.

Input/output ports can be serviced using nomal loop instruction. However, important input/output ports are permitted to interrupt normal program execution. This is a special kind of jump instruction because it occurs at any time and is hardware rather than software initiated. When the return from interrupt statement is executed it would be nice to have the CPU return to exactly the same condition as when the interrupt was encountered. In order to resume normal execution of the program the contents of the accumulators, status registers, and working registers are required, so when an interrupt is encountered the values of all these registers are pushed into the stack, and the SP is decremented with each byte stored. When a return from interrupt statement is encountered the accumulator's status register and working registers are loaded with values as pulled from memory. The value in the stack pointing register is automatically incremented with each byte, to point to the proper location. The stack is usually kept in a special area of the RAM.

10.2.4 Address Bus

The purpose of having an address bus is to give each memory and I/O location of the microcomputer its own unique address. Address buses are usually 2 bytes or 16 bits long. This provides 0 to 65,536 memory address locations. This is the standard length for single-byte 8-bit data microprocessors, the most common type. Microprocessors with a data bus of 2 bytes or 16 bits long usually have a 20-bit address bus. The address bus receives data from the working register module and is usually fed from the addressing function circuit or the increment/decrement circuit.

10.2.5 Data Bus

Data buses are usually 8 bits long. Some microcomputers have 2-byte or 16-bit data buses. Microcomputer systems with 16-bit data buses are beyond the intended scope of this book.

A data bus buffer is provided to isolate the CPU's internal bus from the external data bus. This is accomplished using a three-state (tri-state) buffer. A tri-state buffer is required when the output from two separate devices must be connected to one input. Since tying two outputs together will usually destroy the output circuits, a switch is required to determine which output is connected to the input. The other output is open circuited. This switch is called a tri-state buffer.

10.2.6 Timing Circuit and Control

A microprocessor is provided with three basic timing signals as shown in Figure 10.5. The conventional clock is provided by ϕ_1. The positive pulse of this clock has

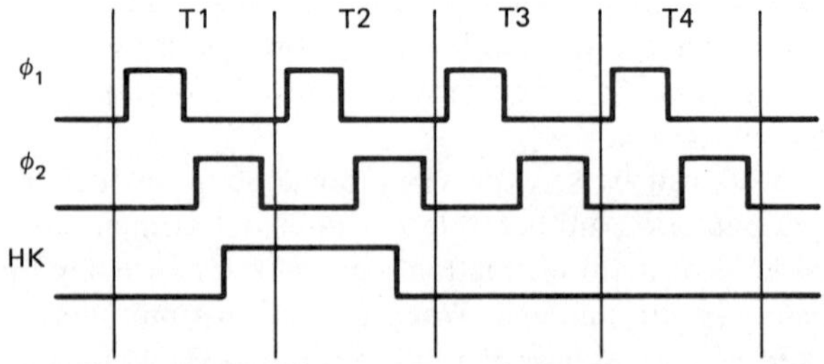

Figure 10.5 Typical system clock arrangement.

a length approximately 40% of the clock period. An additional pulse called ϕ_2 identifies the remaining part of the period. By use of this technique the data and address buses are connected to CPU registers during the ϕ_1 clock or to memory and I/O during the ϕ_2 clock.

The housekeeping (HK) clock is derived from the ϕ_1 and ϕ_2 clock pulses. The HK clock is used to identify the first two states in a cycle. For instruction where housekeeping is not required in these two states, this signal is suppressed.

The CPU control is a sequential circuit. However, it is a sophisticated sequential circuit capable of executing 50 to 100 different instructions. The CPU continuously monitors its own internal status and the status of external components. The status is decoded into a control word. The control word gives the mode of operation and thus determines the exact transition through the state diagram.

As explained above, the CPU loads the address bus and reads from or writes to the data bus. The execution of instructions is controlled by the control bus. The CPU control signals are given in Figure 10.6. The input signals are:

ϕ_1: clock
ϕ_2: clock
INTERRUPT: signal from input device requesting service
HOLD: request to freeze the μP
READY: signal from RAM indicating availability

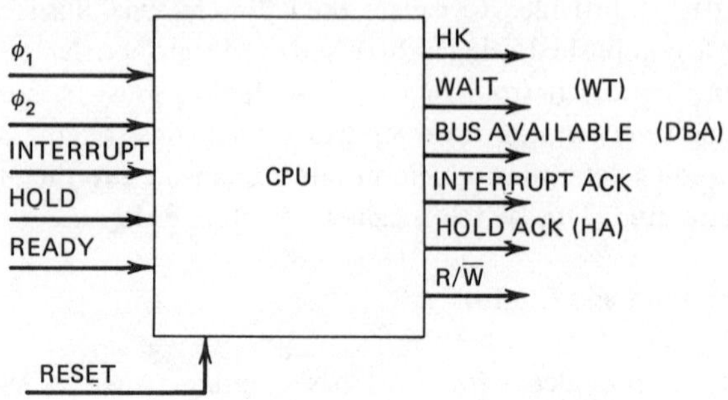

Figure 10.6 Generalized CPU control signals.

The output signals are:

HK: signal to identify states T1 and T2
WAIT: signal to inhibit transfer to data bus
BUS AVAILABLE: signal to enable transfer to data bus
INTERRUPT ACK: CPU response to INTERRUPT request
HOLD ACK: CPU response to HOLD request
R/$\bar{W}$: signal to enable transfer from or to the data bus

10.3 STATE DIAGRAM ILLUSTRATION OF MACHINE CYCLES

It has been mentioned that the number of machine cycles in an instruction cycle may vary from one to five. The number of states in a machine cycle may vary from two to five. Therefore it is obvious that in the execution of any instruction there are a large number of possibilities [30-34]. A detailed study of all the possibilities would achieve little since different manufacturers may tackle the problem differently, and advancements in technology may change the approach. Nonetheless, some studies of how a microprocessor executes an instruction will help the reader understand the following:

- Why different instructions require longer execution times.
- How the control lines on a microprocessor are intended to operate.
- How to choose between alternative methods of programming the same problem.

It is therefore proposed to construct state diagrams to illustrate the general operation machine cycles for the following instructions:

1. Register instructions.
2. Memory instructions.
3. I/O instructions.

A state diagram for a register instruction is given in Figure 10.7. This instruction adds the contents of a register to the accumulator. Since all the information needed to execute the instruction is contained in one address in memory, only one reference to memory is needed; thus only one machine cycle is needed.

The clock ϕ_1 causes a transition to state T1. In state T1 the positive edge of the ϕ_2 clock causes the transition to the housekeeping mode. This, for convenience, can be regarded as an event-driven circuit, and its simple state diagram is presented above state T1. The contents of the program counter are placed on the address line (PC OUT, ADDRESS OUT) and a control word is loaded onto the data bus buffer register (DB → CONTROL). This control word determines the mode of operation (the type of machine cycle) to be executed. The mode here indicates that this is a fetch instruction, which requires a memory read using the data bus.

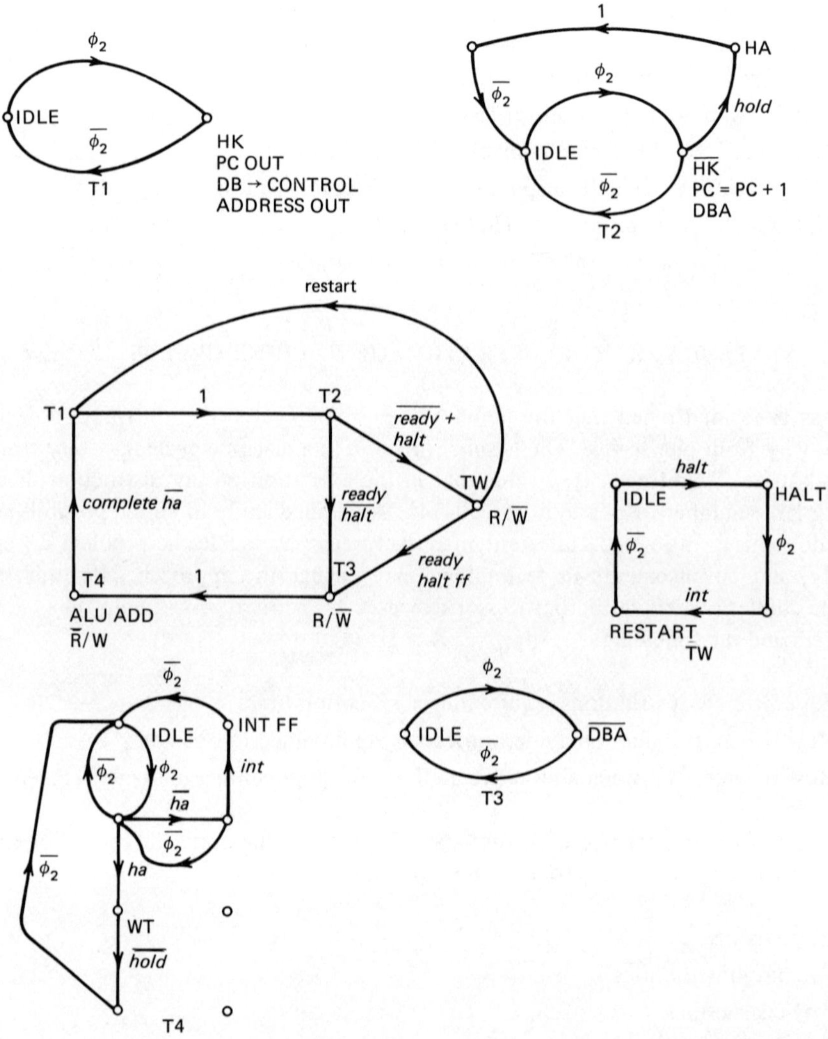

Figure 10.7 Machine cycle 1 example for add register to accumulator; *int* = *interrupt* · *interrupt act.*

On the next clock pulse ϕ_1 there is an automatic transition to state T2. The event-driven state diagram shows the conditions for state T2. On the ϕ_2 clock the housekeeping signal is driven low, the PC is incremented, and the external signal DBA sent to enable data transfer. The external circuits are checked to see if a hold has been received. If it has, an external hold acknowledge (HA) signal is issued. When the state T2 chores are completed there is a transition to state T3 if the memory is available for cycling (*ready*) and if the software instruction *halt* has not been issued. If either of these conditions do occur ($\overline{ready}$ + *halt*) there is a transition to the wait state TW.

Immediately upon entry into state TW an external read signal R/$\overline{W}$ and external wait signal TW are issued. The wait signal inhibits the read signal. In the wait state there is an event-driven circuit. Upon receipt of the *halt* signal a halt flip-flop is set. If an interrupt and an interrupt acknowledge have occurred there are two transitions to the state where a restart signal is issued. The restart signal causes a transition from state TW to state T1 on the next ϕ_1 clock pulse.

If the memory is available for cycling (*ready*) and the halt flip-flop has not been set ($\overline{\text{HALT FF}}$), there is a transition to state T3. In state T3 the external read command R/$\overline{W}$ is issued on the ϕ_1 pulse causing the external R/$\overline{W}$ line to go to read whether or not there was a transition through TW. On occurrence of the ϕ_2 pulse the external enable data line (DBA) is lowered. At this point the fetch instruction is in the CPU decoding circuit.

On the next ϕ_1 pulse there is a natural transition to state T4. In this state the add register to accumulator instruction is executed. The register is loaded into the accumulator register, and the ALU is issued an add instruction. Also the R/$\overline{W}$ line is driven low to the write condition. In this state the last bit of housekeeping is done. If HA is set the circuit goes into a wait mode. If it is not set and an interrupt has occurred, the interrupt flip-flop (INT FF) is set, and an external INTERRUPT ACK signal is issued. If state T4 is the last state to be executed, and in this case it is, and if the HA flip-flop is not set, a transition to state T1 occurs on the next ϕ_1 pulse.

Operation of this machine cycle is also shown in block form in Figure 10.8. On the first clock pulse ① the address of the instruction to be executed is sent from

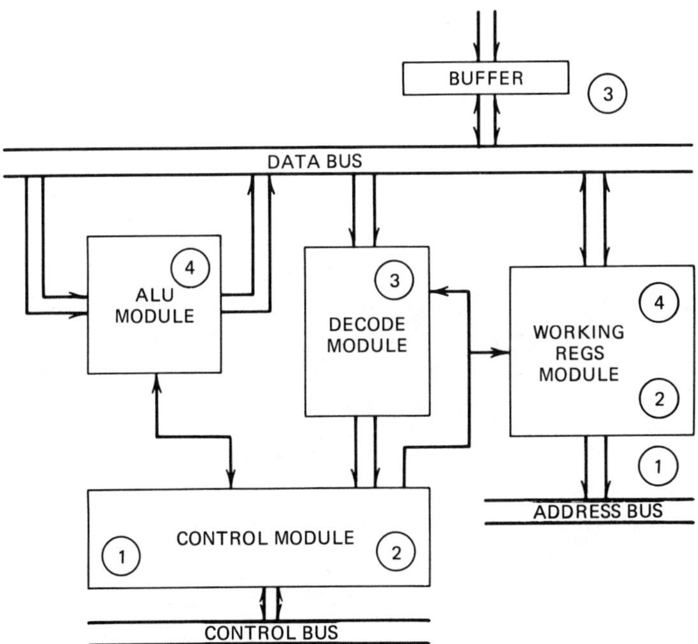

Figure 10.8 Machine cycle 1 block diagram example for add register to accumulator.

the working registers to the address bus, and at the same time the present status of the microprocessor is received in the control module. Typical information that might be received is that an interrupt or a halt is required. However, for this machine cycle the actual information received is that this is a fetch instruction and that a memory read by way of the data bus is required.

On the next clock pulse ② the PC is incremented in the working register module, and the RAM memory is checked to see if it is available for cycling; also the control bus is checked by the control module, and the system status is updated.

On the next clock pulse ③ the instruction is read from ROM through the buffer to the data bus and to the decode module.

On the fourth clock pulse ④ the value of the working register is sent to the ALU module where it is added to the accumulator in the ALU module.

On the next clock pulse the CPU is returned to the initial ① state. This completes the description of the add a register to the accumulator instruction. This instruction cycle is completed in one machine cycle with four states.

An add-memory-location-to-accumulator instruction with an 8-bit address takes two machine cycles. Machine cycle 1 of the add-memory-to-accumulator instruction resembles the machine cycle described above, except that on the fourth clock pulse ④ ALU operation is inhibited.

The state diagram for machine cycle 2 is shown in Fig. 10.9. Starting with a ϕ_1 clock pulse causing a transition into state T1, the positive edge of the ϕ_2 clock causes the address of the memory to be loaded on the address bus. This is also

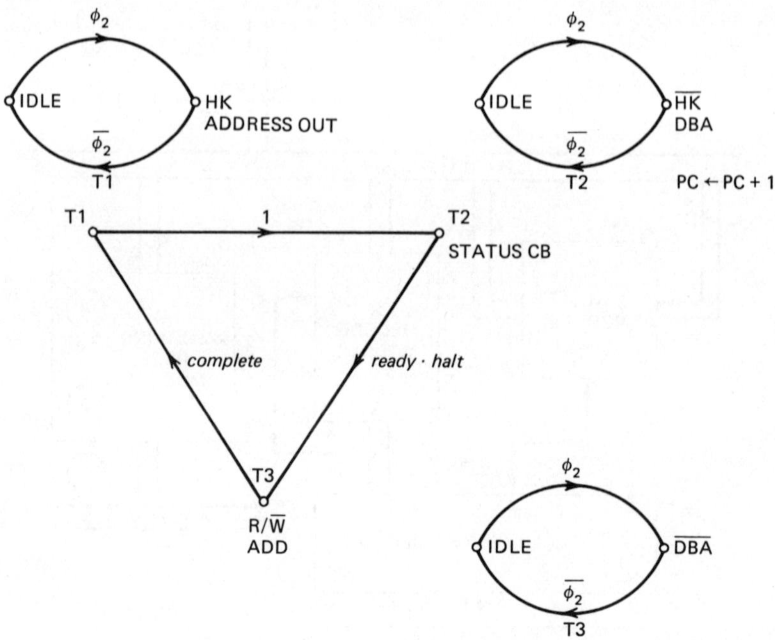

Figure 10.9 Machine cycle 2 example for add memory to accumulator.

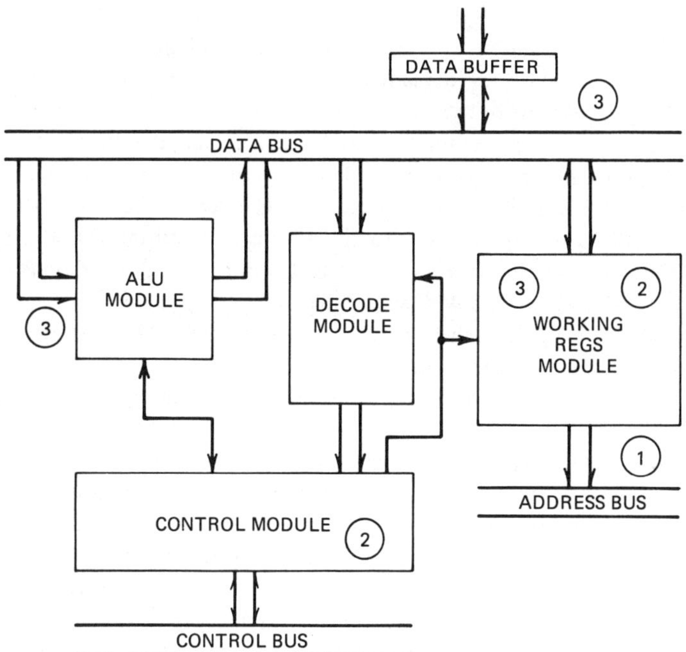

Figure 10.10 Machine cycles 2, 3, and 4. Generalized block diagram.

shown in the block diagram of Figure 10.10. The ① indicates a movement from the working register module to the address bus.

On the next ϕ_1 pulse there is a transition to state T2, ②, as indicated in Figures 10.9 and 10.10. The status of the control bus is checked (STATUS CB), and when clock pulse ϕ_2 occurs the control module enables the data bus (DBA) for data transfer.

On the next ϕ_1 clock, if the memory is ready and a halt has not been received, there is a transition to state T3, ③, where the read ($R/\overline{W}$) and add (ADD) instructions are issued. To achieve this the data is directed to the ALU module and added to the accumulator. On the positive edge of the ϕ_2 clock the enable data line is driven low ($\overline{DBA}$). Thus machine cycle 1 has one read from ROM and machine cycle 2 has one read from ROM and one read from RAM.

Since this completes the instruction add memory to accumulator there is a transition to state T1 on the next ϕ_1 clock. This instruction is executed using machine cycle 1 with four states and machine cycle 2 with three states for a total of two machine cycles and seven states.

An example of an instruction that may take three machine cycles is the load accumulator from I/O. Three cycles are used if the I/O has a 16-bit address. In its operation machine cycle 1 is as given in Figure 10.8. In state T1, ①, the value of the program counter is placed on the address bus. In state T2, ②, the status is checked, and the program counter is incremented by one. In state T3, ③, the instruction is read from ROM through the data buffer and into the decode module. In state T4, ④, a dummy load is done in the ALU module.

Machine cycles 2 and 3 are explained using Figure 10.10. In machine cycle 2 the value of the first address byte is placed on the address bus during state T1, ①. In state T2, ②, the program counter is incremented, and the status is checked. In state T3, ③, the address of the I/O port is sent to the working registers.

Machine cycle 3 completes the instruction. The second address byte is loaded from the working register module to the address bus during ①. During ② the status is checked, and the data bus is enabled. For action ③ the data is exchanged between the I/O port and the ALU accumulator.

Machine cycles 1 and 2 each have one read from ROM. Machine cycle 3 has one read from ROM and one read from I/O.

The execution of this statement required three machine cycles consisting of 10 states.

In summary the number of machine cycles depends upon addressing. Register-to-register instructions require only one machine cycle. Instructions with a single-byte address require two machine cycles. Instructions with a double-byte address require three or more machine cycles. An instruction with a double-byte address often has more than twice the power of an instruction with a single-byte address, and therefore may take more than three machine cycles. The user can connect I/O or memory to single- or double-byte addresses and thereby balance speed against power.

The foregoing description is a generalization of the circuits used by many μP manufacturers. Other manufacturers use enhancements of this design. In one enhancement there is only one clock pulse in each machine cycle. The clock then runs at about one-third speed. All states within a machine cycle are event driven. A further enhancement is to provide improved memory access. This changes the operation and programming in subtle ways. Still another enhancement is to eliminate the ϕ_2 clock.

10.4 INPUT AND OUTPUT (I/O)

An input/output system is required for communication to devices outside the microprocessor. Input and output are very important if the device is to serve a useful function. A microcomputer that cannot receive and/or transmit information serves no useful purpose.

A microcomputer interface is really a buffer or a register to connect a microcomputer CPU to a peripheral device. To output from the CPU the interface register holds data until a peripheral device is ready. In this way the CPU can perform other functions while waiting for the peripheral device to respond. Similarly, on input the interface register holds the data for the CPU, thus permitting the CPU to access that data at its convenience. The main function of the interface is to act as a synchronizing device between the CPU and the peripheral device.

Microcomputer manufacturers market interface adapters to simplify the interfacing task. The two most common interface adapters are the general-purpose parallel interface adapter intended for fast transfer and the serial interface adapter intended for transmission over long distances.

Figure 10.11 shows a block diagram of an input/output port for a microcomputer system. This port must be connected to the data bus for transfer of data and connected to the address bus for identification. In a simple configuration the microprocessor provides the I/O with three signals from the control bus. A clock ϕ is used to synchronize it to other devices in the system. A read/write line $R/\overline{W}$ is used to indicate the direction of data transfer and enable line ENABLE for traffic control. The I/O requests service from the microprocessor using an interrupt request ($\overline{IRQ}$) signal.

Figure 10.12 shows an extended block diagram for the control of a typical I/O port. It is synchronized with the CPU using an interchange on the $\overline{IRQ}$ and ENABLE lines. Similarly, the input/output module is synchronized to the peripheral device using the REQUEST and RESPONSE signals.

A request for data (ENABLE) from the CPU is transferred to the peripheral device on the REQUEST line. When the peripheral device is able to respond it sends a signal on the RESPONSE line. This is transmitted to the CPU by way of the $\overline{IRQ}$ line.

Of course an I/O module is much more sophisticated than this simple structure. A standard package usually has two or three ports. The main problem with an I/O module is that it may have too few pins. Multiplexing within the module allows more effective use of the pins. In this way the number of peripheral devices that can be serviced by one input/output module is increased.

In the early development of microcomputers it was considered important to have the I/O control performed directly by the CPU. This made the I/O somewhat more flexible but placed a burden on the CPU. The structure of a conventional CPU is

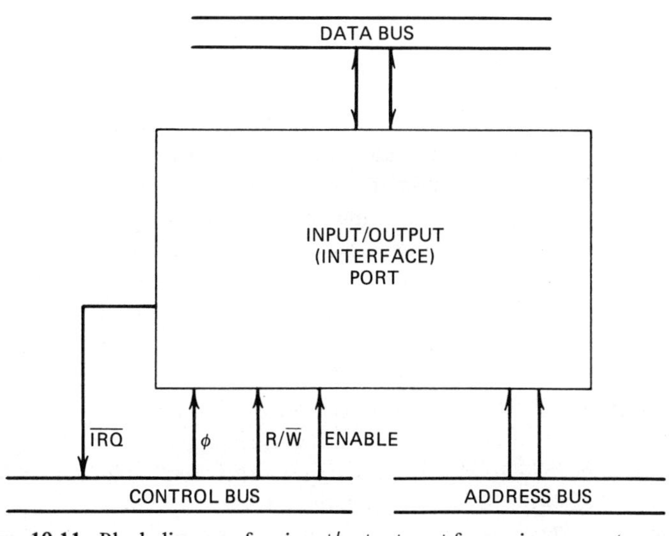

Figure 10.11 Block diagram of an input/output port for a microcomputer system.

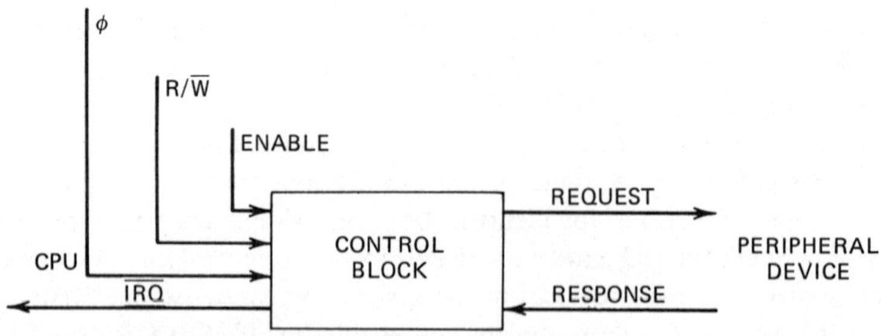

Figure 10.12 I/O control signal.

not designed for I/O tasks, whereas interface adapters are. Interface adapters are made programmable so that the CPU may still have direct I/O control, but the burden of this control is removed. This allows the CPU to be utilized more effectively. Also it leads to two points that the reader must keep in mind about interfaces.

1. The direction of data flow is subject to program control.
2. The operation and direction of the control signals are subject to program control.

Chapter 12 deals with the functions and operations of I/O and interface adapters in a more specific way. The remainder of this chapter gives the frame of reference for the specific and detailed descriptions in Chapter 12 and, in fact, for all remaining chapters in this book.

10.5 MICROCOMPUTER SYSTEM STUDIES

So far in this chapter, all descriptions have been in general terms. Previous chapters describe standard components that do not change from one manufacturer to another. However, microcomputer components are not standard between manufacturers. Microcomputer components really consist of complete circuits. Each circuit is unique to the manufacturer. Indeed, the exact construction and implementation of the circuit may be proprietary information. Therefore it is necessary to study specific manufacturers' microcomputer systems [30–34].

In this book the general operation of the Motorola MC6800 family of components and the Intel 8085A family of components is studied. As an aid to the reader, delimiters are used to identify each manufacturer. □ The reader who wishes to study the Motorola MC6800 family of components should read text between the open squares as you are now doing. □

■ The reader who wishes to study the Intel 8085A family of microprocessor components can read text between the solid squares as you are now doing. ■

The Motorola MC6800 CPU is discussed in this book. Motorola's MC6802, MC68A00, and MC68B00 CPUs have instruction sets compatible to the MC6800. The Intel 8085A CPU is described in this book, and the Intel 8048 and 8080A have compatible instruction sets.

The Motorola MC6800 uses a clock cycle equal to its machine cycle, and the Intel system uses a clock cycle approximately one-third of its machine cycle. To avoid ambiguities the word "cycle," when it appears alone, is to mean machine cycle.

□ 10.5.1 The MC6800 CPU

The symbolic CPU block diagram for the MC6800 CPU is shown in Figure 10.13. The MC6800 has a 16-bit addressing and an 8-bit data bus. It contains the standard decoding module. The ALU has two 8-bit accumulator registers A and B and a status register CC. The working registers in this unit are an index register (IX or X), a program counter (PC), and a stack pointer (SP). All are 16-bit addressing registers.

The CPU control circuit has two clock inputs ϕ_1 and ϕ_2. Its power supply consists of + 5 volts and ground. In addition it has six inputs. These inputs are as follows:

- *Interrupt request ($\overline{IRQ}$).* The interrupt request is provided so that the input/output hardware can request service from the CPU. The CPU can be programmed to disallow this interrupt.
- *Nonmaskable interrupt ($\overline{NMI}$).* This interrupt is similar to the $\overline{\text{IRQ}}$ except that $\overline{\text{NMI}}$ cannot be masked by the programmable interrupt mask.
- *Data bus enable (DBE).* This is a tri-state control signal for the CPU data bus. Normally this signal is derived from the ϕ_2 clock. Its purpose is to guarantee that the output from the CPU and the output from the memory are not tied together at the same time.

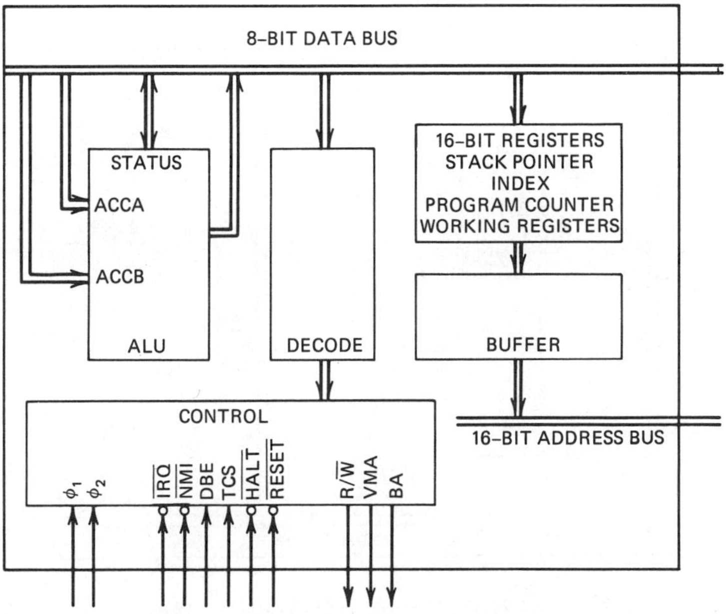

Figure 10.13 MC6800 CPU block diagram.

- *Three-state control (TSC).* The tri-state control affects the address bus. Its purpose is to control the address bus so that no two outputs are tied together. It is used when the CPU and some external hardware are both tied to the address bus.
- *Halt ($\overline{HALT}$).* The halt line is used to stop the CPU. It forces the address bus and data bus into the high-impedance states (that is, it disconnects them).
- *Reset ($\overline{RESET}$).* The reset line is used to initialize the CPU to start program execution.

The CPU has three output commands. These outputs are as follows:

- *Read/write (R/$\overline{W}$).* This signal designates whether the CPU is in the read or write mode. It is also tri-stated open to the high impedance mode by inputs TSC and $\overline{\text{HALT}}$.
- *Valid memory access (VMA).* The VMA line is used to signal memory and I/O that the CPU is performing a read/write operation within a given cycle. This signal must be applied to the chip ENABLE (chip select) input to disable transfer when VMA = 0.
- *Bus available (BA).* If the CPU is stopped for any reason (halt or program instruction), the bus available signal is used to signify that the data bus is available.

Figure 10.14 shows a typical symbolic block Motorola RAM and ROM. It illustrates the method of interconnecting to the CPU address, data, and control bus lines. The remaining control lines are used to interface the I/O devices and to provide start and reset facilities. □

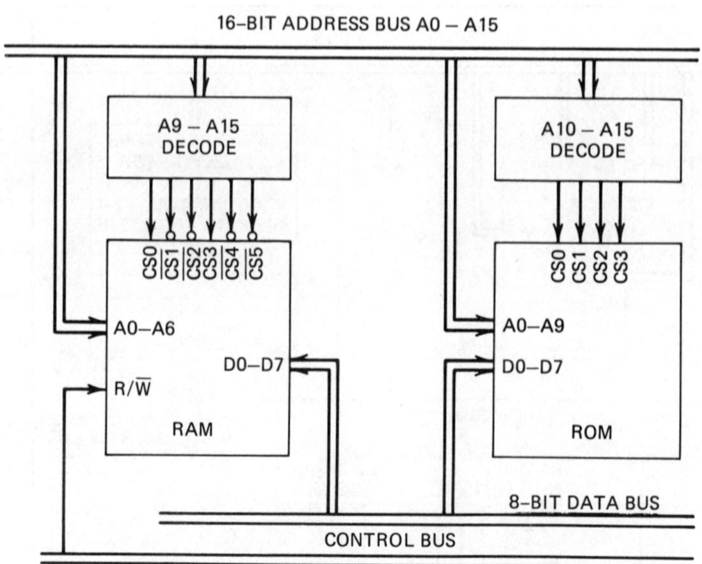

Figure 10.14 Motorola typical RAM and ROM.

■ 10.5.2 The Intel 8085A CPU

A detailed block diagram for an 8085A CPU is shown in Figure 10.15. The 8085A has a 16-bit addressing bus and an 8-bit data bus. However, the data bus and bits A7 to A0 of the address bus use the same bus. This saves pin connections, which are used to extend the control bus.

The CPU contains a standard instruction decode module. The ALU has one 8-bit accumulator A, an 8-bit temporary accumulator, and a status register.

In the working register module there are four 8-bit data registers B, C, D, and E and two 8-bit registers H and L, which function either as data registers or as an address pair of registers. This module also has a stack pointer SP and a program counter PC, which are both 16-bit addressing registers.

The CPU control circuit has two inputs X1 and X2 for a crystal or RC network connected to the internal clock generator. Its power supply consists of + 5 volts and ground.

In addition there are eight inputs. These inputs are as follows:

- *Ready (READY).* Ready is a control signal from memory or a peripheral device that acts as an enable to send or receive data. The CPU will not complete a read or write cycle until the ready signal goes high.
- *Hold (HOLD).* Hold is also a control signal. It indicates that another device requests the use of the address and data buses. The CPU relinquishes the buses at the completion of the current machine cycle. The processor regains control of the buses when the hold signal goes low.

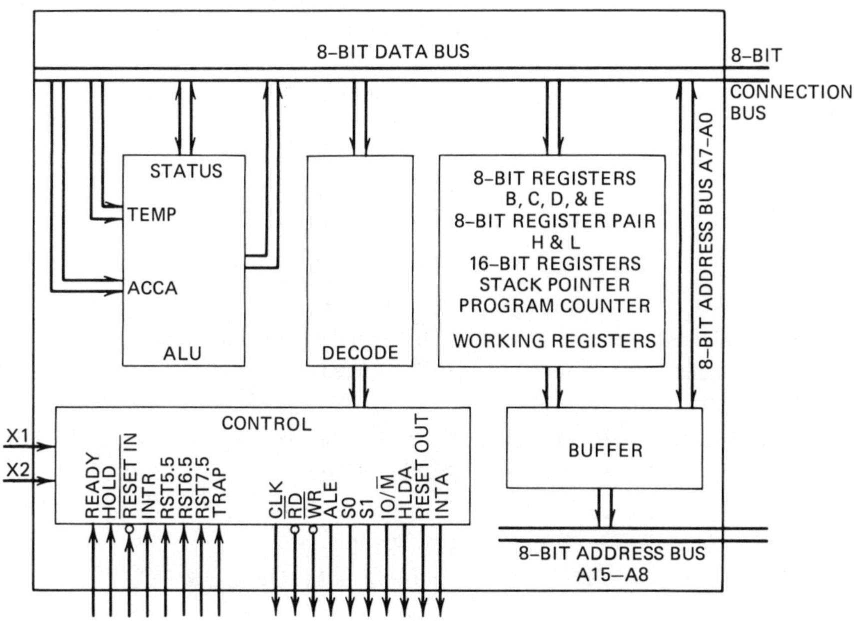

Figure 10.15 8085A CPU block diagram.

- *Reset in ($\overline{RESET\ IN}$).* The $\overline{\text{RESET IN}}$ signal initializes the CPU to start program execution.
- *Interrupt request (INTR).* This is an interrupt signal. The interrupt is enabled and disabled by software.
- *Restart interrupts (RST5.5, RST6.5, and RST7.5).* These three interrupts are used to cause internal restarts to be automatically inserted. The hierarchy of priority for interrupts is RST7.5, RST6.5, RST5.5, and INTR.
- *Trap (TRAP).* The trap interrupt is a nonmaskable restart interrupt, that is, it cannot be masked by the stystem software.

The CPU has 10 output commands. These output commands are as follows:

- *Clock (CLK).* The clock output provides a signal that may be used as a system clock. The period of the clock is twice the input period.
- *Read ($\overline{RD}$).* This signal indicates that the selected memory or I/O device is to be read and the data bus is available for data transfer. It is tri-stated to the high impedance mode by the input HOLD and the halt (HLT) instruction.
- *Write ($\overline{WR}$).* The write signal indicates that the data on the data bus is to be written into the selected memory or I/O location. This signal is driven into the high impedance mode by the input HOLD and the HLT instruction.
- *Address latch enable (ALE).* This output occurs during the first clock pulse in any machine cycle and is used to enable the address to be latched into the proper peripheral. ALE is also used to strobe status.
- *Data bus status (S0, S1).* These two signals can be decoded according to Table 10.1 to determine if the CPU is in the halt, write, read, or fetch mode.

Table 10.1 Control Status

S1	S0	Mode
0	0	Halt
0	1	Write
1	0	Read
1	1	Fetch

- *Input, output, or memory ($IO/\overline{M}$).* This signal is used to indicate whether the read/write signal is to memory or IO. The output is tri-stated on HOLD and on the HLT instruction.
- *Hold acknowledge (HLDA).* This signal is a handshake response to the hold request input. It indicates CPU relinquishing of the bus.
- *Reset out (RESET OUT).* This signal indicates that the CPU is in the middle of a reset. It is intended for system synchronization.
- *Interrupt acknowledge (INTA).* Its purpose is to interrogate I/O to determine an interrupt address.

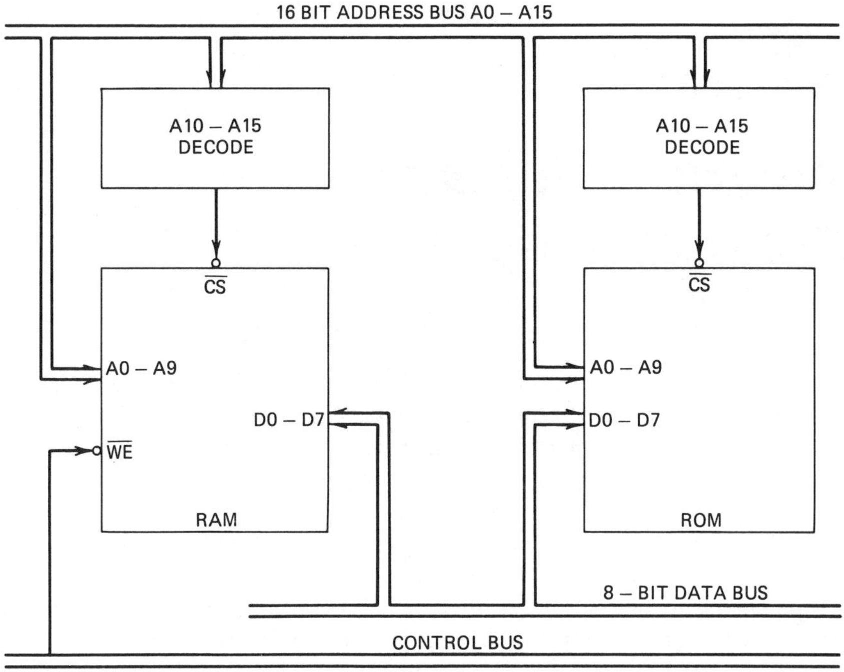

Figure 10.16 Intel typical RAM and ROM.

The CPU address bus, data bus, and control lines are used to interconnect the CPU to other system components.

Figure 10.16 shows a symbolic block diagram for typical Intel RAM and ROM. It illustrates the method of interconnection of the CPU address and data buses and also illustrates the use of the CPU control signals. The read ($\overline{RD}$) and write ($\overline{WR}$) signals control the write enable ($\overline{WE}$). The remainder of the control signals is used to interface I/O devices and to provide start and reset facilities. ■

These signals are introduced in this chapter to provide background information for subsequent chapters, particularly Chapters 12 and 13. The reader will better understand the use of control signals after reading subsequent chapters. As a consequence the reader will better understand this section after reading Chapters 12 and 13. Block diagrams in these subsequent chapters form a natural interface and aid in understanding the use of CPU control signals.

Please do not attempt further processing of these signals in your brain. Acknowledge and interrupt and start your brain working on Chapters 11, 12, and 13.

CHAPTER 11

Programming and Programming Aids

The basic structure of a microcomputer is given in Chapter 10. It illustrates the concept of a CPU as a development from preceding chapters. The concept of programmable circuits is first introduced in Chapter 6. The point is reached where a circuit is not only capable of minor variations in execution, but also is programmable in a truly sophisticated sense. The philosophy of interactive and nested sequential circuits is also developed. This idea is truly powerful in concept and realization.

Different sequential circuits are executed according to the contents of the instruction register. A natural advancement is the development of an instruction set. An instruction set is simply a means of telling the microcomputer what to do by way of the instruction register. The microcomputer is told what to do by a series of instructions (rather than by single instructions) as a matter of efficiency.

A coded instruction leads to the execution of a number of sequential circuits. A series of coded instructions are in themselves a type of master sequential circuit. This master sequential circuit is referred to as a program. A program is executable once it is contained in memory. Since the content of memory is easily altered (that is, at least conceptually) it is referred to as software.

Historically or traditionally digital system designers have tried to choose a combination of sequential circuits and instruction sets that optimize all the considerations of the design. These considerations are versatility, cost, physical size, and utility. As explained in Chapter 10, optimizing the design centers around the two problems of access to the storage bank or memory, and the length (number of bits) of an instruction. The CPU is designed to execute a particular set of instructions under these constraints.

The job of the programmer is to perform a specific operation within the constraints of the instruction set. The programmer follows the same basic steps as the logic designer, which are as follows:

1. Construction of the program flow chart (a programmer's state diagram).
    (a) Determine the inputs and outputs of the system.
    (b) Determine their cause and effect relationship.
    (c) Determine the starting point of the sequence.
    (d) Arrange the inputs in a sequential progression of events.
    (e) Determine when the event is dependent upon an external decision.

(f) Check that inputs that should lead to the same destination do so. This is exactly equivalent to the redundancy checks given in Chapter 3.

(g) The procedure should continue until the program forms a complete loop. The most effective programs are used repeatedly. It is therefore good practice to design all exit points to be nonterminal.

The flow chart is more subtle than a state diagram because inputs and outputs occur in a more subtle form. Also, having gone through the foregoing steps two or three times, they become repetitious, and they may not be placed on paper. Nonetheless the programmer must always be aware of the required procedure.

2. Convert the flow chart to a sequence of instructions (program) as defined by the microcomputer instruction set.
3. The program is entered into the microcomputer memory. This step is usually accomplished in two stages. In the development or temporary stage the program is placed in RAM. Programs in RAM must be executed using a microcomputer system with a special ROM executive program. The fact that the temporary stage receives considerable attention as a teaching and development tool should not confuse the reader as to his ultimate aim. In the permanent or final stage the program is placed in ROM.
4. Write an executive for the final program. The executive consists in general of a few instructions that link the hardware reset function to the first executable program instruction.

The reader is assumed to be proficient in program flow charting. In this chapter programming details such as instruction set format, rules, and characteristics are described. This gives the required background knowledge to accomplish step 2 above. This chapter also includes details of entering a program into RAM. The remainder of steps 3 and 4 is described in Chapters 12 and 13 respectively.

11.1 MNEMONIC PROGRAMMING

A microcomputer can only execute instructions from the instruction register. The series of 1s and 0s that occurs in the instruction register is called a machine code [32, 33]. It is extremely cumbersome to program in machine code. To overcome this difficulty, programming languages have been developed. One type of programming language is called assembly. The programmer writes the source program in assembly language mnemonic form. An assembler program translates the source program into the machine code instructions by way of a one-to-one mapping into machine code instructions.

Figure 11.1 shows a block diagram of the assembler function. Its input is the source program. The output is an object file and an assembly listing. The object file is conveniently understood by microcomputers. The assembly listing is a file conveniently understood by man. The object file is loaded into memory to form the machine instructions that constitute a program. The assembly listing is a super

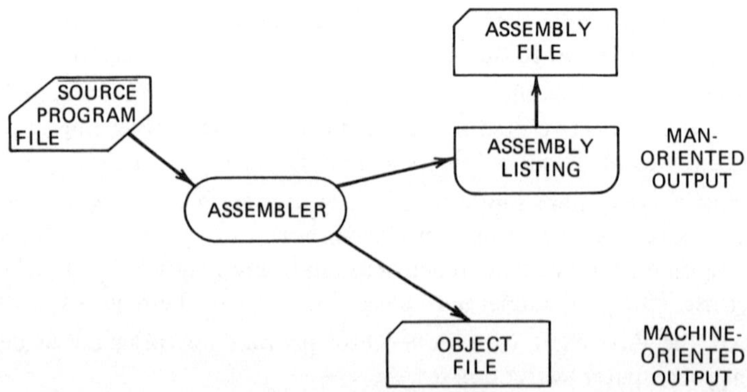

Figure 11.1 Assembler functions.

kind of source program. It is a super version because, along with a simple listing of the instructions and comments (man-oriented part), the listing also includes the locations of each instruction in memory and the machine coding of each mnemonic instruction (machine-oriented part). It makes an excellent working and debugging file. For this reason an assembly listing should always be kept in a file.

11.1.1 Assembler Programs

In general an assembler program requires that the source program file be processed two or more times. Each processing is called a pass. In a two-pass system the following tasks are done.

Pass 1

- Labels are assigned numerical values and placed in a symbol table. The purpose of a two-pass system is to allow the programmer to use labels explained in Sections 11.2.3 and 11.7.
- Syntax is checked and errors are listed.

Pass 2

- Remaining undefined symbols are evaluated or an error message is given.
- The assembly listing is printed and errors associated with any line are noted.

Not all programmers use assembler programs. Some act as their own assembler and map each mnemonic directly into machine code. This practice teaches the programmer to exercise extreme care. It also familiarizes the programmer with machine code. A knowledge of machine code is valuable when debugging. For this reason the reader is advised to practice converting source listings into machine code. Machine codes are included in this book for that purpose. However, use of the assembler program is touted in this book as the most effective means of programming microcomputers.

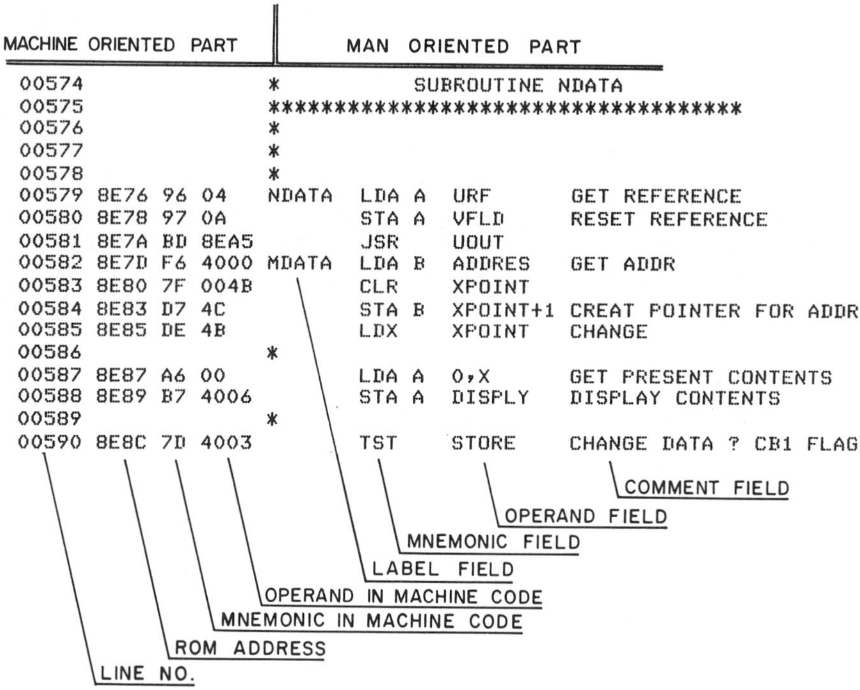

Figure 11.2 Assembler listing format.

11.1.2 Assembler Listing

The assembler listing is a record of the source statement and the object (machine) code for that source statement.

The assembler listing shown in Figure 11.2 illustrates the format. This typical listing is from a Motorola assembler. The overall program is for control of a 13.8 kV alternator for electric power generation. The part of the program illustrated is from the display of control parameters subroutine.

Details required to read an assembler listing in this format are explained in Sections 11.2 and 11.3.

11.2 ASSEMBLER FORMATS

Remembering that the objective of the assembler is to make programming easier, one of the most important functions of the assembler's syntax is to eliminate errors. The assembly instruction is a three- or four-letter mnemonic symbol that is designed to help the programmer remember the definition of the instruction, rather than the hexadecimal two-character code representation of the exact binary number that the machine executes. The assembler program recognizes a certain format. The assembler instruction is written on one line. Each line is divided into four fields. The field sequence cannot be changed, and the correct order is as follows: label, mnemonic, operand, and comment.

11.2.1 The Mnemonic Field

The mnemonic field is considered first because it is the most important field. In the assembler language instruction it is this field that defines an operation of the microcomputer. It specifies the object code to be generated and thus the action to be performed by the CPU. The mnemonic field begins after the label field and a space. The label field can be omitted but the space must occur. In general, the mnemonic field is terminated by one or more spaces.

11.2.2 The Operand Field

The operand field gives information about the recipient of the instruction. For example, it may identify a location in RAM memory, or an I/O port. Instructions that operate on registers have no operand field. In the assembler format the operand field begins after the mnemonic field is terminated by one or more spaces. The mnemonic field is stored in 1 byte of ROM and the operand field, if it exists, requires 1 or 2 bytes. Thus an assembler instruction occupies 1, 2, or 3 sequential bytes of ROM memory.

11.2.3 The Label Field

The label field is provided for the programmer's convenience. If an identifier is placed in the label field it is a means of addressing this instruction. The programmer can access this instruction directly without knowing its actual address in the program. The label field is thus an important asset when programming any type of looping instruction, such as jump, branch, and call. The label field begins immediately after the carriage return, which terminates the previous instruction. Label fields are terminated by a space (Motorola) or colon (Intel).

11.2.4 The Comment Field

The comment field is another field provided for the programmer's convenience. It allows the programmer to write the intent of the instruction in common English. This is a great aid when correcting and for general debugging. Comments in many ways form the primary and most important documentation of the program. It thus becomes an absolutely essential field if the program is to serve more than one user. The comment field begins after the operand field is terminated by a space and is itself terminated by a carriage return (CR).

Intel comment fields must start with a semicolon (;).

The following is an example the source program file set as input to the assembler and the man-oriented part of the assembler listing produced by the assembler. Column numbers show characteristic spacing. In the following (CR) means carriage return.

□ The Motorola assembler does the following formatting:

Source program file

```
NM JSR UOUT (CR)
  CLR XPOINT CLEAR HIGH BYTE (CR)
  STAB XPOINT+1 CREATE POINTER (CR)
```

Assembler listing

```
1       8       15         24
Label   Mnemon  Operand    Comment
NM      JSR     UOUT
        CLR     XPOINT     CLEAR HIGH BYTE
        STAB    XPOINT+1   CREATE POINTER  □
```

■ The Intel assembler does the following formatting:

Source program file

```
  LXI H,BUFR ;LOAD MEMORY ADDRESS (CR)
CI: CALL 0F803H ;GET CHARACTER (CR)
  MOV M,A ;STORE IN MEMORY BUFR+N (CR)
```

Assembler listing

```
1       9       17       25
Label   Mnemon  Operand  Comment
        LXI     H,BUFR   ;LOAD MEMORY ADDRESS
CI:     CALL    0F803H   ;GET CHARACTER
        MOV     M,A      ;STORE IN MEMORY BUFR+N  ■
```

11.3 NOTATIONS AND RULES

Characters used in a source program are a subset of ASCII. Symbol usage makes a program easier to read and easier to write. It also reduces the risk of error during modification and debugging. After symbols have been defined they can be used in the operand field. Symbols are defined automatically when they appear in the label field of an instruction. This is extremely helpful when modifying data that is constantly referred to by other statements. The programmer then need only change the defining statement rather than all statements where the symbol is used in the instruction in operand fields.

Certain words are reserved as predefined symbols. Microcomputer manufacturers also recognize the benefits of symbols and use them in both the mnemonic and operand fields to ease the programmer's chore. These special symbols are given at the end of this section.

Symbols are created using the following rules:

- The first character in the symbol must be alphabetic.
- The remaining symbols may be composed of the letters A through Z, the numbers 0 through 9, and special characters, . (period), _ (underscore), and $ (dollar sign).
- A symbol may contain up to 6 (Motorola) or 8 (Intel) characters. If the symbol is created with more characters than allowed, excess characters are ignored. Predefined symbols and reserved words cannot be redefined.

There are certain symbols that are not recognized by the assembler but are used in abbreviating instruction descriptions.

Symbol	Meaning
←	Transfer to (becomes)
↔	Is exchanged with
T	Test
M	Memory
M ( )	Memory with location in brackets

□ The following are reserved symbols in the Motorola assembler:

- The contents of the registers are specified by the characters corresponding to their name. The register names are: A, B, X, SP, PC, and CC and refer to accumulators A and B, index, stack pointer, program counter, and status registers, respectively. Sixteen-bit registers use the suffix H to denote the higher-order byte and suffix L to denote the lower-order byte.

Certain characters are recognized in a special way by a Motorola assembler. These characters, to a degree, are used in the program examples included in this book. These characters are as follows (the most frequently used definitions are marked with an asterisk):

- Characters used as prefixes:

  # (pounds sign) specifies the immediate mode of addressing*

  $ (dollar sign) specifies a hexadecimal number*

  @ (commercial at) specifies an octal number

  % (percent) specifies a binary number

  ' (apostrophe) specifies an ASCII literal character.

- Characters used as suffixes:

  B (letter B) specifies a binary number

  H (letter H) specifies a hexadecimal number

  O (letter O) specifies an octal number

  Q (letter Q) specifies an octal number

,X (comma and letter X) specifies the indexed mode of addressing. 0, X may be written X.*

A number with no suffix or prefix is decimal. As mentioned above, the space is used as a separating character and the carriage return, written (CR), is used as a statement terminator. □

■ The following are reserved symbols in the Intel assembler:

- The contents of 8-bit registers are specified by the character corresponding to the register name. The register names (r, r1, r2) are A, B, C, D, E, H, and L.
- The contents of 16-bit double registers and register pairs consisting of two 8-bit registers are specified by the two characters corresponding to the register name or register pair (rp). The double register names are PC and SP. These registers use the suffix H to denote the higher-order byte and the suffix L to denote the lower-order byte. The register pair names are BC, DE, and HL.
- The contents of an 8-bit status register are specified by processor status word (PSW).

Certain characters are recognized in a special way by the Intel assembler. To a degree these characters are used in program examples included in this book.

The following characters are used as suffixes:

O (letter O) specifies an octal number

Q (letter Q) specifies an octal number

B (letter B) specifies a binary number

H (letter H) specifies a hexadecimal number.* However, all numbers must begin with a numerical digit. A 0 (number 0) is used to prefix all hexadecimal numbers beginning with the digits A through F. The hexadecimal number 46 is written 46H, whereas the hexadecimal number B8 is written 0B8H. Intel uses a separate mnemonic for immediate addressing.

Numbers without suffixes are in decimal. The most frequently used definitions are marked*.

The following characters designate registers:

- DDD destination
- SSS source

DDD or SSS	Register
111	A
000	B
001	C
010	D
011	E
100	H
101	L

NNN represents a binary number.

The following characters are used as delimiters:

blank	(blank) a field separator
,	(comma) separates operands in the operand field
( )	(double parenthesis) delimits an expression
(CR)	(carriage return) statement terminator
;	(semicolon) delimits (starts) a comment
:	(colon) delimits (ends) a label ■

11.4 MEMORY ADDRESSING

One complete cycle is required to transfer a byte from ROM to the CPU. When a RAM location has a 16-bit address, two transfers are required to identify the address bus. This can be time consuming. For these reasons microprocessor designers have devoted considerable effort to develop improved techniques for data transfer. One technique is to provide additional registers for temporary storage. For example, memory need not be accessed for temporary storage of an addressing variable. Another technique is to provide more effective addressing modes to reduce addressing bus access times.

In general there is a trade-off between working registers and addressing modes. A CPU that has many working registers can survive with fewer addressing modes. A CPU that has many addressing modes can survive with fewer working registers. In fact, the purpose of the more sophisticated addressing mode is to so increase the power of an instruction that the need for working registers is reduced.

The following is a brief explanation of some addressing modes used in microcomputers:

- *Extended addressing mode.* The extended addressing mode is the conventional or standard method described above. The 2-byte address of the operand is sent to the address bus. This addressing mode requires 1 byte for the mnemonic field and 2 bytes for the operand field. These 3 bytes are accessed with three to seven memory cycles.
- □ *Direct addressing mode.* The direct addressing mode reduces the operand size by 1 byte. However, the operand contains only 1 byte and can address locations 0 through 255 only. This places a special importance on the variables chosen for the first 256 locations in memory. This addressing mode requires 2 bytes, one for the mnemonic and one for the operand, and three to five machine cycles are used in the execution of the instruction. In general this mode represents a savings of 1 byte of code and one machine cycle. □
- *Immediate addressing mode.* The immediate addressing mode may be more easily understood if one realizes that it is not really an addressing mode. In this mode, the address (operand) is replaced with a value in the range of 0 to 255. The instruction is executed in two cycles. As an example of the purpose

of this addressing mode consider an instruction "load 100." Normally this is interpreted: load the contents of memory location 100. This mode changes the interpretation to: load the value 100.

- *Inherent mode.* The inherent mode also is not an addressing mode. Some instructions do not access memory or I/O, therefore they have no operand. Examples of such instructions are a register transfer or a clear register within the CPU. Since there is no operand the instruction consists of the 1-byte mnemonic and can be executed in one cycle.

□ • *Indexed addressing mode.* In the indexed addressing mode the operand is written as a positive number. It is interpreted as the sum of the operand and the index register value. The positive number is in the range of 0 to 255. The complete instruction consists of 2 bytes with five to seven memory cycles. □

■ • *Indirect addressing mode.* In the indirect addressing mode the operand is specified as the contents of memory as specified by a register. No offset is permitted. The complete instruction consists of 2 bytes with three memory cycles. ■

Using the indexed or indirect addressing modes relieves the programmer of the necessity of keeping track of the addresses of all variables and parameters stored in data memory. All or part of this is assigned to the pointing registers described above. Although effort is required to set up the pointers, there is a quick reward when a large number of variables are needed or when certain variables are addressed repeatedly.

□ • *Relative addressing mode.* The relative addressing mode is intended to alter the sequence of program execution. It resembles the indexed and indirect addressing modes, which use the same basic principles to address data memory. However, it is unique in that its purpose is to alter the program counter. Thus the operand is used in a way more suited to the purpose. The operand is interpreted as a signed number, and the operand specifies a memory location relative to the program counter's current value. As explained in Chapter 10, the program counter points to the next executable instruction (the current address + 2). Thus the destination of the relative addressing mode must be written - 125 to + 129 memory locations from the instruction itself. The complete instruction consists of 2 bytes with four memory cycles. □

■ Intel uses different mnemonics to designate different modes of addressing. ■

□ Motorola uses a # prefix in the operand field to designate the immediate mode and a suffix X in the operand field to designate the indexed mode. All other modes are mnemonic dependent.

Omission of the # prefix is a common programming error. Writing a 25 instead of #25 causes the instruction to access memory location 25 instead of using the value 25 directly. □

The reader is reminded that the list of prefixes and suffixes that the assembler recognizes is given in Section 11.3.

11.5 PROGRAMMING MODELS

From the programmer's point of view, the CPU has a specific register structure. The programmer is not concerned with the transfer of the program from ROM to the instruction decoder or the operation of the control module. All information about the operation of these two modules is transferred to the status register of the ALU module. This convenience means that the programmer's model consists of the ALU module registers and the working register module.

□ 11.5.1 Motorola's MC6800 Programming Registers

The programming model of the MC6800 is shown in Figure 11.3. The programmable registers consist of two 8-bit accumulators, A and B. The intent of registers A and B is that they be used to execute logic and arithmetic expressions in the ALU module. So these two registers represent the data manipulation registers.

In the working register module there is a program counter, a stack pointer, and an indexing register; each a 16-bit-long addressing register. The stack pointer's purpose is to address locations in RAM. The program counter is intended to control the sequential execution of a program through ROM. The index register's purpose is to address locations in RAM for general data transfer.

From the viewpoint that each of the foregoing registers has a specific purpose, this model represents a minimum register model. To increase the effectiveness of the model Motorola has used the extended, direct, indexed, immediate, inherent, and relative modes of addressing to improve its data handling capabilities.

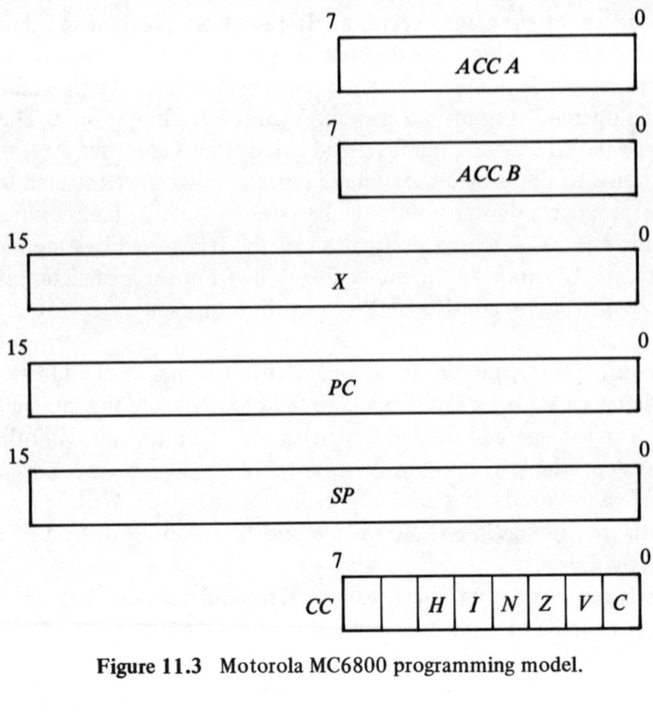

Figure 11.3 Motorola MC6800 programming model.

The status register in the Motorola is also called a condition code register. The status register is an 8-bit register, but bits b7 and b6 are unused. Whenever an instruction is executed in the CPU it automatically sets the appropriate status bit as follows:

H = bit b5. This is the half-carry bit, set whenever there is a carry from the least significant nibble to the most significant nibble within a byte and whenever an add instruction of any type is executed.

I = bit b4. This is the interrupt mask bit, set or reset by software instructions.

N = bit b3. This is the negative status bit, set high if bit b7 of the accumulator output is set; otherwise it is cleared.

Z = bit b2. This is the zero bit, set if memory contents or accumulator contents are 0's; otherwise it is cleared.

V = bit b1. This is the overflow bit, set if the result of an arithmetic operation in the ALU causes overflow.

C = bit b0. This is the carry bit, set if there is a carry as a result of the operation of the ALU. □

■ 11.5.2 Intel 8085 Programming Model

Figure 11.4 shows the Intel 8085A programming model. The 8-bit accumulator register A is associated with the ALU module. There are four 8-bit data registers B, C, D, and E in the working register module. These registers may be regarded as redundant accumulators provided to simplify programming. The working register module has a 16-bit stack pointing register SP to address RAM locations. There is also a 16-bit program counter register PC to control sequential execution of programs through ROM. In addition there is a two 8-bit register pair H and L. The intended purpose of this register pair is for addressing RAM in the manipulation of data.

Intel has introduced versatility in the exact employment of registers B, C, D, E, H, and L. The programmer can use these registers as he sees fit; however using the registers in the manner intended gives a strong and systematic way of programming. The Intel 8085A uses immediate, extended, indirect, and inherent modes of addressing.

The ALU module has an 8-bit status register with three unused bits (D5 = D3 = 0 and D1 = 1). It is called a processor status word (PSW) by Intel. The following are the status bits available to monitor the ALU modules:

Zero: (Z). This bit is set if the result of the ALU operation is zero.

Sign: (S). This bit is set if the most significant bit of the result of accumulator operation is 1.

Parity: (P). This bit is set if an exclusive or operation on the result of the ALU operation is 1.

Carry: (CY). This bit is set if there is a carry or borrow as a result of an add, subtract, or compare instruction execution in the ALU.

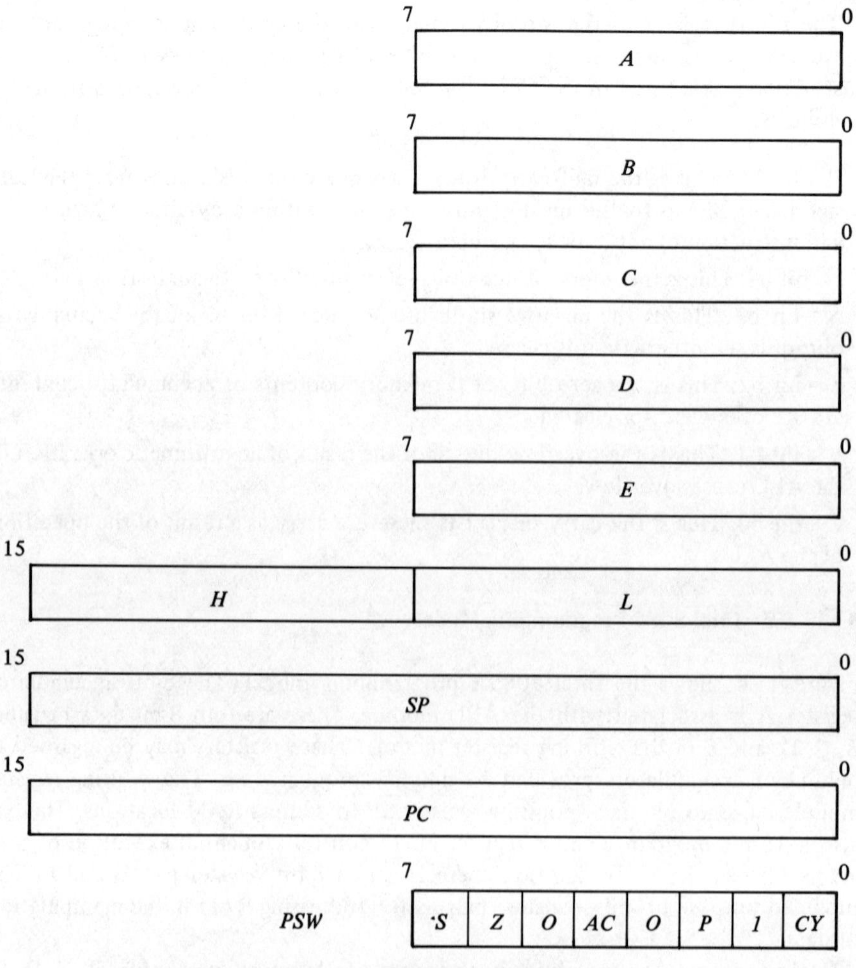

Figure 11.4 Intel 8085A programming model.

Auxiliary Carry: (AC). This bit is set if there is a carry between the least significant and most significant nibble of the byte as a result of an addition, subtraction, increment, decrement, comparison, or logical instruction execution in the ALU. ■

11.6 THE INSTRUCTION SET

The point has already been made that a microcomputer can and does execute many different types of functions. In general a microcomputer has in excess of 70 instructions in its repertoire. As an aid to programming, these 70 plus instructions are classified. This gives the reader a systematic categorization for easy recall.

The instruction set can be divided into three divisions:

1. Instructions that alter accumulators (8-bit registers) and memory.
2. Instructions that alter memory pointers (16-bit registers).
3. Instructions that alter status, special I/O, and miscellaneous.

Division 1 can be given three classifications based on the function performed as follows:

- Data transfer
- Arithmetic
- Logic

Division 2 can be classified according to whether the function is performed on the registers that point at memory data RAM or the register that points at the program ROM as follows:

- Index, indirect, and stack
- Branch and jump

Division 3 is simply classified as

- Status, I/O, and special

These classifications are more rigidly applied to register functions than to actual registers because of the following:

1. Accumulator instructions that use the ALU are obviously arithmetic or logic but also transfer data.
2. There can be a transfer of information from accumulators to pointing registers.
3. Certain registers have a dual role; sometimes they are used as accumulators and sometimes (in pairs) as pointers.

For these reasons classifications used in this book are those of the manufacturer.

The instructions that fall within each classification are given in the remainder of this section. Some examples of their use are included. The intended purpose is to identify the programmer's tools.

A summary of the definitions is given in the appendices. The symbols and instruction set for:

□ The Motorola MC6800 are given in Appendix C. □

■ The Intel 8085 A are given in Appendix D. ■

Also included in each appendix is a quick reference programming guide.

11.6.1 Data Transfer Instructions

This set of instructions is used to transfer data between accumulators or between an accumulator and memory.

□ A summary of Motorola instructions is given in Table 11.1. The description is easily read using the notation from Section 11.3. For example the description $A \leftarrow M$ is read: accumulator A becomes the contents of memory location M. The description $M \leftarrow M + 1$ is read: the contents of memory location M become the contents of M plus 1. The reader is cautioned that M means the contents of memory location M and A means the contents of register A in all modes except the immediate mode. In this mode M means the actual value M (not the contents of M). All novice programmers, even those who have been cautioned, make this fundamental and annoying mistake a number of times.

Motorola data transfer instructions always use the ALU module.

Please note that the immediate addressing mode is available only to the load instruction and that the direct addressing mode is available to the load and store instructions only. For both of these instructions the inherent addressing mode is not available.

The following example of a program:

- Loads a number from memory address NUM
- Divides the number by 2
- Returns the result to memory location NUM

Label	Mnemon	Operand	Comment
DIV	LDAA	NUM	GET NUMBER
	ASRA		DIVID BY TWO
	STAA	NUM	STORE IN MEMORY

In the above program A is loaded with contents of NUM. Assume that the SP is loaded with the address of NUM. The following program is equivalent to the foregoing program but executes faster.

Label	Mnemon	Operand	Comment
DVSP	PULA		GET NUMBER
	ASRA		DIVID BY TWO
	PSHA		STORE IN MEMORY

The SP and PC increment and decrement in the same way. The similarity is that when a byte is moved from the memory (RAM or ROM) the CPU register (SP or PC) is incremented. This similarity helps the programmer to remember the characteristics of the PSH and PUL instructions. However it is more important to remember that since the PC points to the program (ROM with high addresses) and SP points to storage (RAM with low address) normal program execution tends to increase the distance between these pointers. Jump and PUL instructions tend to reset memory pointers toward the starting addresses.

The following is an example of a program that clears five memory locations starting at memory address CLNUM:

Label	Mnemon	Operand	Comment
MEMCL	CLR	CLNUM	FIRST LOCATION CLEAR
	CLR	CLNUM+1	SEC CLEAR
	CLR	CLNUM+2	THIRD CLEAR
	CLR	CLNUM+3	FOURTH CLEAR
	CLR	CLNUM+4	ALL CLEAR

A program that executes a little faster is

Label	Mnemon	Operand	Comment
FMEMCL	CLRA		CLEAR ACCA
	STAA	CLNUM	FIRST LOCATION CLEAR
	STAA	CLNUM+1	SECOND
	STAA	CLNUM+2	THIRD
	STAA	CLNUM+3	FOURTH CLEAR
	STAA	CLNUM+4	ALL CLEAR

Assuming that the SP is loaded with the address CLNUM+4 the following program is still faster:

Label	Mnemon	Operand	Comment
SFCL	CLRA		CLEAR ACC A
	PSHA		CLNUM+4 CLEAR
	PSHA		THREE CLEAR
	PSHA		TWO CLEAR
	PSHA		ONE CLEAR
	PSHA		CLNUM CLEAR □

■ A summary of Intel move, load, and store instructions is given in Table 11.2 for which PSW is unaffected. In the Intel system addressing modes are described as part of the mnemonic. Brackets signify indirect addressing. The description is easily read by using the notation from Section 11.3. For example, the description $M(HL) \leftarrow r$ is read: the contents of memory location pointed at by the register pair HL become the contents of r. Remember that letters signify the contents of a register. The term "data" is used when actual values are given. In the Intel literature conventional addressing using a 2-byte operand is described as direct addressing. Intel data transfer instructions do not use the ALU module.

□ **Table 11.1 Motorola Data Transfer Instructions**

Operation	Mnemonic	Description	Addressing Mode: Immed	Direct	Index	Extend	Inher	Codes	Cycles	Bytes	Status: 5 H	4 I	3 N	2 Z	1 V	0 C
CLEAR	CLR	M(X) ← #00			√			6F	7	2						
		M ← #00				√		7F	6	3		0	0	1	0	0
	CLRA	A ← #00					√	4F	2	1						
	CLRB	B ← #00						5F	2	1						
DECREMENT	DEC	M(X) ← M(X) − 1			√			6A	7	2						
		M ← M − 1				√		7A	6	3			T	T	T	
	DECA	A ← A − 1					√	4A	2	1						
	DECB	B ← B − 1					√	5A	2	1						
INCREMENT	INC	M(X) ← M(X) + 1			√			6A	7	2						
		M ← M + 1				√		7A	6	3						
	INCA	A ← A + 1					√	4A	2	1			T	T	T	
	INCB	B ← B + 1					√	5A	2	1						
LOAD ACCUMULATOR	LDAA	A ← #M	√					86	2	2						
		A ← M		√				96	3	2						
		A ← M(X)			√			A6	5	2						
		A ← M				√		B6	4	3			T	T	0	
	LDAB	B ← #M	√					C6	2	2						
		B ← M		√				D6	3	2						
		B ← M(X)			√			E6	5	2						
		B ← M				√		F6	4	3						
STORE ACCUMULATOR	STAA	M ← A		√				97	4	2						
		M(X) ← A			√			A7	6	2						
		M ← A				√		B7	5	3			T	T	T	T
	STAB	M ← B		√				D7	4	2						
		M(X) ← B			√			E7	6	2						
		M ← B				√		F7	5	3						

TRANSFER ACCUMULATOR	TAB	B ← A			√	16	2	1	T T 0
	TBA	A ← B			√	17	2	1	
ROTATE LEFT	ROL	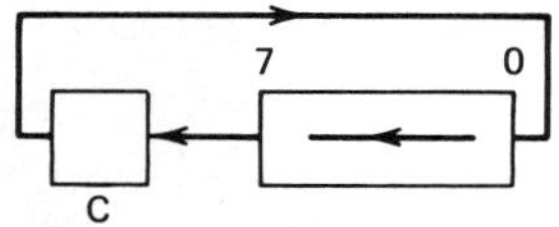	√			69	7	2	
				√		79	6	3	T T T b7
	ROLA				√	49	2	1	
	ROLB				√	59	2	1	
ROTATE RIGHT	ROR	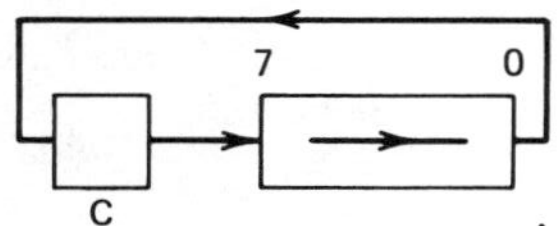	√			66	7	2	
				√		76	6	3	T T T b0
	RORA				√	46	2	1	
	RORB				√	56	2	1	
ARITHMETIC SHIFT LEFT	ASL	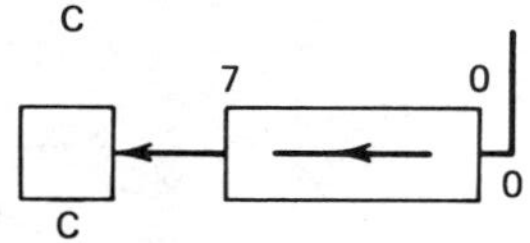	√			68	7	2	
				√		78	6	3	T T T b7
	ASLA				√	48	2	1	
	ASLB				√	58	2	1	
ARITHMETIC SHIFT RIGHT	ASR	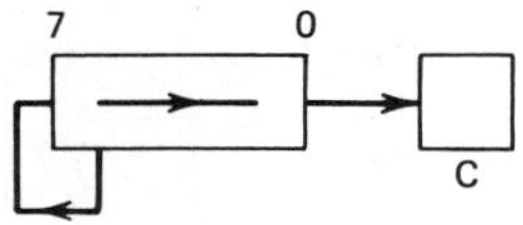	√			67	7	2	
				√		77	6	3	T T T b0
	ASRA				√	47	2	1	
	ASRB				√	57	2	1	
LOGICAL SHIFT RIGHT	LSR	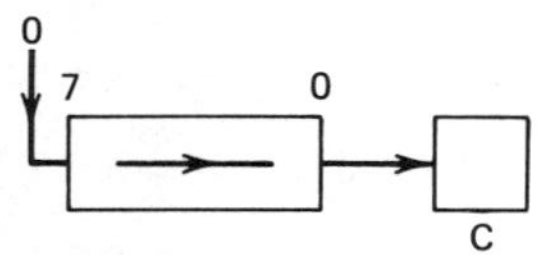	√			64	7	2	
				√		74	6	3	0 T T b0
	LSRA				√	44	2	1	
	LSRB				√	54	2	1	
PUSH DATA ONTO STACK	PSHA	M(SP) ← A SP ← SP − 1			√	36	4	1	
	PSHB	M(SP) ← B SP ← SP − 1			√	37	4	1	
PULL DATA	PULA	SP ← SP + 1 A ← M(SP)			√	32	4	1	
	PULB	SP ← SP + 1 B ← M(SP)			√	33	4	1	

■ Table 11.2 Intel Data Transfer Group

Operation	Mnemonic and Operand	Description	Add Mode: Immediate	Add Mode: Direct	Add Mode: Indirect	Add Mode: Inherent	Cycles	Bytes	Code
MOVE	MOV r1, r2	r1 ← r2				√	1	1	01DDDSSS
	MOV r, M	r ← M(HL)		√			2	1	01DDD110
	MOV M, r	M(HL) ← r		√			2	1	01110SSS
	MVI r, data	r ← byte 2	√				2	2	00DDD110
	MVI m, data	M(HL) ← byte 2	√				3	2	36
LOAD REGISTER PAIR	LXI B	B ← byte 3 C ← byte 2	√				3	3	01
	LXI D	D ← byte 3 E ← byte 2	√				3	3	11
	LXI H	H ← byte 3 L ← byte 2	√				3	3	21
LOAD ACCUMULATOR	LDA M	A ← M		√			4	3	3A
	LDAX B	A ← M(BC)			√		2	1	0A
	LDAX D	A ← M(DE)			√		2	1	1A
STORE ACCUMULATOR	STA M	M ← A		√			4	3	32
	STA B	M(BC) ← A			√		2	1	02
	STA D	M(DE) ← A			√		2	1	12
LOAD H and L	LHLD	L ← M H ← M + 1		√			5	3	2A
STORE H and L	SHLD	M ← L (M + 1) ← H		√			5	3	22
EXCHANGE HL WITH DE	XCHG	H ↔ D L ↔ E				√	1	1	EB

S	Z	0	AC	0	P	1	CY

Status Register

Following are examples of programs that illustrate the way to write mnemonics and operands as well as illustrate the interpretation of the descriptions.

Three ways to load register A with the contents of memory location 1000 are as follows:

1. Assume register pair HL contains 1000 hex and use indirect mode of addressing

   ```
   MOV    A, M
   ```

   The description M means "the contents of memory location pointed at by register pair HL."
2. Use extended mode addressing

   ```
   LDA    1000H
   ```

   The description 1000H means "the contents of memory location 1000."
3. If you happen to know that location 1000 contains hexadecimal number F6 memory addressing can be avoided using the immediate mode

   ```
   MVI    A,0F6H
   ```

The following is an example of a program that clears five memory locations starting at memory location CLNUM:

Label	Mnemon	Operand	Comment
MEMCL:	MVI	A,00	;CLEAR ACCA
	STA	CLNUM	;CLEAR FIRST
	STA	CLNUM+1	;2ND
	STA	CLNUM+2	;3RD
	STA	CLNUM+3	;4TH
	STA	CLNUM+4	;5TH

A program that executes a little faster is

Label	Mnemon	Operand	Comment
FMCL:	MVI	A,00	;CLEAR ACCA
	MOV	A,CLNUM	;CLEAR FIRST
	MOV	A,CLNUM+1	;2ND
	MOV	A,CLNUM+2	;3RD
	MOV	A,CLNUM+3	;4TH
	MOV	A,CLNUM+4	;5TH ■

11.6.2 Arithmetic Instructions

This set of instructions uses the arithmetic part of the ALU module for individual alteration of any accumulator. Each accumulator is altered using an additional input from either another accumulator or a memory location's contents.

□ Table 11.3 Motorola Arithmetic Instructions

Operation	Mnemonic	Description	Addressing Mode: Immed	Direct	Index	Extend	Inher	Code	Cycles	Bytes	Status: 5 H	4 I	3 N	2 Z	1 V	0 C
ADD	ADDA	A ← #M + A	√					8B	2	2						
		A ← M + A		√				9B	3	2						
		A ← M(X) + A			√			AB	5	2						
		A ← M + A				√		BB	4	3	T		T	T	T	T
	ADDB	B ← #M + B	√					CB	2	2						
		B ← M + B		√				DB	3	2						
		B ← M(X) + B			√			EB	5	2						
		B ← M + B				√		FB	4	3						
	ABA	A ← A + B					√	1B	2	1						
ADD WITH CARRY	ADCA	A ← A + #M + C	√					89	2	2						
		A ← A + M + C		√				99	3	2						
		A ← A + M(X) + C			√			A9	5	2						
		A ← A + M + C				√		B9	4	3	T		T	T	T	T
	ADCB	B ← B + #M + C	√					C9	2	2						
		B ← B + M + C		√				D9	3	2						
		B ← B + M(X) + C			√			E9	5	2						
		B ← B + M + C				√		F9	4	3						
2's COMPLEMENT	NEG	M ← 00 − M(X)			√			60	7	2						
		M ← 00 − M				√		70	6	3						
	NEGA	A ← 00 − A					√	40	2	1			T	T	T	T
	NEGB	B ← 00 − B					√	50	2	1						
DECIMAL ADJUST ADD	DAA	[a]					√	19	2	1			T	T	T	T
SUBTRACT	SUBA	A ← A − #M	√					80	2	2						
		A ← A − M		√				90	3	2						
		A ← A − M(X)			√			A0	5	2						
		A ← A − M				√		B0	4	3						
	SUBB	B ← B − #M	√					C0	2	2			T	T	T	T
		B ← B − M		√				D0	3	2						
		B ← B − M(X)			√			E0	5	2						
		B ← B − M				√		F0	4	3						
	SBA	A ← A − B					√	10	2	1						
SUBTRACT WITH CARRY	SBCA	A ← A − #M − C	√					82	2	2						
		A ← A − M − C		√				92	3	2						
		A ← A − M(X) − C			√			A2	5	2						
		A ← A − M − C				√		B2	4	2			T	T	T	T
	SBCB	B ← B − #M − C	√					C2	2	2						
		B ← B − M − C		√				D2	3	2						
		B ← B − M(X) − C			√			E2	5	2						
		B ← B − M − C				√		F2	4	3						

[a]Converts to a BCD format after addition. It regards each byte as two 4-bit BCD numbers.

□ A summary of Motorola instructions is given in Table 11.3.

The following is an example of a program that adds two numbers from memory locations, NUMI and NUMI+1, and returns the result to memory location NUMI:

Label	Mnemon	Operand	Comment
NADD	LDAA	NUMI	GET FIRST NO
	ADDA	NUMI+1	ADD SECOND NO
	STAA	NUMI	RETURN TO MEM

Under the assumption that the address of NUMI is stored in index register X the following is an alternate way to write the above program:

Label	Mnemon	Operand	Comment
N2ADD	LDAA	X	GET FIRST NO
	ADDA	I, X	ADD SECOND NO
	STAA	X	RETURN TO MEM □

■ A summary of Intel single-byte add and subtract instructions is given in Table 11.4. All indirect addressing uses the HL register pair and all PSW test bits are subject to flag conditions.

A summary of the Intel double-byte add instructions is given in Table 11.5. Only the CY flag is subject to change in the PSW.

A summary of the Intel single-byte increment and decrement instructions with PSW tests is given in Table 11.6.

A summary of the Intel double-byte (word) increment and decrement instructions is given in Table 11.7. The status register is unaffected.

The following is an example of a program that adds two numbers from memory locations, NUMI and NUMI+1, and returns the result to memory location NUMI.

Label	Mnemon	Operand	Comment
NIAD:	LXI	H, NUMI+1	;LOAD HL WITH ADDRESS OF NUMI+1
	MOV	A,M	;LOAD A WITH CONTENT OF NUMI+1
	DCX	H	;CHANGE ADDRESS TO NUMI
	ADD	M	;ADD CONTENT OF NUMI
	MOV	M,A	;STORE RESULT IN NUMI ■

11.6.3 Logic Instructions

This set of instructions uses the logic part of the ALU module for individual alteration of any accumulator. Each accumulator is altered using an additional input from either another accumulator or a memory location's contents.

□ A summary of Motorola logic operator instructions is given in Table 11.8.

■ **Table 11.4 Intel Add and Subtract Instructions for Register A**

Operation	Mnemonic and Operand	Description	Add Mode: Immediate	Direct	Indirect	Inherent	Cycles	Bytes	Code
ADD	ADD r	A ← A + r				√	1	1	10000SSS
	ADD M	A ← A + M(HL)			√		2	1	86
	ADI data	A ← A + byte 2	√				2	2	C6
ADD WITH CARRY	ADC r	A ← A + r + CY				√	1	1	10001SSS
	ADC M	A ← A + M(HL) + CY			√		2	1	8E
	ACI data	A ← A + byte 2 + CY	√				2	2	CE
SUBTRACT	SUB r	A ← A − r				√	1	1	10010SSS
	SUB M	A ← A − M(HL)			√		2	1	96
	SUI data	A ← A − byte 2	√				2	2	D6
SUBTRACT WITH BORROW	SBB r	A ← A − r − CY				√	1	1	10011SSS
	SBB M	A ← A − M(HL) − CY			√		2	1	9E
	SBI data	A ← A − byte 2 − CY	√				2	2	DE
DECIMAL ADJUST ACCUMULATOR	DDA	Converts the contents of A from HEX to BCD				√	1	1	27

7							0
S	Z	0	AC	0	P	1	CY
T	T		T		T		T

Status Register

■ **Table 11.5 Intel Double-Byte (Word) Add Instructions**

Operation	Mnemonic and Operand	Description	Add Mode				Cycles	Bytes	Code
			Immediate	Direct	Indirect	Inherent			
ADD REGISTER PAIR TO HL	DAD B	HL ← HL + BC				√	3	1	09
	DAD D	HL ← HL + DE				√	3	1	19
	DAD H	HL ← HL + HL				√	3	1	29
	DAD SP	HL ← HL + SP				√	3	1	39

S	Z	0	AC	0	P	1	CY
							T

Status Register

■ **Table 11.6 Intel Single-Byte Increment and Decrement Instructions**

Operation	Mnemonic and Operand	Description	Add Mode: Immediate	Direct	Indirect	Inherent	Cycles	Bytes	Code
INCREMENT	INR r	$r \leftarrow r + 1$				√	1	1	00DDD100
	INR M	$M(HL) \leftarrow M(HL) + 1$			√		3	1	34
DECREMENT	DCR r	$r \leftarrow r - 1$				√	1	1	00DDD101
	DCR M	$M(HL) \leftarrow M(HL) - 1$			√		3	1	35

S	Z	0	AC	0	P	1	CY
T	T		T		T		

Status Register

■ **Table 11.7 Intel Double-Byte Increment and Decrement Instructions**

Operation	Mnemonic and Operand	Description	Add Mode: Immediate	Direct	Indirect	Inherent	Cycles	Bytes	Code
INCREMENT	INX B	BC ← BC + 1				√	1	1	03
	INX D	DE ← DE + 1				√	1	1	13
	INX H	HL ← HL + 1				√	1	1	23
DECREMENT	DCX B	BC ← BC − 1				√	1	1	0B
	DCX D	DE ← DE − 1				√	1	1	1B
	DCX H	HL ← HL − 1				√	1	1	2B

S	Z	0	AC	0	P	1	CY

Status Register

□ Table 11.8 Motorola Logic Operator Instructions

Operation	Mnemonic	Description	Addressing Mode: Immed	Direct	Index	Extend	Inher	Code	Cycles	Bytes	Status: 5 H	4 I	3 N	2 Z	1 V	0 C
AND	ANDA	A ← #M · A	✓					84	2	2						
		A ← M · A		✓				94	3	2						
		A ← M(X) · A			✓			A4	5	2						
		A ← M · A				✓		B4	4	3			T	T	0	
	ANDB	B ← #M · B	✓					C4	2	2						
		B ← M · B		✓				D4	3	2						
		B ← M(X) · B			✓			E4	5	2						
		B ← M · B				✓		F4	4	3						
1's COMPLEMENT	COM	$M(X) \leftarrow \overline{M}(X)$			✓			63	7	2						
		$M \leftarrow \overline{M}$				✓		73	6	3			T	T	0	1
	COMA	$A \leftarrow \overline{A}$					✓	43	2	1						
	COMB	$B \leftarrow \overline{B}$					✓	53	2	1						
EXCLUSIVE OR	EORA	A ← #M ⊕ A	✓					88	2	2						
		A ← M ⊕ A		✓				98	3	2						
		A ← M(X) ⊕ A			✓			A8	5	2						
		A ← M ⊕ A				✓		B8	4	3			T	T	0	
	EORB	B ← #M ⊕ B	✓					C8	2	2						
		B ← M ⊕ B		✓				D8	3	2						
		B ← M(X) ⊕ B			✓			E8	5	2						
		B ← M ⊕ B				✓		F8	4	3						
INCLUSIVE OR	ORAA	A ← #M + A	✓					8A	2	2						
		A ← M + A		✓				9A	3	2						
		A ← M(X) + A			✓			AA	5	2						
		A ← M + A				✓		BA	4	3			T	T	0	
	ORAB	B ← #M + B	✓					CA	2	2						
		B ← M + B		✓				DA	3	2						
		B ← M(X) + B			✓			EA	5	2						
		B ← M + B				✓		FA	4	3						

A summary of the Motorola data test instructions is given in Table 11.9. □

■ A summary of the Intel logic operator and compare instructions is given in Table 11.10. All PSW bits are tested.

A summary of the Intel rotate, complement, and set carry instructions is given in Table 11.11. The instructions affect the CY bit in the PSW. For the rotate left instruction CY ← A7. For the rotate right instruction CY ← A0.

The Intel complement accumulator instruction is given in Table 11.12. This instruction does not change PSW flags. ■

11.6.4 Index, Indirect and Stack Instructions

To assist the programmer in the task of keeping track of data memory (RAM) address pointers are provided. Of course, keeping track means that the programmer includes within his program a number of instructions that modify the pointers. This section gives the main data storage pointer modification instructions.

□ A summary of Motorola index register and stack instructions is given in Table 11.13.

The reader is again cautioned about the meaning of descriptive symbols. The symbol $X \leftarrow X + 1$ means add one to the contents of X and transfer the result to X. The symbol $XL \leftarrow (M + 1)$ means transfer the contents of $(M + 1)$ to XL. XL is, of course, the lower order byte of X.

Operation of the index register is easily understood. The programmer uses this register as a data pointer. It can be incremented or decremented to move sequentially through a data bank. Nonsequential moves are made using the load index register instruction. When instruction execution time is important, time can be saved using the offset feature. Using the offset requires a little more effort on the part of the programmer.

All novice programmers have a problem understanding the features of the stack. In fact the stack is equipped with such powerful and subtle instructions that many, not so novice, programmers get into stack-related problems.

The stack is unique and important in that it provides a super fast way to transfer the contents of registers and data memory. This transfer occurs

- Automatically whenever a jump or branch to subroutine or return instruction is used, as explained in Section 10.2.3.
- Automatically whenever an interrupt is encountered, as explained in Section 10.2.3.
- Under instruction control whenever a push data onto stack or pull data from stack instruction is used as explained in Section 11.6.1.

Because the stack can be manipulated automatically by interrupts and under program control, mutually exclusive and thus correct operation is guaranteed by the following rules:

1. Always disable interrupts before any stack instruction. Enable the interrupt after the stack instruction. Instructions for interrupt control are given in Section 11.6.6.

□ **Table 11.9 Motorola Data Test Instructions**

Operation	Mnemonic	Description	Addressing Mode: Immed	Direct	Index	Extend	Inher	Code	Cycles	Bytes	Status: 5 H	4 I	3 N	2 Z	1 V	0 C
BIT TEST	BITA	#M · A	√					85	2	2						
		M · A		√				95	3	2						
		M(X) · A			√			A5	5	2						
		M · A				√		B5	4	3			T	T	0	
	BITB	#M · B	√					C5	2	2						
		M · B		√				D5	3	2						
		M(X) · B			√			E5	5	2						
		M · B				√		F5	4	3						
COMPARE	CMPA	A – #M	√					81	2	2						
		A – M		√				91	3	2						
		A – M(X)			√			A1	5	2						
		A – M				√		B1	4	3			T	T	T	T
	CMPB	B – #M	√					C1	2	2						
		B – M		√				D1	3	2						
		B – M(X)			√			E1	5	2						
		B – M				√		F1	4	3						
COMPARE ACCUMULATORS	CBA	A – B					√	11	2	1			T	T	T	T
TEST	TST	M(X) – 00			√			6D	7	2						
ZERO OR		M – 00				√		7D	6	3			T	T	0	0
MINUS	TSTA	A – 00					√	4D	2	1						
	TSTB	B – 00					√	5D	2	1						

■ **Table 11.10 Intel Logic Operator and Compare Instructions**

Operation	Mnemonic and Operand	Description	Add Mode: Immediate	Direct	Indirect	Inherent	Cycles	Bytes	Code
AND	ANA r	$A \leftarrow A \cdot r$				√	1	1	10100SSS
	ANA M	$A \leftarrow A \cdot M(HL)$			√		2	1	A6
	ANI data	$A \leftarrow A \cdot$ byte 2	√				2	2	E6
EXCLUSIVE OR	XRA r	$A \leftarrow A \oplus r$				√	1	1	10101SSS
	XRA M	$A \leftarrow A \oplus M(HL)$			√		2	1	AE
	XRI data	$A \leftarrow A \oplus$ byte 2	√				2	2	EE
INCLUSIVE OR	ORA r	$A \leftarrow A + r$				√	1	1	10110SSS
	ORA M	$A \leftarrow A + M(HL)$			√		2	1	B6
	ORI data	$A \leftarrow A +$ byte 2	√				2	2	F6
COMPARE	CMP r	$A - r$				√	1	1	10111SSS
	CMP M	$A - M(HL)$			√		2	1	BE
	CPI data	$A -$ byte 2	√				2	2	FE

S	Z	0	AC	0	P	1	CY
T	T		T		T		T

Status Register

■ **Table 11.11 Intel Rotate, Complement, and Set Carry Instructions**

Operation	Mnemonic and Operand	Description	Add Mode: Immediate	Direct	Indirect	Inherent	Cycles	Bytes	Code
ROTATE A LEFT	RLC	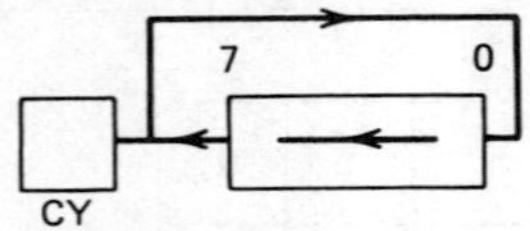				√	1	1	07
ROTATE A RIGHT	RRC	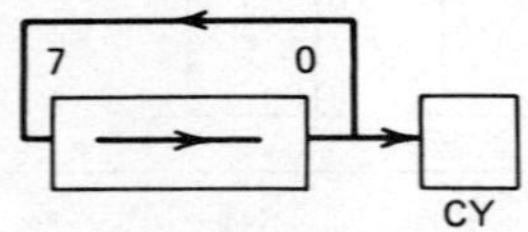				√	1	1	0F
ROTATE A LEFT THRU CARRY	RAL	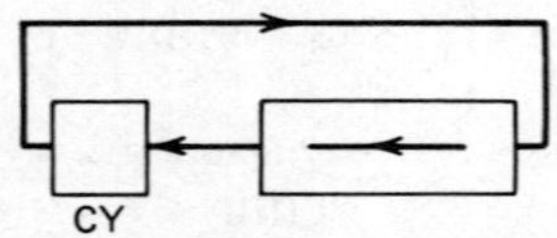				√	1	1	17
ROTATE A RIGHT THRU CARRY	RAR	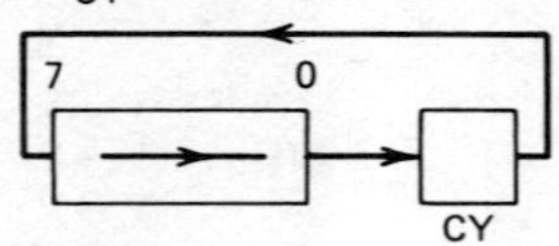				√	1	1	1F
COMPLEMENT CARRY	CMC	$CY \leftarrow \overline{CY}$				√	1	1	3F
SET CARRY	STC	$CY \leftarrow 1$				√	1	1	37

S	Z	0	AC	0	P	1	CY
							T

Status Register

■ **Table 11.12 Intel Complement Accumulator Instruction**

Operation	Mnemonic and Operand	Description	Add Mode				Cycles	Bytes	Code
			Immediate	Direct	Indirect	Inherent			
COMPLEMENT ACCUMULATOR	CMA	$A \leftarrow \overline{A}$				√	1	1	2F

S	Z	0	AC	0	P	1	CY

Status Register

□ Table 11.13 Motorola Index Register and Stack Instructions

Operation	Mnemonic	Description	Addressing Mode: Immed	Direct	Index	Extend	Inher	Code	Cycles	Bytes	Status: 5 H	4 I	3 N	2 Z	1 V	0 C
COMPARE INDEX REGISTER	CPX	XH – #M XL – #M + 1	√					8C	3	3						
		XH – M XL – (M + 1)		√				9C	4	2			T	T	T	
		XH – M(X) XL – M(X + 1)			√			AC	6	2						
		XH – M XL – (M + 1)				√		BC	5	3						
DECREMENT INDEX REGISTER	DEX	X ← X – 1					√	09	4	1				T		
DECREMENT STACK POINTER	DES	SP ← SP – 1					√	34	4	1						
INCREMENT INDEX REGISTER	INX	X ← X + 1					√	08	4	1				T		
INCREMENT STACK POINTER	INS	SP ← SP + 1					√	31	4	1						
LOAD INDEX REGISTER	LDX	XH ← #M XL ← #M + 1	√					CE	3	3						
		XH ← M XL ← (M + 1)		√				DE	4	2			T	T	T	
		XH ← M(X)			√			EE	6	2						

		XL ←M(X + 1)											
		XH ←M XL ←(M + 1)				√		FE	5	3			
LOAD STACK POINTER	LDS	SPH ←#M SPL ←#M + 1	√					8E	3	3			
		SPH ←M SPL ←(M + 1)		√				9E	4	2	T	T	T
		SPH ←M(X) SPL ←M(X + 1)			√			AE	6	2			
		SPH ←M SPL ←(M + 1)				√		BE	5	3			
STORE INDEX REGISTER	STX	M ←XH (M + 1) ←XL		√				DF	5	2			
		M(X) ←XH M(X + 1) ←XL			√			EF	7	2	T	T	T
		M ←XH (M + 1) ←XL				√		FF	6	3			
STORE	STS	M ←SPH (M + 1) ←SPL		√				9F	5	2			
		M(X) ←SPH M(X + 1) ←SPL			√			AF	7	2	T	T	T
		M ←SPH (M + 1) ←SPL				√		BF	6	3			
TRANSFER FROM INDEX TO STACK	TXS	SP ←X – 1					√	35	4	1			
TRANSFER FROM STACK TO INDEX	TSX	X ←SP + 1					√	30	4	1			

2. Set the stack pointer to a high location in data memory at the beginning of any program using interrupts. Remember you will have to apply rule 1 to manipulate the stack pointer.

These are easy rules to forget; but failure to comply means that a return from interrupt instruction will generate unpredictable actions. □

■ A summary of Intel stack-related instructions is given in Table 11.14. PSW flags are unaltered. The reader is referred back to the data transfer group of instructions and, in particular, Tables 11.5 and 11.7. Those instructions have a dual purpose. They

- alter data
- alter pointers.

The symbol SP + 2 means add 2 to the contents of SP. The symbol M(SP + 2) means the contents of the memory location whose address is 2 greater than SP.

The Intel stack-related instructions are super fast ways of transferring the contents of a register pair and data memory. It is fast because:

- Words are transferred.
- Indirect address, which is automatically incremented or decremented to the correct value, is used.

Stack-related instructions are ideal for pushing the contents of accumulators and pointers into memory before servicing an interrupt and for popping the content of memory back to accumulators and pointers after servicing the interrupt. The CPU working registers are thus returned to their pre-interrupt condition, and normal program execution continues.

A summary of the Intel pop processor status word instruction is given in Table 11.15. By definition this instruction alters all PSW flags.

The POP PSW instruction is ideal for restoring the CPU to full operating condition.

Immediately after an interrupt a PUSH PSW instruction saves the status. After the interrupt has been serviced a POP PSW immediately before a return from interrupt restores the status register to its normal program condition. The sequence is illustrated as follows:

```
interrupt
         PUSH PSW
             .
             .
             .
         services routine
             .
             .
             .
         POP PSW
return from interrupt.
```

■

■ **Table 11.14 Intel Stack-Related Instructions**

Operation	Mnemonic and Operand	Description	Add Mode: Direct	Add Mode: Indirect	Add Mode: Inherent	Cycles	Bytes	Code
PUSH ACCUMULATORS ONTO STACK	PUSH B	M(SP − 1) ← B M(SP − 2) ← C SP ← SP − 2		√		3	1	C5
	PUSH D	M(SP − 1) ← D M(SP − 2) ← E SP ← SP − 2		√		3	1	D5
	PUSH H	M(SP − 1) ← H M(SP − 2) ← L SP ← SP − 2		√		3	1	E5
PUSH PSW ONTO STACK	PUSH PSW	M(SP − 1) ← A M(SP − 2) ← PSW SP ← SP − 2		√		3	1	F5
POP FROM STACK TO ACCUMULATORS	POP B	C ← M(SP) B ← M(SP + 1) SP ← SP + 2		√		3	1	C1
	POP D	E ← M(SP) D ← M(SP + 1) SP ← SP + 2		√		3	1	D1
	POP H	L ← M(SP) H ← M(SP + 1) SP ← SP + 2		√		3	1	E1

S	Z	0	AC	0	P	1	CY

Status Register

■ **Table 11.15 Intel Pop Processor Status Word Instruction**

Operation	Mnemonic and Operand	Description	Add Mode				Cycles	Bytes	Code
			Immediate	Direct	Indirect	Inherent			
POP STATUS FROM STACK	POP PSW	PSW ← M(SP) A ← SP + 1 SP ← SP + 2				√	3	1	F1

S	Z	0	AC	0	P	1	CY
T	T		T		T		T

Status Register

11.6.5 Branch and Jump Instructions

A program is basically a file of sequential instructions, which is stored in ROM if it is permanent and stored in RAM if it is under development. Normally the program counter is incremented to follow the sequential progression of the program. However there must be a mechanism to load the program counter to facilitate major and minor looping as required in the control of any process.

This section gives the program counter modification instructions.

□ A summary of Motorola's jump and branch instructions is given in Table 11.16.

The branch instruction uses the relative addressing mode, which offsets the PC by the contents of the operand. However, as explained in Chapter 10, the PC is also altered by the execution of a machine cycle. Executing a branch instruction takes two machine cycles and this causes the PC to be incremented by 2. So the offset is the address of the destination instruction minus the address of the branch instruction minus 2.

The easy way to determine the offset is to label the destination instruction and write an instruction, branch-to-label. The assembler will determine the required value.

Motorola provides three hardware interrupts. These hardware interrupts may be said to resemble instructions in the following ways:

- The PC is altered.
- Registers are stored in memory.
- Registers are restored from memory using equivalent software instructions.
- There are software interrupts that initiate the same sequence of operations as the hardware interrupts.

A summary of Motorola's hardware interrupts is given in Table 11.17. All these operations are followed by an automatic jump instruction.

A summary of Motorola's interrupt instructions is given in Table 11.18. All the operations are followed by an automatic jump.

Software interrupts can be used in the debugging stage to test service subroutine before or independently of the construction of interrupt hardware. Wait for interrupt can be used to synchronize the execution of the main program and service subroutines. □

■ A summary of Intel's branch instructions is given in Table 11.19.

The branch instructions use the immediate mode of addressing. The PC is altered by the operand, which occurs in the last 2 bytes of a 3-byte instruction. Byte 3 gives the *most* significant 8 bits of the address, and byte 2 gives the *least* significant 8 bits. However, it is much wiser to specify the address as a label or an expression and use an assembler.

Intel 8085 provides four hardware interrupts. These hardware interrupts may be said to resemble instructions in the following ways:

- The value of PC is stored in the stack. The PC is altered.
- The PC is restored from the stack using an equivalent software instruction.

Table 11.16 Motorola's Jump and Branch Instructions

Operation	Mnemonic	Test	Description	Addressing Mode			Codes	Cycles	Bytes
				Rel	Index	Extend			
BRANCH ALWAYS	BRA	1	$PC \leftarrow PC + 2 + REL$	√			20	4	2
BRANCH IF CARRY CLEAR	BCC	$C = 0$	$PC \leftarrow PC + 2 + REL$	√			24	4	2
BRANCH IF CARRY SET	BCS	$C = 1$	$PC \leftarrow PC + 2 + REL$	√			25	4	2
BRANCH IF ZERO SET	BEQ	$Z = 1$	$PC \leftarrow PC + 2 + REL$	√			27	4	2
BRANCH IF ⩾ ZERO	BGE	$N \oplus V = 0$	$PC \leftarrow PC + 2 + REL$	√			2C	4	2
BRANCH IF POSITIVE	BGT	$Z + (N \oplus V) = 0$	$PC \leftarrow PC + 2 + REL$	√			2E	4	2
BRANCH IF HIGHER	BHI	$C \cdot Z = 0$	$PC \leftarrow PC + 2 + REL$	√			22	4	2
BRANCH IF ⩽ ZERO	BLE	$Z + (N \oplus V) = 1$	$PC \leftarrow PC + 2 + REL$	√			27	4	2
BRANCH IF LOWER OR SAME	BLS	$C + Z = 1$	$PC \leftarrow PC + 2 + REL$	√			23	4	2
BRANCH IF < ZERO	BLT	$N \oplus V = 1$	$PC \leftarrow PC + 2 + REL$	√			20	4	2

BRANCH IF MINUS	BMI	N = 1	PC ← PC + 2 + REL	√			28	4	2
BRANCH IF NOT EQUAL ZERO	BNE	Z = 0	PC ← PC + 2 + REL	√			26	4	2
BRANCH IF OVERFLOW CLEAR	BVC	V = 0	PC ← PC + 2 + REL	√			28	4	2
BRANCH IF OVERFLOW SET	BVS	V = 1	PC ← PC + 2 + REL	√			29	4	2
BRANCH IF PLUS	BPL	N = 0	PC ← PC + 2 + REL	√			2A	4	2
BRANCH TO SUBROUTINE	BSR	1	M(SP) ← PCL + 2 M(SP – 1) ← PCH + 2 SP ← SP – 2 PC ← PC + 2 + REL		√		80	8	2
JUMP	JMP	1	PC ← X		√		6E	4	2
			PCH ← M(PC + 1) PCL ← M(PC + 2)			√	7E	3	3
JUMP TO SUBROUTINE	JSR	1	M(SP) ← PCL + 2 M(SP – 1) ← PCH + 2 PC ← X		√		AD	8	2
			M(SP) ← PCL + 3 M(SP – 1) ← PCH + 3 PCH ← M(PC + 1) PCL ← M(PC + 2)			√	BD	9	3

□ **Table 11.17 Motorola Hardware Interrupts**

Operation	Mnemonic	Description	Status 5 H	4 I	3 N	2 Z	1 V	0 C
RESET (ACTIVE LOW)	$\overline{\text{RES}}$	PCH ← FFFE PCL ← FFFF		1				
NON-MASKABLE INTERRUPT (ACTIVE LOW)	$\overline{\text{NMI}}$	M(SP) ← PCL M(SP – 1) ← PCH M(SP – 2) ← XL M(SP – 3) ← XH M(SP – 4) ← A M(SP – 5) ← B M(SP – 6) ← CC SP ← SP – 7 PCH ← FFFC PCL ← FFFD		1				
INTERRUPT REQUEST (ACTIVE LOW)	$\overline{\text{IRQ}}$	M(SP) ← PCL M(SP – 1) ← PCH M(SP – 2) ← XL M(SP – 3) ← XH M(SP – 4) ← A M(SP – 5) ← B M(SP – 6) ← CC SP ← SP – 7 PCH ← FFF8 PCL ← FFF9		[a]				

[a] If set, IRQ is ignored; if not set, IRQ is executed and status bit 4 is set during the execution of a service routine.

A summary of Intel's hardware interrupts is given in Table 11.20. If the interrupt enable bit is reset, an automatic restart is executed, and the hardware instructions follow. The trap interrupt masks the enable bit and always executes. ■

11.6.6 Status, I/O, and Special Instructions

There are a number of instructions that are used in initialization, as well as resetting and restarting the microcomputer. This section gives status and I/O modification instructions.

□ A summary of the Motorola status and interrupt control instructions is given in Table 11.21.

Motorola classifies the interrupt mask as status. Thus bit b4 refers to ALU module status register. □

■ A summary of the Intel I/O group of instructions is given in Table 11.22.

It is good practice to disable all other interrupts when servicing an accepted interrupt. The practice should also be used when using IN and OUT instructions

□ Table 11.18 Motorola Interrupt Instructions

Operation	Mnemonic	Description	Inherent Add Mode	Code	Cycles	Bytes	Status 5 H	4 I	3 N	2 Z	1 V	0 C
NO OPERATION	NOP	PC ← PC + 1	√	01	2	1						
RETURN FROM INTERRUPT	RTI	CC ← M(SP + 1) B ← M(SP + 2) A ← M(SP + 3) XH ← M(SP + 4) XL ← M(SP + 5) PCH ← M(SP + 6) PCL ← M(SP + 7) SP ← SP + 7	√	3B	10	1						
RETURN FROM SUBROUTINE	RTS	PCH ← M(SP + 1) PCL ← M(SP + 2)	√	39	5	1						
SOFTWARE INTERRUPT	SWI	PC ← PC + 1 M(SP) ← PCL M(SP − 1) ← PCH M(SP − 2) ← XL M(SP − 3) ← XH M(SP − 4) ← A M(SP − 5) ← B M(SP − 6) ← CC SP ← SP − 7 PCH ← FFFA PCL ← FFFB	√	3F	12	1						
WAIT FOR INTERRUPT	WAI	PC ← PC + 1 M(SP) ← PCL M(SP − 1) ← PCH M(SP − 2) ← XL M(SP − 3) ← XH M(SP − 4) ← A M(SP − 5) ← B M(SP − 6) ← CC WHEN INTERRUPT ARRIVES STATUS BIT 4 IS SET PCH ← FFF8 PCL ← FFF9		3E	9	1		T				

■ **Table 11.19 Intel Branch Instructions**

Operation	Mnemonic	Test	Description	Add Mode			Cycles	Bytes	Code
				Direct	Indirect	Inherent			
JUMP	JMP	1	PCH ← byte 3	✓			3	3	C3
	JC	CY = 1	PCL ← byte 2	✓			3	3	DA
	JNC	CY = 0	for all jumps	✓			3	3	D2
	JZ	Z = 1		✓			3	3	CA
	JNZ	Z = 0		✓			3	3	C2
	JP	S = 1		✓			3	3	F2
	JM	S = 0		✓			3	3	FA
	JPE	P = 1		✓			3	3	EA
	JPO	P = 0		✓			3	3	E2
CALL	CALL	1	M(SP – 1) ← PCH	✓			5	3	CD
	CC	CY = 1	M(SP – 2) ← PCL	✓			5	3	DC
	CNC	CY = 0	SP ← SP – 2	✓			5	3	D4
	CZ	Z = 1	PCH ← byte 3	✓			5	3	CA
	CNZ	Z = 0	PCL ← byte 2	✓			5	3	C4
	CP	S = 1	for all calls	✓			5	3	F4
	CM	S = 0		✓			5	3	FC
	CPE	P = 1		✓			5	3	EC
	CPO	P = 0		✓			5	3	E4
RETURN	RET	1	PCL ← M(SP)		✓		3	1	C9
	RC	CY = 1	PCH ← M(SP + 1)		✓		3	1	D8
	RNC	CY = 0	SP ← SP + 2		✓		3	1	D0
	RZ	Z = 1			✓		3	1	C8
	RNZ	Z = 0	for all returns		✓		3	1	C0
	RP	S = 1			✓		3	1	F0
	RM	S = 0			✓		3	1	F8
	RPE	P = 1			✓		3	1	E8
	RDO	P = 0			✓		3	1	E0
JUMP INDIRECT MOVE HL TO PC	PCHL	1	PC ← HL			✓	1	1	E9

■ **Table 11.20 Intel Hardware Interrupts**

Operation	Mnemonic and Operand	Description
NONMASKABLE INTERRUPT	TRAP	PC ← data data = 24H
MASKABLE INTERRUPT	RST 5.5	PC ← data data = 2CH
MASKABLE INTERRUPT	RST 6.5	PC ← data data = 34H
MASKABLE INTERRUPT	RST 7.5	PC ← data data = 3CH

that contain code words for programming peripherals. Peripheral programming is explained in Chapter 12. ■

In this section the reader has been introduced to very important and fundamental instruction sets. The treatment in this book is necessarily introductory in nature and is intended to form the groundwork for further and more specialized study [32, 33].

The remainder of the book is more global or system operation oriented.

11.7 ASSEMBLER DIRECTIVES

The instruction set described in Section 11.6 can be termed *executable instructions.* The word executable is intended to mean that the instructions can be expressed in machine code form. The assembler program that performs this one to one mapping must be given *directives* that essentially describe the configuration of the assembler program. A directive is an instruction that is not translated into machine code, but that resembles an executable instruction in format. However directives are capable of performing other tasks for the programmer. The following is a list of some tasks assigned to the assembler and executed using directives:

1. *Identify statements to be assembled.* Starting the assembler is assigned to system commands, which is the subject of the next section. An assembly is stopped when a *termination directive* is encountered.
2. *Define all data symbols.* An efficient programmer uses symbols for data. Data symbols are used in immediate addressing mode instructions. When a symbol is used for data, the data must be defined using a *symbol directive.*
3. *Define all known variables.* Programmers use variables in symbol form for addressing data memory (RAM). Known variables are used in direct, extended, indexed, and indirect addressing modes. When a known variable is represented by a symbol it is assigned to data memory. The symbol must be defined using a *data definition directive.*

□ **Table 11.21 Motorola Status Control Instructions**

Operation	Mnemonic	Description	Inherent Add Mode	Code	Cycles	Bytes	Status 5 H	4 I	3 N	2 Z	1 V	0 C
CLEAR CARRY	CLC	C ← 0	√	0C	2	1						0
CLEAR INTERRUPT MASK	CLI	I ← 0	√	0E	2	1		0				
CLEAR OVERFLOW	CLV	V ← 0	√	0A	2	1					0	
SET CARRY	SEC	C ← 1	√	0D	2	1						1
SET INTERRUPT MASK	SEI	I ← 1	√	0F	2	1		1				
SET OVERFLOW	SEV	V ← 1	√	0B	2	1					1	
LOAD CC WITH ACCA	TAP	CC ← A	√	06	2	1	T		T	T	T	T
LOAD ACCA WITH CC	TPA	A ← CC	√	07	2	1						

■ Table 11.22 Intel I/O and Special Instructions

Operation	Mnemonic and Operand	Description	Add Mode: Direct	Indirect	Inherent	Cycles	Bytes	Code
INPUT	IN port	A ← port	√			3	2	DB
OUTPUT	OUT port	port ← A	√			3	2	D3
ENABLE INTERRUPTS	EI	I ← 0			√	1	1	FB
DISABLE INTERRUPTS	DI	I ← 1			√	1	1	F3
READ INTERRUPT MASK	RIM	A ← flags[a]			√	1	1	20
SET INTERRUPT MASK	SIM	flags ← A[b]			√	1	1	30
HALT	HLT	PC ← PC			√	1	1	76
NO OPERATION	NOP	PC ← PC + 1			√	1	1	00

All mnemonics copyright Intel Corporation, 1978.

[a] Read interrupt mask to register A (RIM in Table 11.22) uses the following format:

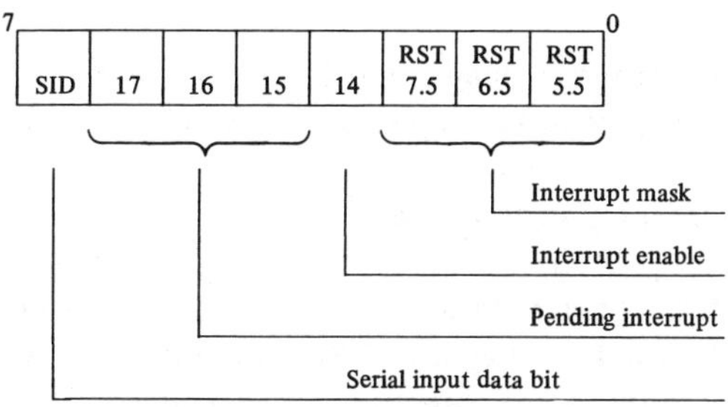

[b] Set interrupt mask from register A (SIM in Table 11.22) uses the following format:

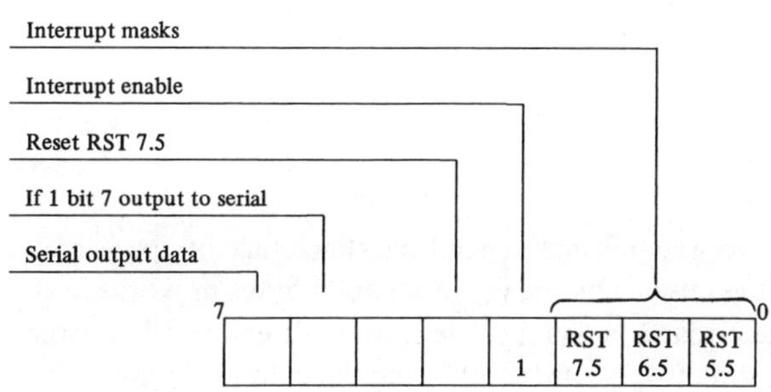

It is sometimes convenient to store a string of characters in ASCII codes. This is done to form output messages. Data definition directive can be used for this additional purpose.

4. *Define all unknown variables.* When a program is executed the value of an unknown variable(s) is determined. Such variables are used in direct, extended, indirect, and indexed addressing modes. Often it is convenient to store such variables in data memory. A *memory reservation directive* is used to define variables that require memory storage locations.

5. *Assign program and data locations.* The assembler automatically replaces address symbols with a determined specific memory location. In addition, it is necessary for the programmer to assign instructions to specific locations in memory, and often the first line of data is assigned a memory location to prevent overwriting. Assignment of memory addresses is done using a *location control directive.*

6. *Assist in modular programming.* The best way to write a complex program is to dissect it into manageable parts and write an executive to control the parts. This concept was introduced for sequential circuits in Section 6.12. It becomes necessary for separate programs to transfer information between themselves. The programmer transfers such information using *program linkage directives.*

7. *Control of the formatting of the assembler listing.* The format of the assembler listing may be changed to

- improve appearance
- improve comprehensibility
- shorten computation time

Formatting of the assembler listing is accomplished use *listing format control directives.*

Directives are written in mnemonic form using ASCII characters. The format, symbol, and meaning of specific directives is given in the following subsections.

11.7.1 Directive Notation and Rules

In general the format and notation rules given in Sections 11.2 and 11.3 for writing instructions are used to write directives. However there are some exceptions and some additional rules. Some directives must omit one or more of the following fields:

- label
- operand
- comment

□ In addition to the normal separating characters of space and carriage return Motorola directives use a comma (,) to separate bytes or words in the operand. □

■ Intel differentiates a symbol definition directive label from an instruction label and other directive labels by replacing the colon delimiter with a space. ■

In addition both Motorola and Intel accept the following operators in the operand field:

Operator	Meaning
*	multiplication
/	division
+	addition
−	subtraction

The memory location counter is represented by the symbol LC. Some assemblers for Motorola and Intel instruction permit additional operators in the operand field.

□ 11.7.2 Motorola Assembler Directives

The following is a set of Motorola assembler directives arranged according to task. The directives given are summarized in Appendix E. The meaning, special formatting requirements, and an illustrative example are given with each mnemonic.

1. *Termination directive*

   END–end of program

   - Last statement of source program, requests new file
   - No label permitted

Label	Mnemon	Operand	Description
	END		Return to files

2. *Symbol directive*

   EQU–equate symbol

   - Assign value to data symbol
   - Normal format

Label	Mnemon	Operand	Descriptive
PIAPA	EQU	$4005	PIAPA ← 4005
HOS	EQU	$A01	HOS ← 0A01

   Consider the statement

Label	Mnemon	Operand	Comment
	LDAA	$4005	A ← 4005

   This statement should be replaced with

Label	Mnemon	Operand	Comment
PIAPA	EQU	$4005	gives PIA address
	LDAA	PIAPA	A ← 4005

   Use of the symbol instead of the value as the load instruction operand is more convenient and leads to fewer errors. Subsequent changes in a PIA address means one directive rather than a number of instructions will require change.

3. *Data definition directives*

   (a) FCB–form constant byte

   - Used to place data bytes in memory
   - Normal format

Label	Mnemon	Operand	Description
	FCB	$1B	M(LC) ← 1B LC ← LC + 1

Label	Mnemon	Operand	Description
NUM	FCB	,$0F,23	NUM ← LC M(LC) ← 00 M(LC+1) ← 0F M(LC+2) ← 17 LC ← LC + 3

Assuming the address location counter is at 102 the following occurs:

LOCATION	DATA
102	1B
103	00
104	0F
105	17

The contents of memory location as indicated by the assembler location counter takes the value 1B. Similarly NUM takes the new value of the location counter. Hexadecimal values 00, 0F, 17 are loaded into the next three locations (17 hex = 23 decimal). Commas are used to separate bytes of memory in the operand field.

(b) FDB–form double constant byte

- Used to place data words in memory
- Normal format

Label	Mnemon	Operand	Description
NUMB	FDB	$1B	NUMB ← LC M(LC) ← 001B LC ← LC + 2
	FDB	,$0F,$17	M(LC) ← 0000 M(LC+2) ← 000F M(LC+4) ← 0017 LC ← LC + 6

Assuming the address location is at 106 the following occurs:

LOCATION	DATA
106	001B
108	0000
10A	000F
10C	0017

Again NUMB takes the value of the assembler location counter and the contents of memory location NUMB (a word) takes the value 001B. Similarly, the contents of the next three words in memory are 0000, 000F, and 0017. Commas are used to separate words of memory in the operand field.

(c) FCC–form constant characters

- Translates characters into their 7 bit ASCII code as shown in Table B.3
- Normal format

Label	Mnemon	Operand	Description
MSG1	FCC	/LOVE/	MSG1 ← LC
			M(LC) ← 4C
			M(LC + 1) ← 4F
			M(LC + 2) ← 56
			M(LC + 3) ← 45
			LC ← LC + 4
MSG2	FCC	6,LOVE	MSG2 ← LC
			M(LC) ← 4C
			M(LC + 1) ← 4F
			M(LC + 2) ← 56
			M(LC + 3) ← 45
			M(LC + 4) ← 20
			M(LC + 5) ← 20
			LC ← LC + 6

The description can be written as M(LC) ← TEXT in ASCII. Assuming the address location counter is at 0A02 the following occurs:

LOCATION	DATA
0A02	4C
0A03	4F
0A04	56
0A05	45
0A06	4C
0A07	4F
0A08	56
0A09	45
0A0A	20
0A0B	20

A string of ASCII characters can be used to store a message in a man-oriented language such as English. The operand can be written as

- Text between identical delimiters. Each delimiter should be a single character. If numbers are used as delimiters the text must not begin with a comma.
- Text following count and separator (,). Spaces are valid characters.

4. *Memory reservation directive*

RMB–reserve memory bytes

- Used to save space for variables (move the location counter in data memory)
- Normal format

Label	Mnemon	Operand	Description
	RMB	4	LC ← LC + 4

Label	Mnemon	Operand	Description
CLNUM	RMB	5	CLNUM ← LC LC ← LC + 5
TABLE	RMB	$20	TABLE ← LC LC ← LC + 20

The label if it occurs is given the value of the location counter. The location counter is incremented by the value of the operand.

5. *Location control directive*

ORG–Origin

- Used to move the location counter in program memory
- No label permitted

Label	Mnemon	Operator	Description
	ORG	$100	LC ← 0100
	ORG	$A30F	LC ← A30F

It makes sense to use directives like this

Label	Mnemon	Operator	Description
BEGIN	EQU	$A30F	BEGIN ← A30F
	ORG	BEGIN	LC ← BEGIN

Modification and debugging becomes easier.

6. *Program linkage directives*

NAM–name

- Used to name a program or provide a top of the page heading
- No label or comment field permitted

Mnemon	Operand
NAM	DUMMY

7. *Listing format control directives*

(a) OPT–option

- Control format of assembler output
- No label permitted

Label	Mnemon	Operand	Description
	OPT	0	An object tape is made

There are some useful options for formatting and control of the assembly listing. These options are as follows with the necessary part of the symbol in capitals and the optional part in lowercase.

Symbol	Description
Generate	Give code generated by FCC directive (selected by default)
List	Lists the assembly in long format (selected by default)

Symbol	Description
Otape = (FILENAME)	Object program is saved in an object file tape named (FILENAME). Overwrites any existing file of the same type and the same name
Symbol	Symbol table is given in listing

(b) PAGE—advance paper to top of next page

- Paper is advanced to top of next page
- No label or operand permitted

Label	Mnemon	Operand	Description
	PAGE		Advance paper

(c) SPC—space N lines

- Space listing number of lines given in operand
- No label permitted

Label	Mnemon	Operand	Description
	SPC	N	listing line ← listing line + N

This directive is terminated if the listing is incremented to greater than the total page lines. □

■ 11.7.3 Intel Assembler Directives

The following are Intel assembler directives arranged according to task. The directives are summarized in Appendix F. The meaning, special formatting requirements and an illustrative example are given after each mnemonic.

1. *Termination directive*

   END—end of program

   - The last statement of source program. PC takes value of operand.
   - Normal formatting permitted.

Label	Mnemon	Operand	Description
	END	N	;PC ← N, LAST STATEMENT

   The operand is usually left blank.

2. *Symbol directives*

   (a) EQU—equate symbol

   - Assigns value to data symbol.
   - Normal format.

Label	Mnemon	Operand	Description
PPIUA	EQU	2FH	;PPIUA ← 2F
FULL	EQU	0FFH	;FULL ← FF

This directive is often used to increase the flexibility of immediate addressing mode instructions as follows:

Label	Mnemon	Operand	Description
PPIUA	EQU	2FH	;PPIUA ← 2F
	MVI	A,PPIUA	;A ← 2F

Subsequent changes in the value of PPIUA require only the change of the EQU directive, which occurs once. The MVI instruction, which usually occurs a large number of times, is not altered.

(b) SET–set symbol

- Assign value to data symbol, and value can be changed dynamically as the program executes.
- Normal format.

Label	Mnemon	Operand	Description
NUMI:	SET	10H	;NUMI ← 10
	ADI	NUMI	;A ← A + 10
NUMI:	SET	20H	;NUMI ← 20
	ADI	NUMI	;A ← A + 20

3. *Data definition directives*

(a) DB–data byte

- Used to place data byte in memory.
- Normal format.

Label	Mnemon	Operand	Description
	DB	1BH	;M(LC) ← 1B
			;LC ← LC + 1
NUMI:	DB	0FH,23,	;NUMI ← LC
			;M(LC) ← 0F
			;M(LC + 1) ← 17
			;M(LC + 2) ← 00
			;LC ← LC + 3
MSG:	DB	'LOVE'	;MSG ← LC
			;M(LC) ← 4C
			;M(LC + 1) ← 4F
			;M(LC + 2) ← 56
			;M(LC + 3) ← 45
			;LC ← LC + 4

The last description can be abbreviated to M(LC) ← 'TEXT' IN ASCII. Assuming the address location counter is at 0A0B the following occurs.

LOCATION	DATA
0A0B	1B
0A0C	0F
0A0D	17
0A0E	00

LOCATION	DATA
0A0F	4C
0A10	4F
0A11	56
0A12	45

The data 1B is loaded into memory location given by the location counter. The location counter is incremented, and NUMI takes the value of the value of the location counter. The data is loaded into the next three memory locations, and the counter is incremented. The counter is incremented with each load, and liberty is taken in this description: MSG takes the value of the location counter, and the ASCII equivalent of love 4C4F5645 is loaded into memory, and the location counter is simultaneously incremented.

(b) DW–data word

- Used to place data words in memory.
- Normal format.

Label	Mnemon	Operand	Description
NUMI:	DW	1BH	;NUMI ← LC ;M(LC) ← 1B00 ;LC ← LC + 2
	DW	0FH,17H,'AB',	;M(LC) ← 0F00 ;M(LC + 2) ← 1700 ;M(LC + 4) ← 4241 ;M(LC + 6) ← 0000 ;LC ← LC + 8

The label NUMI is given the value of the location counter. The operand is loaded into the word memory location indicated by the location in the reverse order (001B becomes 1B00). This is typical format for addresses. Items can be listed and separated by commas as shown in the second directive. The four words 000F, 0017, 0'A', and 0000 are stored as 0F00, 1700, 4100, and 0000 respectively. Bytes are reversed, and 'A' causes the ASCII code for A to be stored in memory.

4. *Memory reservation directive*

DS–data storage

- Used to save space for variables (move the location counter through data memory).
- Normal format.

Label	Mnemon	Operand	Description
	DS	2	;LC ← LC + 2
N2ADD:	DS	6	;N2ADD ← LC ;LC ← LC + 6
TABLE 1:	DS	20	;TABLE 1 ← LC ;LC ← LC + 20

If a label occurs it is given the value of the location counter. The location counter is incremented by the value of the operand.

5. *Location control directives*

ORG–Origin

- Used to move the location counter.
- Normal format.

Label	Mnemon	Operand	Description
PAG1:	ORG	0FFH	;PAG1 ← LC
			;LC ← FF

The label is set to the present value of the location counter. The location counter is set to the value of the operand. It is more effective to write the above directive as follows:

Label	Mnemon	Operand	Description
START	EQU	0FFH	;START ← FF
PAG1:	ORG	START	;PAG1 ← LC
			;LC ← FF

This makes modification and debugging easier.

6. *Program linkage directive*

NAME–name

- Used to name a program for system executive statements. System executive statements is the topic of the next section.
- Normal format.

Label	Mnemon	Operand	Description ■
	NAME	PROG1	

The foregoing directives form a minimum set and are considered to be fundamental. They are summarized in Appendices E and F. Many tasks can be done by the assembler program, and therefore the list of available directives is large.

Assembling a program is one of many system functions that a microcomputer system can perform. The next section deals with some of the number of system functions available.

11.8 SYSTEMS AND SYSTEM EXECUTIVE STATEMENTS

A system that can assemble instructions and directives into a file of machine code bytes has the following hardware components:

- Microcomputer with large memory capacity.
- Diskettes for permanent storage of files and programming aid programs.

- Console for system control using commands and system file display.
- Printer for permanent copies.

Such a system is capable of performing other functions. Manufacturers are forever increasing the functional capabilities of microcomputer systems. Examples of functions that can be provided in a microcomputer system are ROM (etc.) programmers and circuit (hardware) emulators. Such functions require additional hardware and are beyond the scope of this book.

However, as an aid to programming ROM it is possible to run the program in RAM to verify correct operation using a monitor module. A convenient way to correct errors in the source file is to use an editor module.

A minimum system required for program development has the following software programs under the command of an executive module:

- Editor module (used to create a source file).
- Assembler module (used to create an object file).
- Monitor module (used to run programs).

Programs are developed by an iterative process following the repetitive steps shown in Figure 11.5. Somewhere in this process the programmer must do some system and program housekeeping chores. The system is equipped with the two following programs for these functions:

- Print module
- Directory module

The print module is used to print (or display) source and assembly listings. The directory is used to print (or display) all files on diskettes.

The following subsections describe the interconnection of all the fundamental modules and the initialization of the executive module. A summary of programming commands for the subordinate modules is given in Appendices G and H.

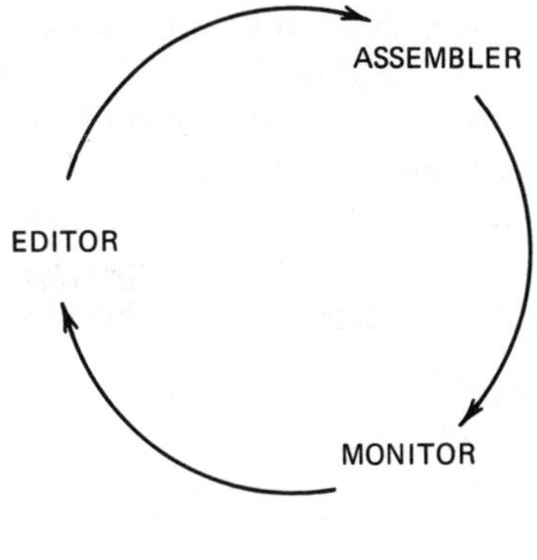

Figure 11.5 System software packages.

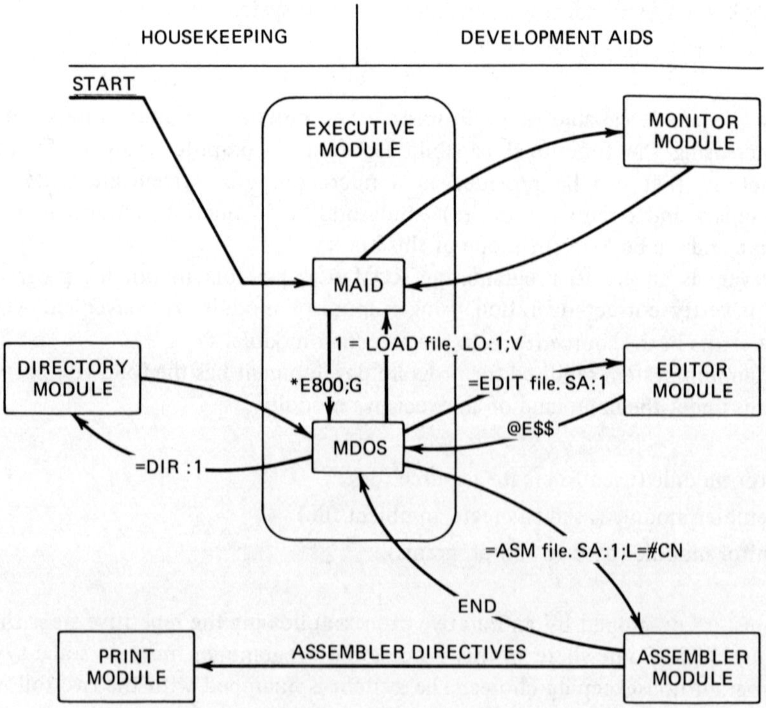

Figure 11.6 Motorola system software modules.

□ 11.8.1 Motorola Fundamental Module Configuration

A minimum modular configuration of a Motorola system is shown in Figure 11.6.

This development system is started manually, as described in Appendix G, which transfers control to the system console. The start-up sequence calls the hardware-introduced executive MAID.

MAID is in essence a monitor module. This executive is identified by the * prompt it uses at the beginning of each line. It can run your program for you if you supply the address of the first line of program.

MAID can also run the MDOS system program. The first line of MDOS is at E800. This address and the comand ;Go calls MDOS. The command is written E800;G.

The MDOS program identifies itself with an = prompt at the beginning of each line. The following are definitions of commands that call modules. Uppercase denotes compulsory characters; lowercase represents a name provided by the programmer.

An MDOS return to MAID command is

= LOAD file. LO :1; V (CR)

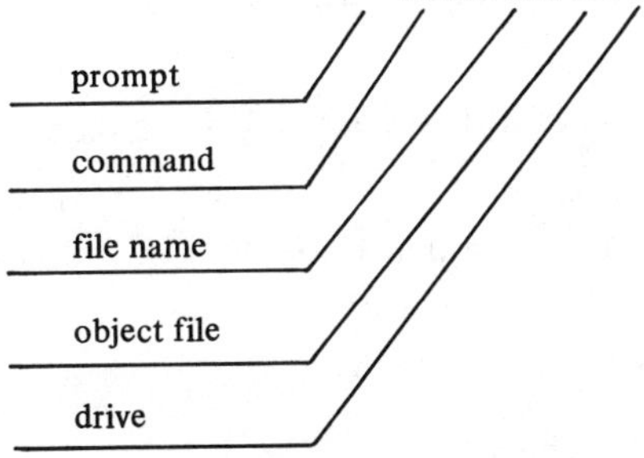

Another technique, and probably the most common, is for the operator to press reset and enter MAID on the console.

From MDOS the editor is called using the command

= EDIT file. SA :1 (CR)

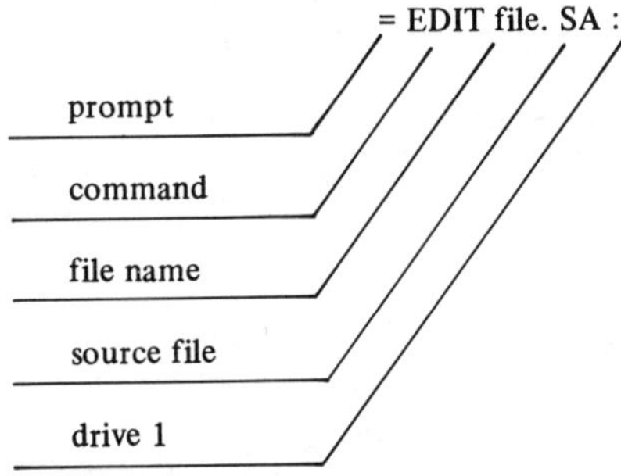

The editor program identifies itself with a @ prompt. An editor return to MDOS command is

@ E$$

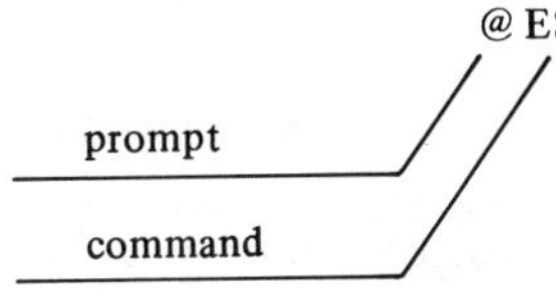

The first editor command needed is A as given in Appendix G.

From MDOS the assembler is called using the command

= ASM file. SA :1; L = #CN (CR)

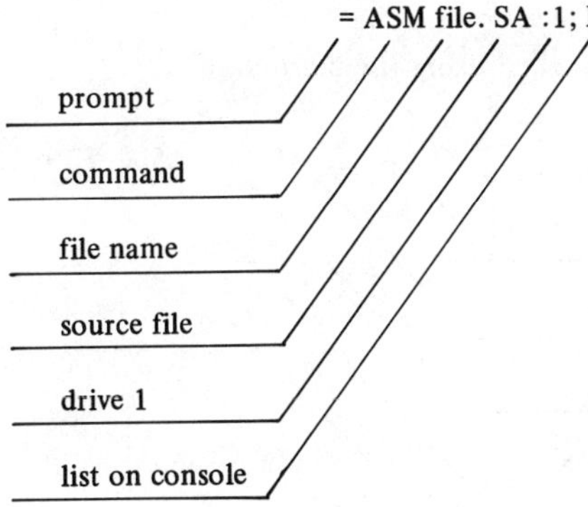

The assembler should create an object file. However if it creates a paper tape listing use the command

= EXBIN file.LX: 1(CR)

to create an object file (file.LO).

Assembler format is given in Sections 11.6 and 11.7. An assembler return to MDOS is not a command but the directive

END

The description is given in Appendix E.

From MDOS the directory is called using the command

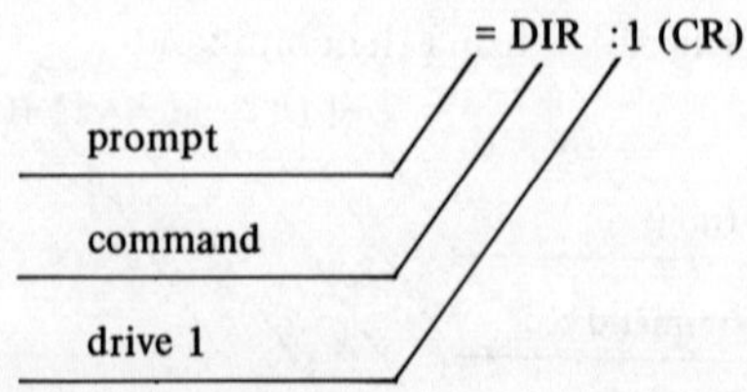

A directory return to MDOS is automatic.

A summary of some of the executive, monitor, editor, and directory commands is given in Appendix G. Your system will vary. Consult your system expert for a more accurate and complete description of your system. □

■ 11.8.2 Intel Fundamental Module Configuration

A minimum module configuration of an Intel development system is shown in Figure 11.7

This development system is started manually, as described in Appendix H, which transfers control to the system console. The start-up sequence calls the hardware-introduced executive ISIS-II. ISIS-II can call any of the submodules.

The ISIS-II program identifies itself with the – prompt at the beginning of each line.

From ISIS-II the editor is called using the command

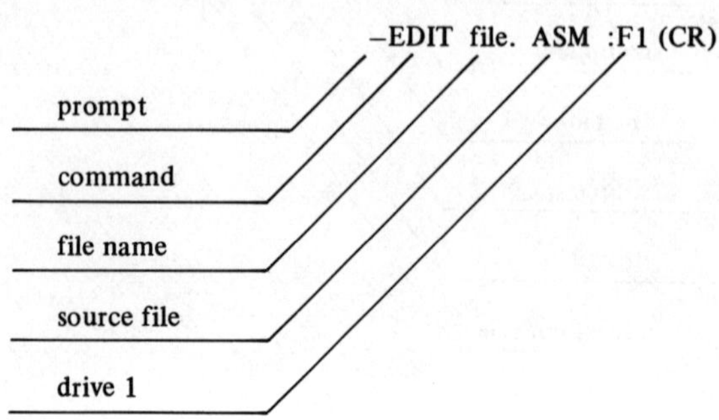

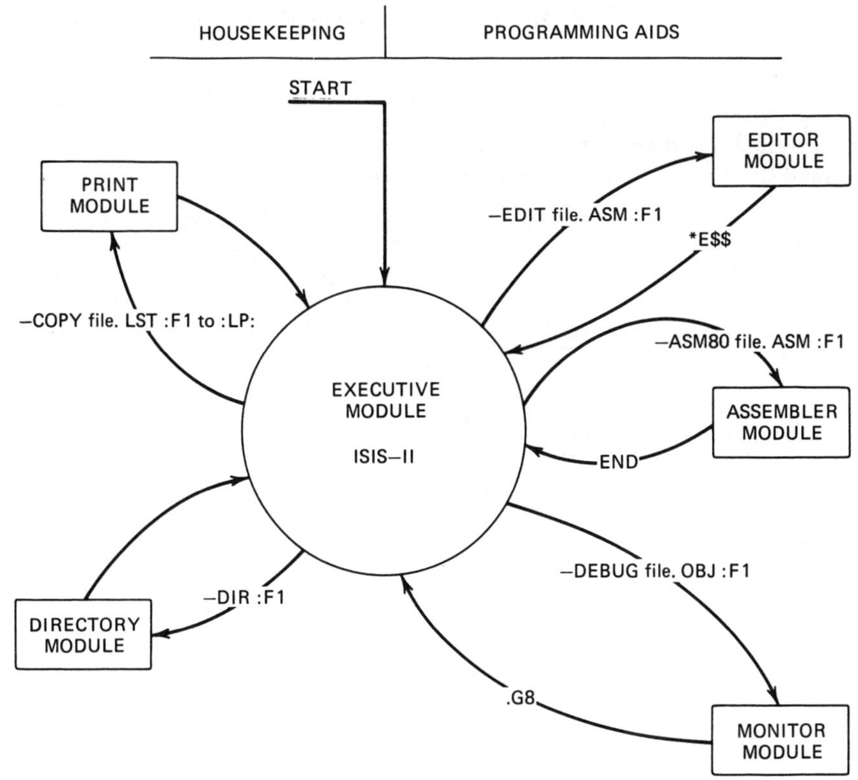

Figure 11.7 Intel system software modules. Reprinted by permission of Intel Corporation.

The editor program identifies itself with a * prompt. An editor return to ISIS-II command is

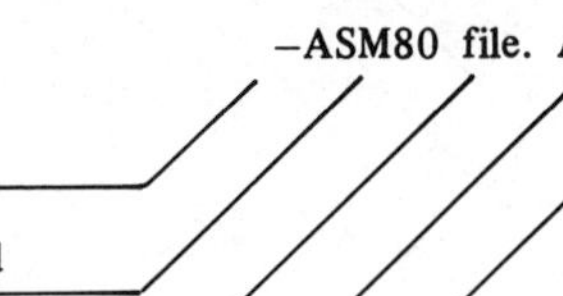

From ISIS-II the assembler is called using the command

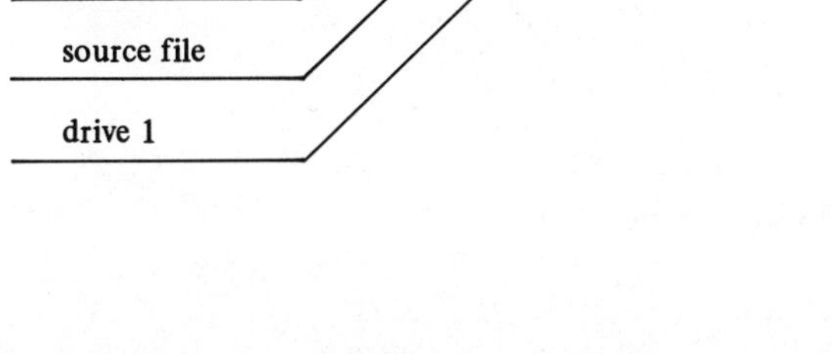

Assembler format is given in Appendix D. An assembler return to ISIS-II is not a command but the directive

END

The description is given in Appendix F.

From ISIS-II the monitor is called using the command

–DEBUG file. OBJ :F1 (CR)

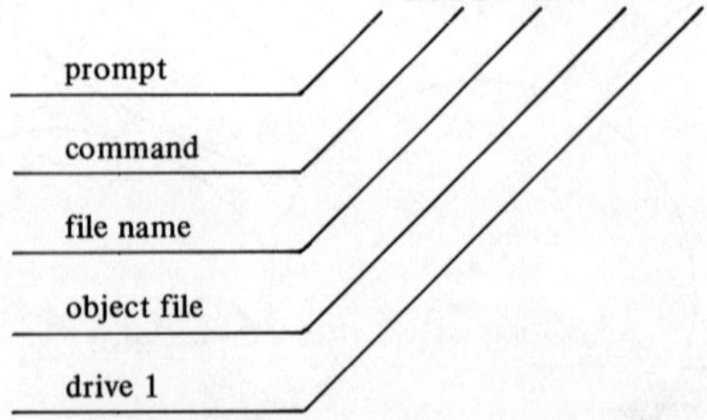

The debug program identifies itself with a · prompt. After debugging your program at return to ISIS-II command is

• G8 (CR)

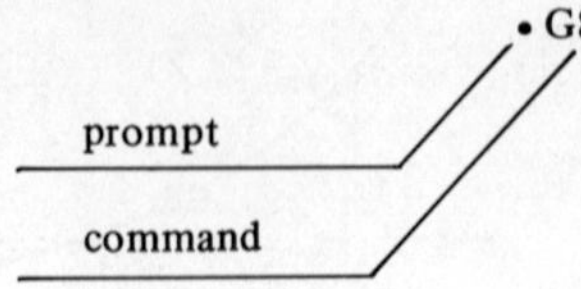

From ISIS-II the directory is called using the command

–DIR :F1 (CR)

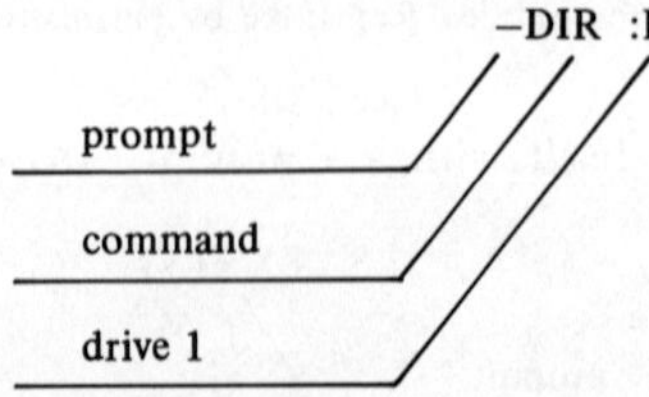

A directory return to ISIS-II is automatic.

From ISIS-II the print module is called using the command

–COPY file.ASM :F1 TO :LP:(CR)

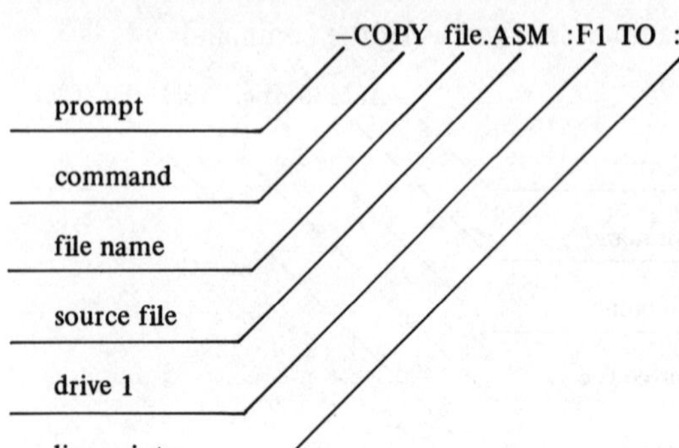

A print return to ISIS-II is automatic.

A summary of some of the executive, editor, monitor, directory, and print commands are given in Appendix H. Your system will vary. Consult your system expert for a more accurate and complete description of your system. ■

The modules and commands described form a fundamental set. There are many more. The interested reader is referred to manufacturers' manuals given in the References.

11.9 PROBLEMS

P11.1 Write a program that generates a variable time delay of 1, 5, 10, or 20 ms. The flow chart is shown in Figure P11.1. Counter number 1 should cause a 100-μs total delay.

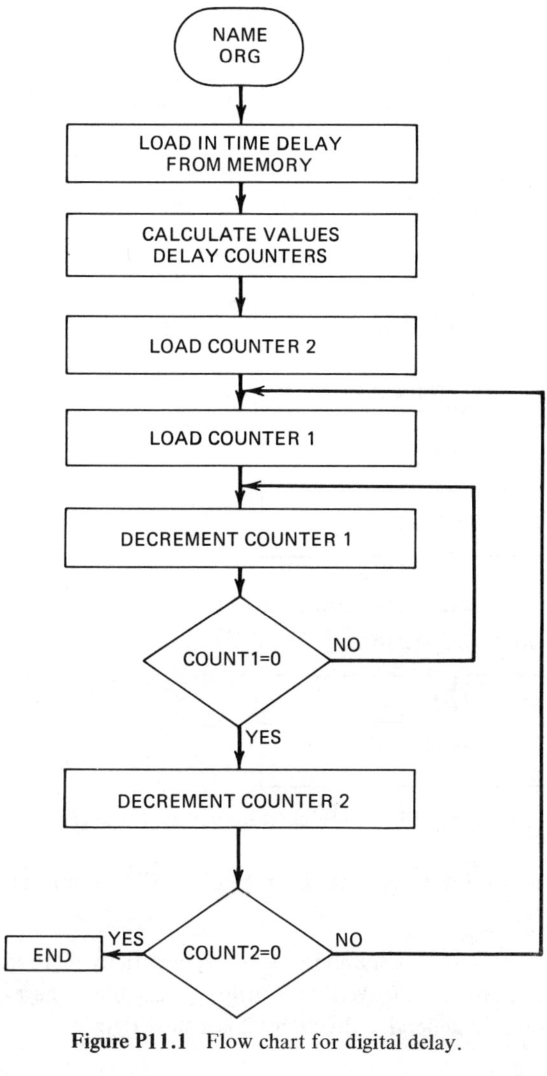

Figure P11.1 Flow chart for digital delay.

P11.2 Write a subroutine to subtract one 16-bit word from another 16-bit word.

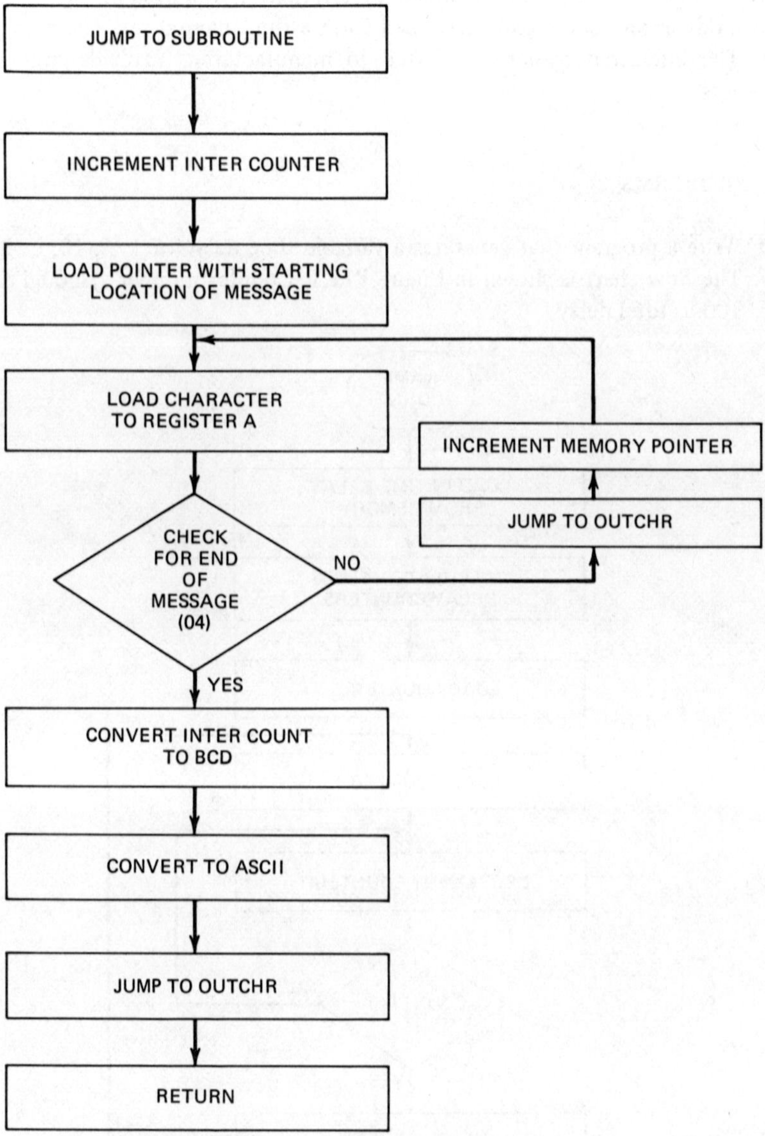

Figure P11.2 Flow chart for interrupt routines.

P11.3 Write a subroutine to convert from hex to BCD. Store the three characters in memory.

P11.4 There are sixteen 8-bit numbers stored sequentially in 16 memory locations starting at address AVE. Write a routine to calculate the average using double precision. In general, a divide by 2 is a shift right.

P11.5 Write the following interrupt service routine:

(a) Cause the nonmaskable interrupt location to execute a jump to the subroutine.

(b) Increment an interrupt counter.

(c) Cause the message LOVE and the value of the interrupt counter to be displayed.

The display subroutine (which someone else has written) is labeled OUTCHR. It outputs an ASCII character from the A register. The flow chart is shown in Figure P11.2.

P11.6 Write the following assembly program:

(a) Set the location counter to RAM and reserve 15 consecutive memory locations for a table.

(b) Set the location counter to virtual ROM and write a program to transfer from a ROM table to the 15 RAM reserve locations.

(c) Set the location counter to virtual ROM and load "HAVE A GOOD DAY" into memory.

CHAPTER 12

Input, Output Techniques

Microprocessors are sophisticated logic processors. The main function of any processor is to transform data or information. A microcomputer system receives a given input and massages or transforms this input to an output useful to a particular application. Because of this type of application the input/output function of a microcomputer system is important. Microcomputer systems are designed for ease in adaptability to a wide variety of interfaces. This chapter describes the features of some of the I/O devices marketed by microcomputer manufacturers that simplify the interfacing problem.

The fact that it is not difficult to design special-purpose logic circuits has been illustrated in this book. Defining comprehensive design criteria is difficult. It is important that one of the design criteria be that systems can be extended in an evolutionary way into more complex and sophisticated circuits. This leads to the basic philosophy behind a microcomputer system design. This philosophy is one of evolutionary advance in which existing developments are not discarded. An expandable system is modular in its components. This is a less wasteful process.

In any microcomputer system the software development cost far outweighs the cost of hardware. Because I/O interface adapters usually have a degree of programmability, they can be a great aid in freeing computational time and effort by offloading the CPU of ill-suited tasks and increasing the system capacity. Standard interface adapters allow the CPU to treat I/O functions exactly like memory functions, and this simplifies interface design. Also, the interface adapter does the synchronizing task, which eliminates problems between high-speed CPUs and low-speed peripheral devices.

Information that must be transferred between the CPU and the I/O device has two general classifications

- Control and status signals
- Data to be processed by the CPU

Data movement between memory and peripheral devices can be done in two ways:

1. Transfer through the CPU by way of the data bus. By use of this technique the transfers are program controlled. Movement is accomplished by the execution of the data transfer instructions as described in Chapter 11.
2. Transfer bypassing the CPU. Data is now defined as anything stored in an auxiliary mass storage device or in RAM. However, from a practical point of view

RAM is constructed in identifiable bytes. Mass storage is not. Data in mass storage is transferred to RAM for execution. Bypassing the CPU saves time. However, it is still necessary that the CPU supervisory signals be provided.

The general philosophy is to program the instruction sequence to issue the proper control and status signals and to issue data masquerading as control and status signals. The CPU is the master device, and it must be programmed to initialize interface adapters according to the following sequence of events:

1. *Disable interrupts before and enable interrupts after I/O control instructions.* This step is easily missed. It is the equivalent of closing the loop in a state diagram. If omitted the interface adapters are not returned to the idle mode. Therefore, they cannot receive new instructions and continue to execute tasks according to previous values. These previous values may be random or may be from old instructions. They are sure to cause unwanted signals. Some of these signals may be catastrophic in nature.
2. *The CPU must issue operating instructions to each interface adapter.* The CPU transfers data from the memory to registers in the interface adapter. The interface adapter decodes this as an instruction and determines its tasks and modes of operation. The CPU then sends a start command to the interface adapter by way of a control line.
3. *The interface adapter sends flags to the CPU whenever it completes a task.* The CPU generally acknowledges the interface adapter. The CPU may mask the flag if it is engaged in another task and does not wish to be interrupted. This is a required prerogative in all hierarchies.

There are many types of interface adapters, and to cover the scope and intent of all such devices in a book such as this would not serve any useful purpose. However, two interface adapters that are powerful, versatile, and common enough to require special attention follow:

- General-purpose parallel adapters
- General-purpose serial adapters

The operation and programming of all interface adapters is sufficiently similar so that the reader should be well versed in these techniques after completing the examples [31, 32].

12.1 GENERAL-PURPOSE PARALLEL ADAPTER

The function of an adapter is to

- Provide a buffered connection from the data bus to the I/O device.
- Convert the multitude (five or six) CPU control signals to two interface control signals (request and response).

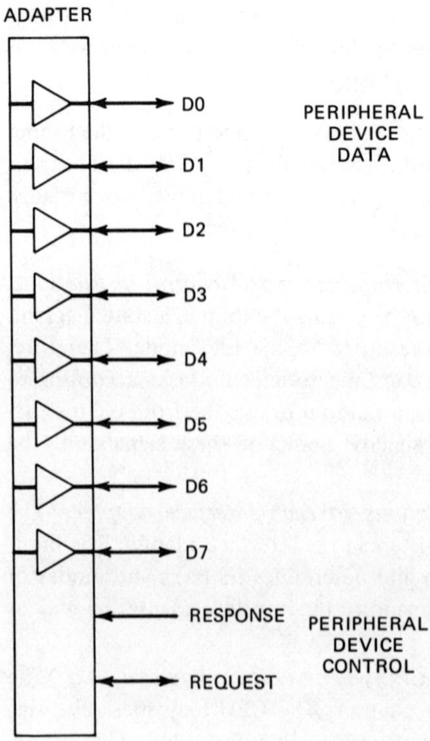

Figure 12.1 Adapter with buffer data lines and two-line control.

Figure 12.1 shows a typical adapter interconnection model. This figure illustrates the simplification of I/O using adapters. The data lines are programmable for direction (input or output). The control line request is programmable for direction. Both control lines (request and response) are programmable for

- Transition. The control may detect high-to-low or low-to-high signal level changes.
- Duration. The control may detect a level change or a pulse.
- Mask. The control may be enabled or disabled.

Figure 12.2 gives the programming model for the adapter. Writing into the D register sets the direction of data flow between the adapter and peripheral device. Writing into the C register sets the control mode. The following sections define the codes that must be written into each register.

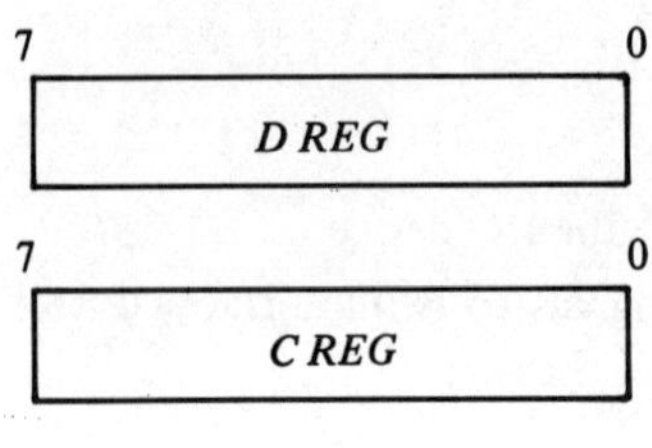

Figure 12.2 Adapter programming model.

Conceptually the foregoing model is simple and correct and really describes the task to be done.

There are some cute tricks used to overcome a physical problem and to demonstrate manufacturers' individuality. The physical problem to be overcome is limited availability of pin connections. Thus, eight lines are used to access two 8-bit registers. This is accomplished by choosing one bit (say b6) of the C (control) register to act as a binary switch. The procedure is as follows:

1. Write a byte with b6 = 1 onto the eight lines. The write instruction connects the lines to the D register.
2. Write the desired contents of the D register onto the eight lines. The D register is loaded, and the trailing edge of the write instruction control pulse connects the lines to the C register.
3. Write the desired contents of the C register onto the eight lines.

The C register can be modified at any time. The D register is modified by repeating all the above steps.

Individual manufacturer preferences are described in the following sections and are summarized in Appendices I and J.

□ 12.1.1 Motorola Peripheral Interface Adapter

The Motorola Peripheral Interface Adapter (PIA) is a general-purpose parallel device. The user must program the PIA registers. Three programming functions are shown in Figure 12.3 and are executed using the following steps:

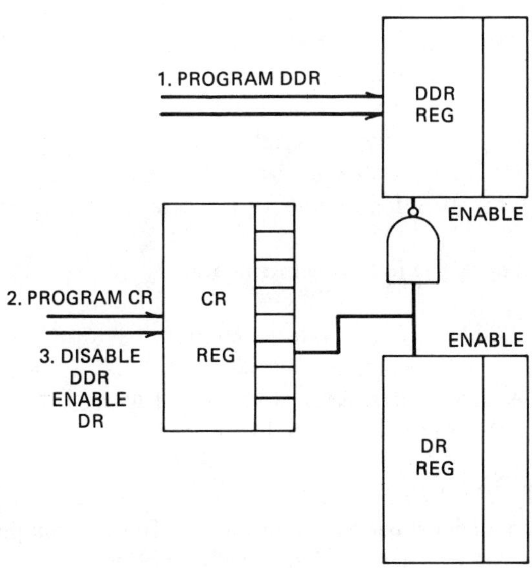

Figure 12.3 PIA registers.

1. Program the direction of data flow on the data register (DR) to either input or output. A data direction register (DDR) in the PIA is used for this task.
2. Program the operation of the peripheral interface control lines. A control register (CR) in the PIA is used. Each control line must be programmed to
   (a) Enable the control line.
   (b) Set the transition of the control line to either high-to-low or low-to-high.
   (c) Set the signal direction either out to the peripheral device or from the peripheral device. Not all control lines can be programmed for signal direction.
3. Disable the data direction register input so that it cannot be changed inadvertently under normal operation, and enable the data buffer on the PIA to receive data.

Information to accomplish step 1 is contained in 1 byte, and information to accomplish both steps 2 and 3 is contained in 1 byte. If the DDR register is at one address and the CR register is at the next address (and this is the recommended configuration), the two instructions used to program a Motorola PIA are as follows:

```
LDX   CW1   Set desired DDR and CR
STX   DDR   Send to registers
```

The LDX operator loads all the values for steps 1, 2, and 3 into the 2-byte X register. The operand CW1 (control word 1) contains a word to program both DDR and CR and to disable the DDR and enable the data buffer.

The STX operator stores the values to the PIA. The first byte goes to operand or location DDR. The second byte goes to location DDR + 1 or CR.

A $\overline{\text{RES}}$ (reset) signal drives all registers in the PIA to 0. The $\overline{\text{RES}}$ signal is issued to all PIAs any time a microcomputer system is turned on.

Because the peripheral interface is subject to program control, it is possible to change its operation dynamically in the middle of a program. If step 3 is reversed, that is enable the DDR and disable the data buffer, the procedure described can be repeated to change the direction of flow of the data and the operation of the control lines.

Step 3 is reversed using the following instruction:

```
CLR   CR    Reset registers
```

This operator clears the CR register. The control word, CW1, is decoded from DDR and CR. Their descriptions are as follows:

Data Direction Register (DDR). The DDR is used to establish the direction (input or output) of each individual peripheral buffer line. This is accomplished by writing to the DDR. Each line is controlled individually. Writing a zero or one causes the corresponding line to function as an input or output respectively. For example, the hexadecimal byte 00 causes all eight lines to be input. The hexadecimal byte FF causes all eight lines to be output, and the byte 0F causes lines b7 to b4 to be input and b3 to b0 to be output.

Control Register (CR). Figure 12.4 shows the control register configuration for the Motorola PIA. The operation and programming of this device is as follows:

1. *Operation.* CA1 is an adapter input signal from the peripheral device. $\overline{\text{IRQ1}}$ is the output from the PIA to the CPU.

CA2 represents a control line between the adapter and the peripheral device. This line can be programmed to be either input or output. $\overline{\text{IRQ2}}$ is the output signal to the CPU.

2. *Programming.* The programming of signal CA1 is shown in Table 12.1. Bit 1 sets the level control. For bit b1 = 1 the flag is set on a transition from low to high and if bit b1 = 0 the flag is set on a transition from high to low. These flags are automatically reset when the corresponding data register is read. Bit b0 controls the enabling of the flag. When b0 = 1, the flag is enabled. When b0 = 0, the flag is disabled. CA1 is always an input to the PIA. It cannot be programmed for input/output. The reader is asked to recall problem P6.6 in which a circuit that responds to this type of programming is designed.

Table 12.2 shows the programming of signal CA2 for input. The reader is referred to problem P6.7 for a circuit that responds to this type of programming. Bit b5 controls the CA2 direction. When bit b5 = 0, CA2 is programmed for input. Bit b4 is the CA2 level control. When b4 = 1 the flag is set on a low-to-high transition, and when b4 = 0 the flag is set on a high-to-low transition. Bit b3 controls the enabling of flag 2. When b3 = 1 this flag is enabled. When b3 = 0 this flag is disabled.

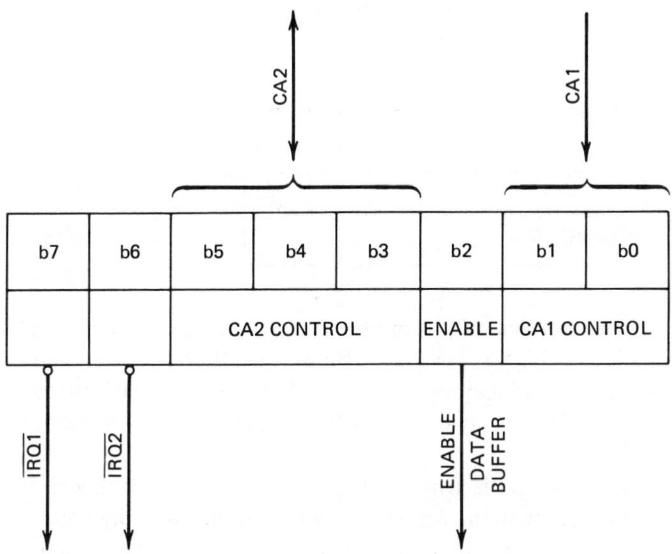

Figure 12.4 Control register.

Table 12.1 Input Programming for CA1

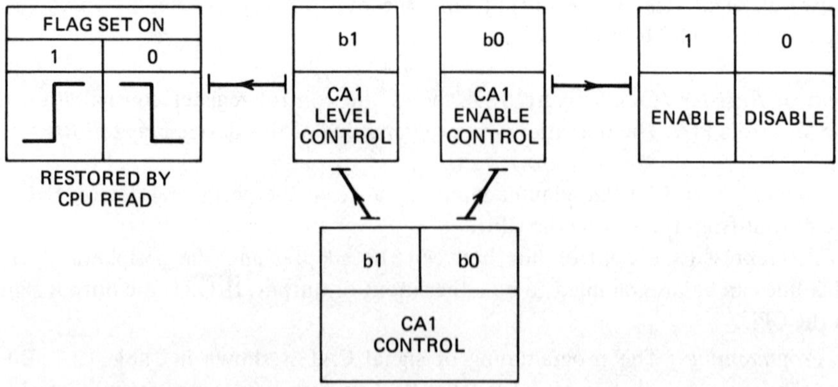

Table 12.2 Input Programming for CA2

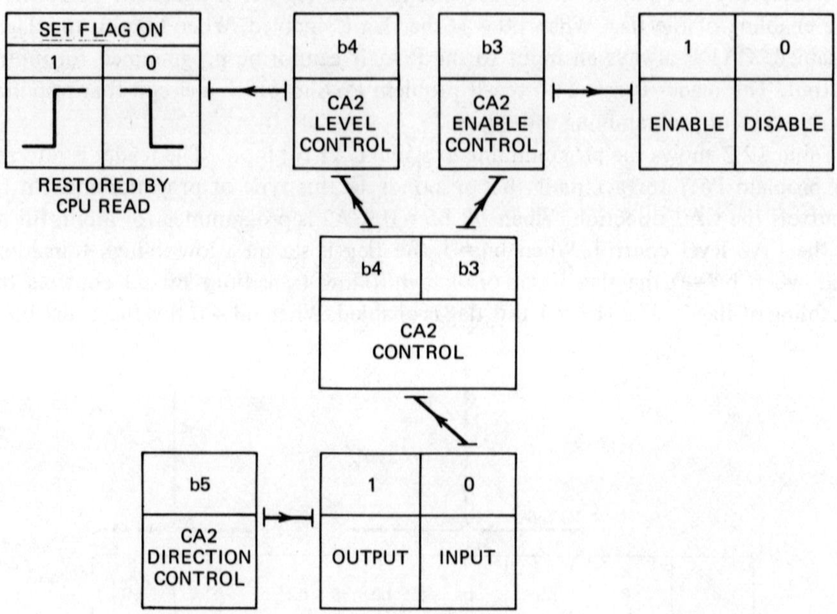

Table 12.3 illustrates the output programming for CA2. As above, bit b5 controls the CA2 direction. If b5 = 1, CA2 is set for output. Bit b4 controls the CA2 enable (note that the roles of b4 and b3 are reversed between input and output). When bit b4 = 1 the level of CA2 changes on a CPU write. When bit b4 = 0 the level of CA2 changes on a CPU read.

On a CPU read, CA2 goes from high to low. If bit b3 = 0 this transition is restored by a CA1 transmission, which causes CA2 to be normally high. If bit b3 = 1 this transition is restored by chip enable, which causes CA2 to be a pulse.

When bit b4 = 1, a CPU write, CA2 follows the value of bit b3.

Table 12.3 Output Programming for CA2

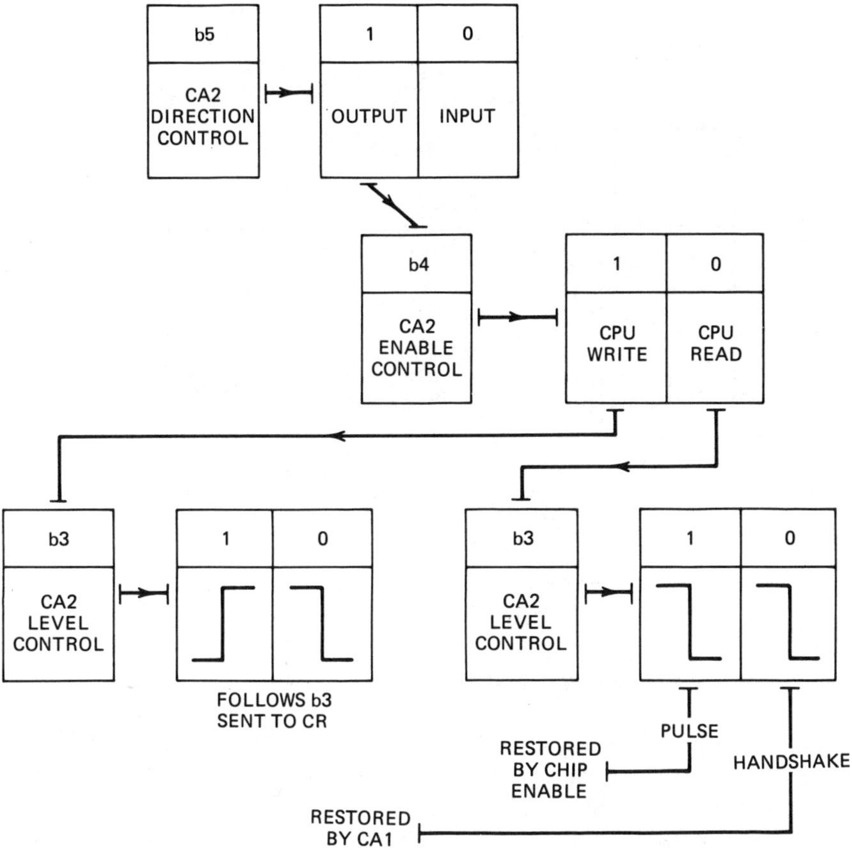

Tables 12.1 to 12.3 are summarized in Appendix F. Figure 12.5 shows the block diagram of the Motorola MC6820 PIA. There are two ports A and B available for connection to the peripheral device. On the CPU side the PIA has eight lines connected to the data bus, five lines to the address bus, and five lines connected to the control bus. The five control lines are:

$\overline{\text{IRQA}}$: an interrupt request flag on register A.

$\overline{\text{IRQB}}$: an interrupt request flag on register B.

R/$\overline{\text{W}}$: signals the device whether it is to be in the read or write mode.

E: chip enable signal is used for timing to control the transfer of data and the timing of all peripheral control signals. It is normally connected to the ϕ_2 clock and is thus the ϕ_2 clock.

$\overline{\text{RES}}$: the chip reset.

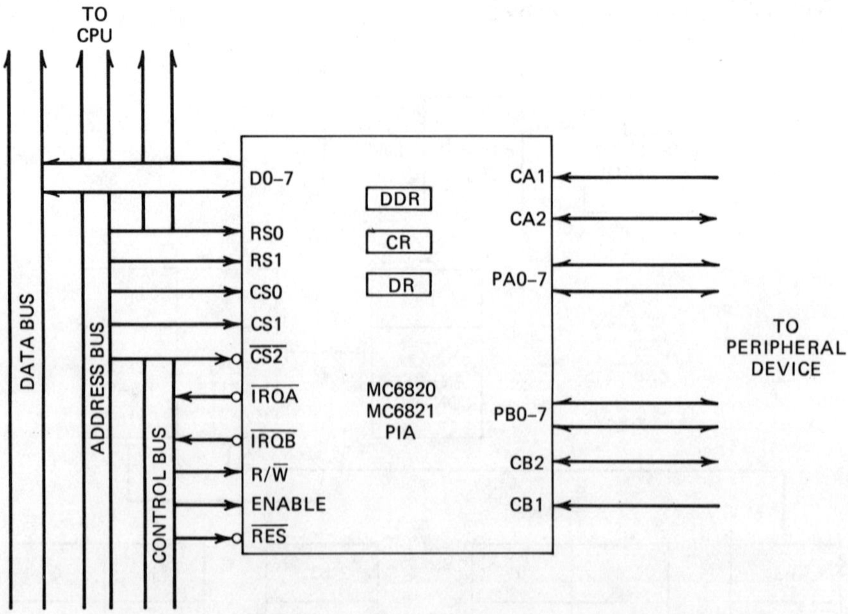

Figure 12.5 Block diagram of Motorola MC6820 and MC6821 PIA.

There are two ways in which Port B differs from Port A.

1. The output programming of CB2 has a slightly different enable control feature. It can only be enabled for CPU write. Table 12.4 shows three blocks for bit b4 control redrawn from Table 12.3 to illustrate this feature. All other programming features of Tables 12.1 to 12.3 are identical (except, of course, Bs should be substituted whenever As occur).
2. Port B has better output drive capacity than Port A.

Considering the two preceding points it seems obvious that one sensible way to use a PIA is to program Port A for input and Port B for output. This is accom-

Table 12.4 Output Programming for Bit b4 of CB2 Control

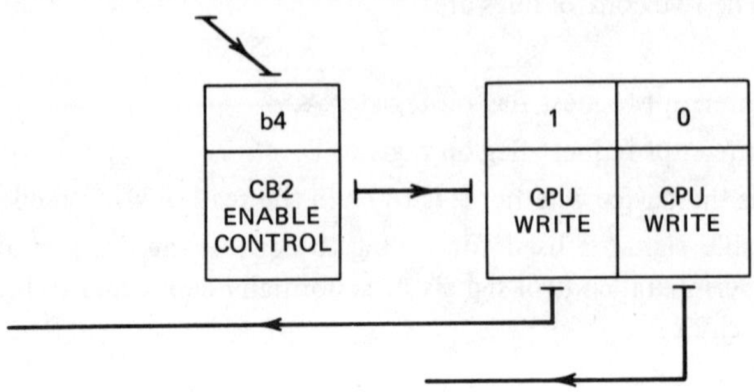

plished by the following initialization program, which also assigns CA2 and CB2 the output direction and the level as normally high.

```
LDX    #$0027    Set values DDRA and CRA
STX    PIAPA     Write to register
LDX    #$FF27    Set values DDRB and CRB
STX    PIAPB     Write to register
```

The suffixes A and B are used to represent the addresses of Port A and Port B of the PIA respectively.

The MC6821 is an enhancement of the MC6820 with improved security, signal, load, and power characteristics. A quick reference summary is given in Appendix I. □

■ 12.1.2 Intel Programmable Peripheral Interface Unit

The programmable peripheral interface unit (PPIU) block diagram is shown in Figure 12.6. The PPIU has eight lines connected to the CPU data bus. It has three lines connected to the address bus. One of these lines, $\overline{CS}$, is a chip select; the other two lines, A0 and A1, select one of the four registers (CONTROL, PA, PB, or PC) in the PPIU. There are two lines ($\overline{RD}$, read and $\overline{WR}$, write) connected to the control bus.

There are 24 port lines. These are divided into three groups of eight. Each group is connected to a port (A, B, or C). Intel provides four ways to use these lines.

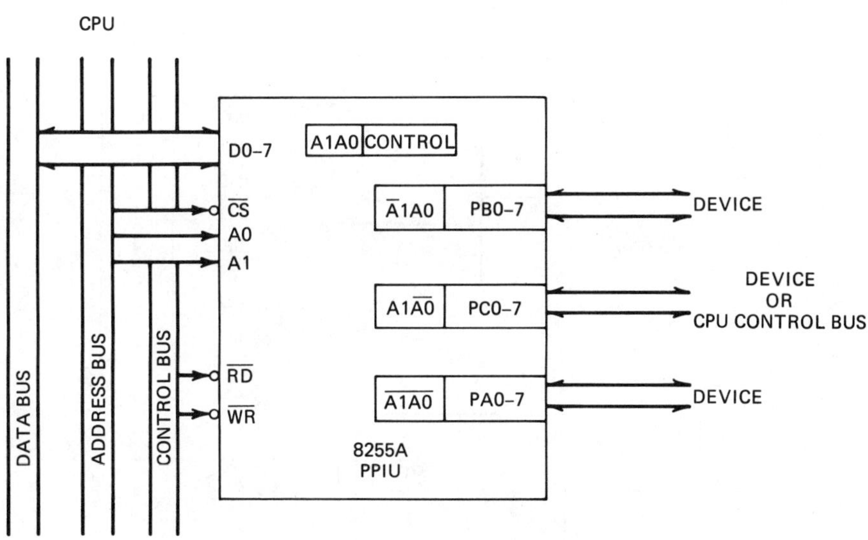

Figure 12.6 Block diagram of Intel 8255A and 8255A-5.

The ways are listed in increasing degrees of sophistication as follows:

- Bit set/reset. Lines can be controlled individually. However this technique is limited to Port C. This can be termed program controlled handshake.
- Mode 0. When no handshaking is required all 24 lines can be used as data lines.
- Mode 1. In this mode Port C lines are used to provide automatic handshake for data Ports A and B.
- Mode 2. In this mode bidirectional data transfer is provided on Port A only. Port C is used to provide automatic handshake with unidirectional control for Port B and bidirectional control for Port A.

Set/Reset Control. Individual bits in Port C are controlled by using the instruction saving set/reset control illustrated in Table 12.5. Bit D7 = 0 selects the set/reset.

Table 12.5 Bit Set/Reset Format

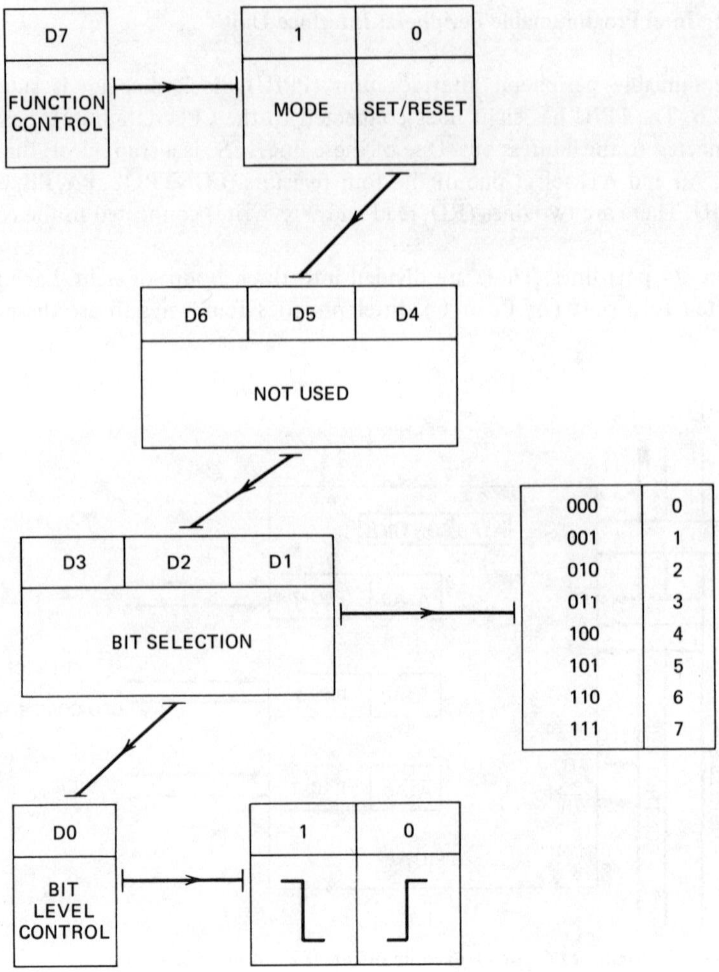

Bits D6, D5, and D4 are not used; thus D6 = D5 = D4 = 0. Bits D3, D2, and D1 select the bit on Port C to be altered. Bit D0 gives the level control. If D0 = 0, the selected bit on Port C goes low. If D0 = 1, the selected bit on Port C goes high.

To cause bit PC6 on Port C to go high the following instructions are used:

```
MVI    A,0DH
OUT    PPIUC
```

The MVI instruction moves the data 0D into register A. Since 0D = 00001101, bit D7 = 0, which selects the bit/reset format. Bits D3, D2, D1 = 110, which selects bit PC6 on Port C, and bit D0 = 1 causes the select bit to go high.

The OUT instruction writes the contents of register A to PPIUC. PPIUC is a symbol for the address of the control register of the PPIU.

Mode 0. In Mode 0 all 24 port lines are used for data. They are programmable for data direction in groups of two 8-bit ports (A and B) and two 4-bit ports (Upper C and Lower C). Table 12.6 shows the Mode 0 control word format.

To cause Port A to be input, Port B output, Port C upper to be input, and Port C lower to be output, the following instructions are used:

```
MVI    A,98H
OUT    PPIUC
```

The MVI instruction moves 98 (10011000) into register A. Mode format is selected since bit D7 = 1. D6, D5 = 00 and D2 = 0 select Mode 0. D4 = 1 selects Port A as input. D3 = 1 selects Port C lower as output. D1 = 0 selects Port B as output, and D0 = 0 selects Port C lower as output.

The OUT instruction causes register A to be written to PPIUC, which is the control register of the PPIU.

Modes 1 and 2. Figure 12.7 shows the division of Port C lines for the control of Ports A and B. Whenever Mode 1, the strobed input/output mode, and Mode 2, the bidirectional bus mode, are selected lines PC0–2 are used for the control of Port B, and lines PC3–7 are used for the control of Port A.

Figure 12.8 shows a simplification of Port C lines for Modes 1 and 2. Port B control operation is the same for both modes. A request for service from the CPU is sent to PC1 on gate 1204. PC1 acts as an enable signal $(\overline{\text{inhibit}})$. PC1 is also sent to the peripheral device as a request for service. When the peripheral device responds by way of PC2 on gate 1204 which is enabled, PC0 goes high. PC0 is sent to the CPU as an interrupt request or flag. Transmission of the interrupt request by PC0 can be inhibited by a software command.

The operation of Port A is determined by the mode control. If Port A is selected for Mode 1 output a request for service from the computer enables gate 1201 by causing PC7 to go high. PC7 is also sent to the peripheral device as a request for

Table 12.6 Mode 0 Control Word Format

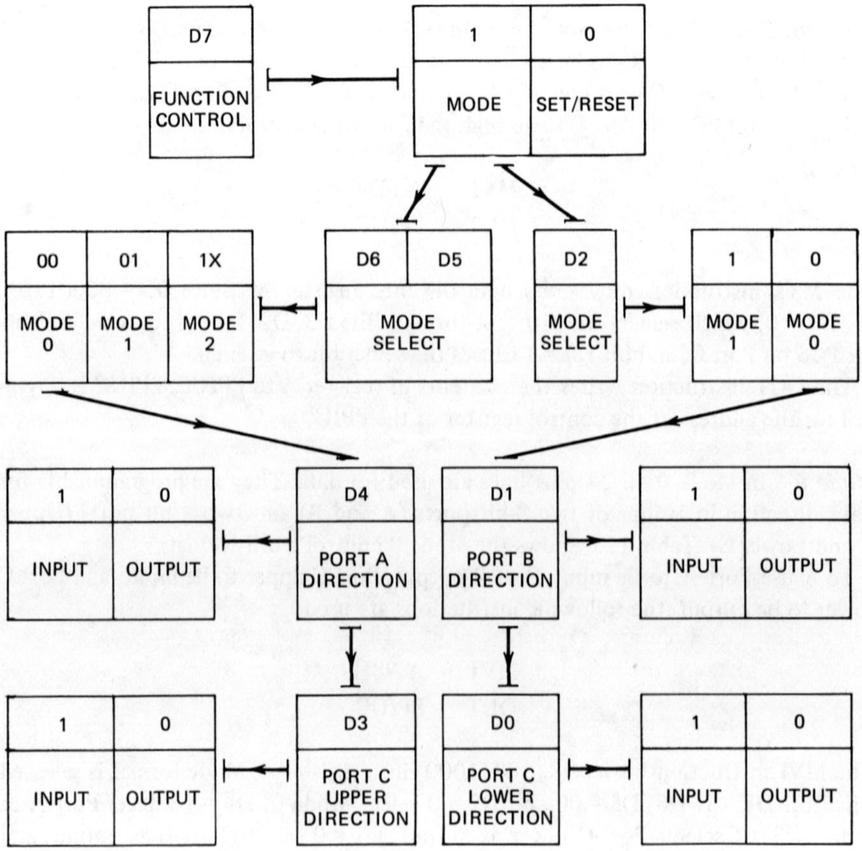

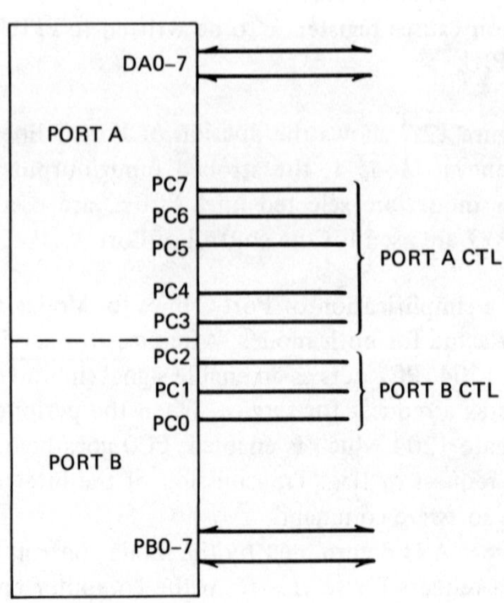

Figure 12.7 Modes 1 and 2 control.

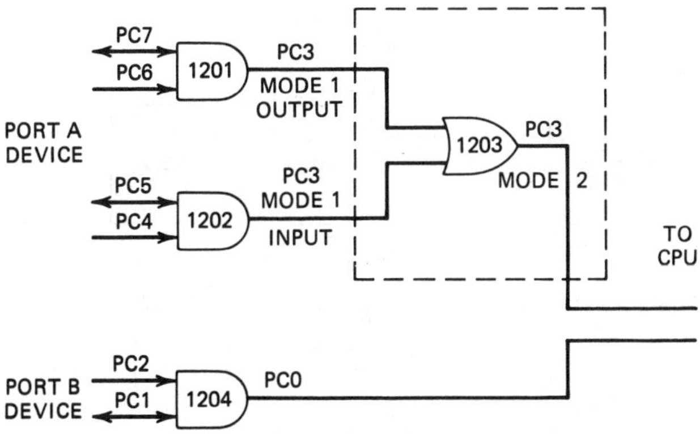

Figure 12.8 Modes 1 and 2 handshake convention.

service. The peripheral device responds on PC6. PC3 is used as an interrupt request to the CPU. When Mode 1 input is selected, request for service from the CPU enables 1202 on line PC5. PC5 is in turn connected to the peripheral device and forms a request signal. The peripheral device responds on PC4, which generates interrupt request to the computer on PC3. In Mode 1, Port A must be selected for input or output. As a result, the unused lines PC4 and PC5 or PC6 and PC7 are available for data I/O. Problem P6.8 gives a typical state diagram for this circuit operation.

In Mode 2, Port A is capable of both input and output operation, and a typical state diagram is given in problem P6.9. In this mode the above input/output signals are routed to an or gate (1203) to form the interrupt request PC3. All four inputs PC4 to PC7 are required, but there is more flexibility on the use of Port A.

Table 12.7 shows the mode control word format for Modes 1 and 2. Bit D7 = 1. Bits D6 and D5 are used to select Port A mode. In Mode 1, bit D4 is used to program the direction of data flow. Bit D3 is used to select the direction for unused Port C lines. D2 is used to select Port B mode. Bit D1 is used for Port B direction, and bit D0 is unused.

In Mode 2 the direction select bits D4 and D3 are unused because this mode is bidirectional.

An acceptable configuration for a PPIU is to have Port A in Mode 2 and Port B in Mode 1 output. The following instructions can be used to initialize the PPIU:

```
MVI    A,0FCH
OUT    PPIUC
```

The MVI operator moves the data FC into register A. Table 12.6 can be used to verify that FC is indeed Port A Mode 2 and Port B Mode 1 output.

The OUT operator sends the contents of register A to PPIUC. PPIUC is again a symbol to represent the address of the control register of the PPIU.

Tables 12.5 to 12.7 are summarized in Appendix J as a quick reference guide. ■

Table 12.7 Modes 1 and 2 Control Word Format

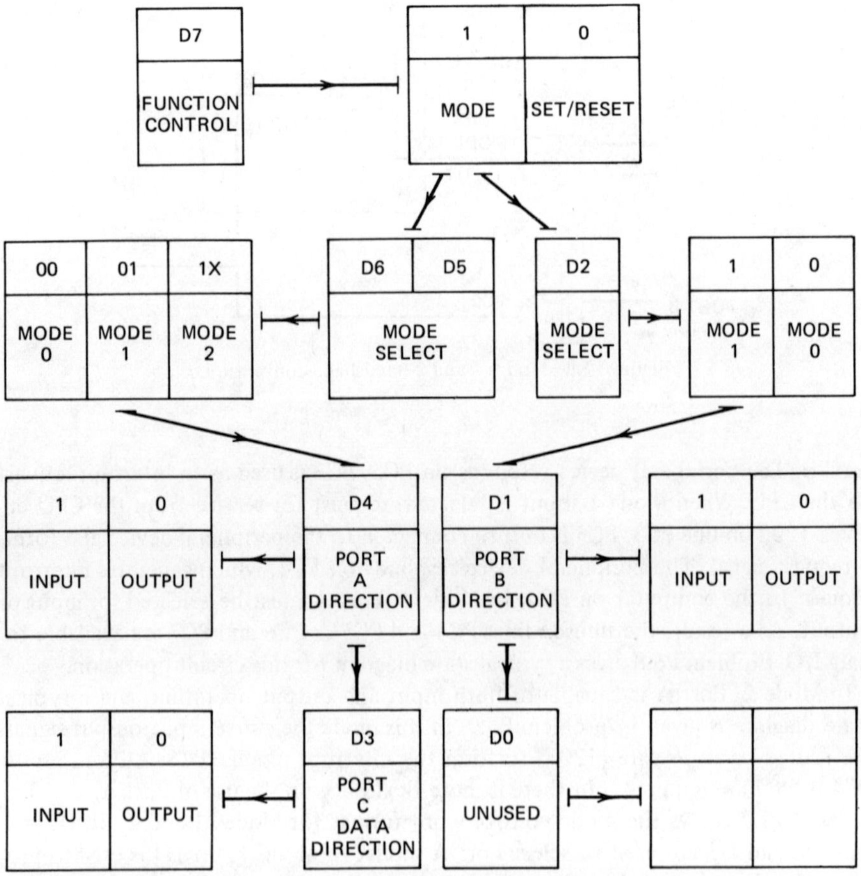

12.2 MICROCOMPUTER SYSTEM COMMUNICATIONS

Telegraph and morse code represent the early forms of effective long-distance communication. Evolution brought about the teletype and the American Standard Code for Information Interchange (ASCII) given in Appendix B. Computer and microcomputer manufacturers follow this standard. For obvious economical reasons, long-distance communication uses serial transmission of information over a single line rather than parallel transmission requiring many lines. The function of a communication interface is to multiplex from many lines to a single line.

Information may be sent asynchronously using timing and framing techniques as described in Section 7.4. In the synchronous mode data is transmitted in synchronism with a clock pulse. To increase the flexibility of operation, a communication interface has the following programmable features:

- Operation with 5, 6, 7, or 8-bit characters.
- Clock rates programmable in multiples of the baud rate (bits per second).

- Operation with shorter stop bits.
- Selection of the baud rates.
- Internal or external synchronization.

In the following methods of programming the communication interfaces are described.

□ 12.2.1 Motorola Communication Interface

The Motorola MC6850 asynchronous communication interface adapter (ACIA) is Motorola's parallel-to-serial converter for asynchronous serial data format. It is used to control such peripherals as modems, CRTs, teletypes, and printer-readers. A block diagram of the interface is shown in Figure 12.9.

There are eight bidirectional data lines, D0 to D7, that connect the interface to the data bus. There are four addressing lines of which three are chip select lines (CS0, CS1, and $\overline{\text{CS2}}$) and an additional register select (RS) which is used to select one of two register pairs within the ACIA: There are three control lines connected to the CPU bus. These are as follows:

- $\overline{\text{IRQ}}$, interrupt request.
- R/$\overline{\text{W}}$, read/write, which selects one of the registers in a register pair.
- E, enable.

On the peripheral side of the ACIA there are two serial data lines as follows:

- Tx Data, output, which is a serial transmit.
- Rx Data, input, which is the serial receive line.

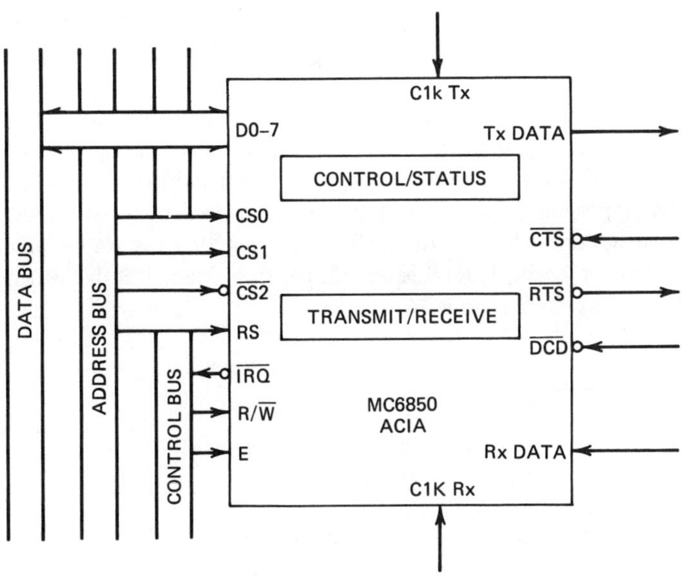

Figure 12.9 Block diagram of Motorola MC6850 ACIA.

There are two clock inputs as follows:

- Clk Rx, receive side clock.
- Clk Tx, transmit side clock.

There are three control signals used for handshake. These control signals are the following:

- $\overline{\text{CTS}}$, clear to send.
- $\overline{\text{DCD}}$, data carry detect.
- $\overline{\text{RTS}}$, request to send.

The ACIA has two control registers called the *control/status* pair and two data registers called the *transmit/receive* pair. RS = 0 selects the control/status pair, and RS = 1 selects the transmit/receiver pair. R/$\overline{\text{W}}$ = 1 selects the status register from the control/status pair and the receive register from the transmit/receive pair. R/$\overline{\text{W}}$ = 0 selects the control register from the control/status pair and the transmit register from the transmit/receive pair. The control register instruction format is shown in Table 12.7.

Bit b7 determines if the control output from the ACIA to the CPU ($\overline{\text{IRQ}}$) is used during the *receive* portion of operation. For the receive portion $\overline{\text{IRQ}}$ can be enabled or disabled using a receive interrupt enable (RIE). RIE = b7 because RIE follows the values of bit b7.

Bits b6 and b5 determine the status of the two control outputs $\overline{\text{RTS}}$ and $\overline{\text{IRQ}}$ during the *transmit* portion of operation. For the transmit portion $\overline{\text{RTS}}$ is controlled directly, and $\overline{\text{IRQ}}$ can be enabled or disabled using a transmit interrupt enable (TIE). The values of $\overline{\text{RTS}}$ and TIE are set/reset by bits b6 and b5 as shown in the table. Bits b6, b5 = 11 cause spaces (rather than marks) to be sent out on Tx Data. A burst of spaces is interpreted as a break character.

Bits b4, b3, and b2 select the word length, parity, and stop bits. The chart given in Table 12.8 is self-explanatory. Bits b1 and b0 are used as a general master reset device and to select the ratio between the clock speed and the baud rate.

The ACIA is programmed by loading the ACIA control register (ACIAC). The following instructions initialize the ACIA assuming that the following conditions are needed. IRQ is disabled, $\overline{\text{RTS}}$ is issued, parity is even, 1 stop bit is used, word length is 8 bits, and clock ratio is 16.

```
LDAA    #03      A = 00000011
STAA    ACIAC    Reset
LDAA    #$19     A = 00011001
STAA    ACIAC    Set control.
```

The operator LDAA loads register A with the hexadecimal byte 03 (the code for master reset).

The operator STAA sends the contents of register A to the ACIA control register.

Table 12.8 Control Register Instruction Format

b7	1	0
RECEIVE INTERR ENABLE CONTROL	ENABLE $\overline{IRQ}$	DISABLE $\overline{IRQ}$

RIE — receive interrupt enable

TIE — transmit interrupt enable

b6 b5	00	01	10	11
TRANSMIT CONTROL	$\overline{RTS}$ = 0 TIE = 0	$\overline{RTS}$ = 0 TIE = 1	$\overline{RTS}$ = 1 TIE = 0	$\overline{RTS}$ = 0 TIE = 0 SPACES

b4 b3 b2 — WORD LENGTH, PARITY, STOP BITS

VALUE	PARITY	STOP BITS	WORD LENGTH
000	EVEN	2	7
001	ODD	2	7
010	EVEN	1	7
011	ODD	1	7
100	NONE	2	8
101	NONE	1	8
110	EVEN	1	8
111	ODD	1	8

b1 b0	00	01	10	11
MASTER RESET AND COUNT RATIO	÷ 1	÷ 16	÷ 64	MASTER RESET

The operator LDAA loads register A with 0001 1001. This is the code for the desired operation.

The operator STAA sends the contents of register A to the ACIA control register.

The CPU interrogates the ACIA by reading the status register. The ACIA status register format is given in Table 12.9. Bit b7 is an interrupt request bit. When the CPU is interrupted, this bit is set. Bit b6 is a parity error. Bit b5 is an overrun error. Bits b5 and b0 are set on $\overline{E}$ and read data, and reset on the next $\overline{E}$ and read data. Bit b4 indicates a framing error. Bit b4 is set on internal Rx Data and reset on the next Rx Data. Bit b3 indicates $\overline{CTS}$ has been received. Bit b2 gives the status of

Table 12.9 Status Register Byte Format

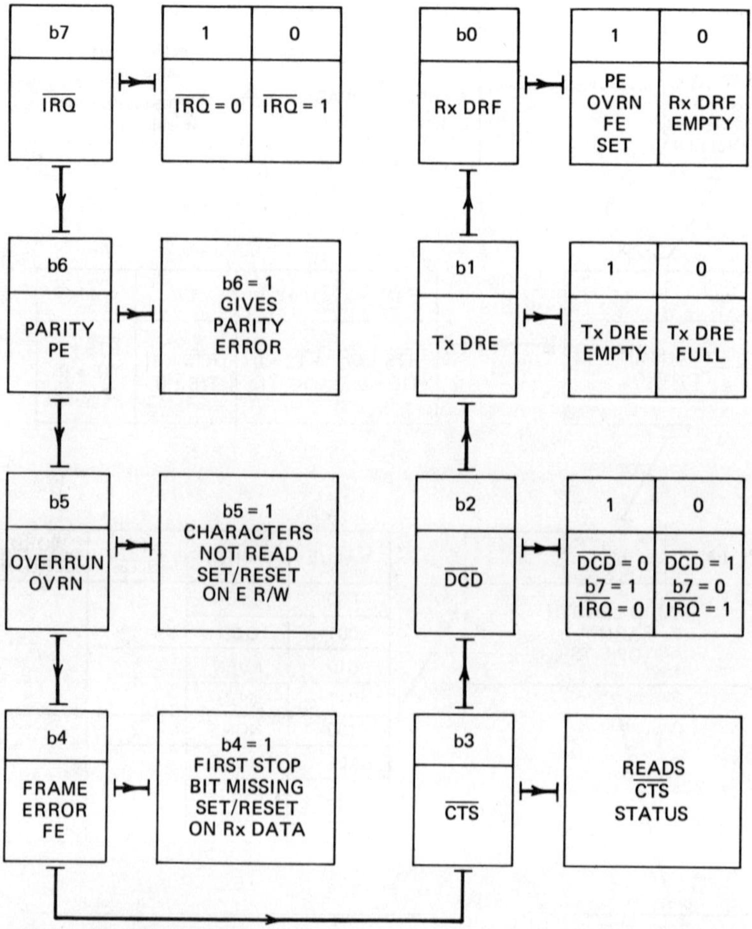

DRF/E Data register FULL/EMPTY
DCD Data carrier detect
CTS Clear to send

the data carrier detect. $\overline{DCD}$ = 1 causes b7 = 1 and $\overline{IRQ}$ = 0. Reading the status and the Rx Data registers or master reset causes bit b2 = 0 and bit b7 = 0. Bit b1 gives the status of the transmitter data register. Tx DRE causes bit b1 = 1. $\bar{E}$ and R/$\overline{W}$ = 0 causes b1 = 0, as does $\overline{CTS}$ = 1. Bit b0 gives the status of the receive data register. $\bar{E}$ X R/$\overline{W}$ causes b0 = 0 as does $\overline{DCD}$ = 1.

The following instructions can be used to check the status of the ACIA:

```
LDAA   ACIAC    Get status
ANDA   #$0C     Check b3 and b2
```

The LDAA operator loads the contents of the ACIA status register into register A. Remember that a load (read) accesses the status register. A store (write) accesses the control register.

The ANDA operator compares the hexadecimal value 0C (00001100 or $\overline{CTS}$ and $\overline{DCD}$) with the ACIA status register. If these two are unequal the zero status flag in the ALU status register of the CPU is cleared. The value of the flag can be used in a conditional branching instruction. □

■ 12.2.2 Intel Universal Interface

The 8251A is also called a universal synchronous/asynchronous receiver/transmitter (USART). A block diagram for this chip is shown in Figure 12.10. The USART has two control registers called the control/status pair. The CPU *writes* control information to the control/status register and *reads* status information from the control/status register. Also the USART has two data registers called the receive/transmit pair. The CPU *writes* data to the transmit register and *reads* data from the receive register. There are eight data lines D0 to D7 connected to the data bus and one chip select line $\overline{CS}$ connected to the address line.

The control lines are as follows:

- Reset.
- Clock lines used in the normal way.
- C/D (control/data) line is used to select the control/status register ($C/\overline{D} = 1$) or the data registers ($C/\overline{D} = 0$).

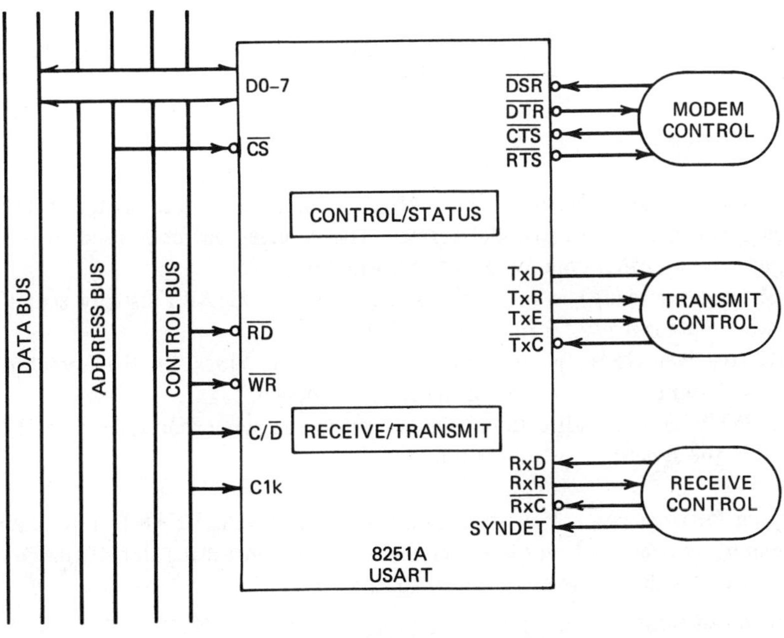

Figure 12.10 8251A USART block diagram.

- $\overline{\text{WR}}$ *write* line initiates transfer from the data bus to the interface registers.
- $\overline{\text{RD}}$ *read* line initiates transfer from the interface registers to the data bus.

$\overline{\text{WR}} = 1$ and $\overline{\text{RD}} = 1$ isolates the interface registers and the data bus.

Additional control lines are shown in Figure 12.10. The USART has modem control, transmit control, and receive control. Their descriptions follow.

Modem Control. The modem control simplifies the interface to any modem. The signals used to interface to a modem are the following:

- $\overline{\text{DSR}}$ (data set ready). The $\overline{\text{DSR}}$ input signal from the modem to the USART is used to indicate when the modem is ready.
- $\overline{\text{DTR}}$ (data terminal ready). The $\overline{\text{DTR}}$ is an output signal from the USART to the modem and may be regarded as a handshake routine with $\overline{\text{DSR}}$.
- $\overline{\text{RTS}}$ (request to send). The $\overline{\text{RTS}}$ is an output signal from the modem used as a request to send.
- $\overline{\text{CTS}}$ (clear to send). The $\overline{\text{CTS}}$ input signal is used to enable the USART to transmit serial data.

Transmitter Buffer and Control. The transmitter buffer accepts data from the parallel data bus and converts it to a serial bit stream. The transmitter buffer also inserts the appropriate characters or bytes and outputs its stream of data on the TxD output line. It transmits data in response to the enable and $\overline{\text{CTS}}$ signal.

TxR (transmitter ready) is an output signal usually sent to the CPU to indicate when the transmitter is ready to accept the character. The TxR is reset by the leading edge of $\overline{\text{WR}}$.

TxE (transmitter empty) is an output signal from the USART. It can be used to signal the CPU that the transmission is complete.

$\overline{\text{TxC}}$ (transmitter clock) is an input to the USART. In asynchronous operation, the baud rate is a fraction of $\overline{\text{TxC}}$ frequency.

Receiver Buffer and Control. The receiver accepts serial data on the RxD input line and converts it to a parallel format. The receiver buffer is used to strip off framing bits and other communication information.

RxR (receiver ready) is an output signal from the USART usually sent to the CPU when data is ready for input to the CPU.

$\overline{\text{RxC}}$ (receiver clock) is the input for the receiver clock. In the asynchronous mode the baud rate is a fraction of the $\overline{\text{RxC}}$ frequency.

The SYNDET is a bidirectional signal used in the synchronous mode and is described in the synchronous subsection.

Interface Programming. For asynchronous operation the USART is programmed by sending two control words in sequence to the control register of the control/status register pair. The two control words are

1. Mode instruction
2. Command instruction

Since the two control words must be loaded into the same register, a little care must be taken. The control word for the mode instruction can be loaded into the control register of the control status pair after a reset only. The second control word or command instruction can be loaded any time $C/\overline{D} = 1$ except immediately after a reset instruction. Thus control word 2, or the command instruction, can be changed after the mode instruction as illustrated in Figure 7.13.

Table 12.10 shows mode control instruction format. Bits D1 and D0 are used to assign a classification task and a selection task. If bits D1 D0 = 00, the mode is classified as synchronous. If D1 D0 = 01, 10, or 11, the mode is classified as asynchronous. For the asynchronous mode, bits D7 D6 are used to assign frame control. D7 D6 = 00 is not used, D7 D6 = 01 assigns 1 stop bit, D7 D6 = 10 assigns $1\frac{1}{2}$ stop bits, and D7 D6 = 11 assigns 2 stop bits. Bits D5 and D4 are used to assign parity control, D5 D4 = 00 and 10 indicate no parity. D5 D4 = 01 gives odd bit parity, and D5 D4 = 11 gives even bit parity. Parity is used only between the USART and the teletype or CRT. The data bus receives no indication of parity.

Bits D3 D2 are used to assign character length. D3 D2 = 00 gives a length of 5 bits, D3 D2 = 01 gives a length of 6 bits, D3 D2 = 10 gives a length of 7 bits and D3 D2 = 11 gives the standard 8-bit character length. Bits D1 D0 are also used in a selection mode. D1 D0 = 00 is not permitted by the classification task. D1 D0 = 01 gives equal clock rate and baud rate. D1 D0 = 10 makes the clock rate to baud rate ratio 16, and D1 D0 = 11 makes this ratio 64.

The command instruction controls the actual operation of the USART. Table 12.11 shows the 8251A command instruction format. D7 = 1 enables a search for a synchronizing character. In asynchronous operation this bit is ignored. Bit D6 = 1 is an internal reset that returns the programming to the mode instruction. Bit D5 = 1 is used to generate a request to send ($\overline{RTS}$) to the modem control. Bit D4 is used to reset any error indication in the status register of the control/status register pair. Bit D3 = 1 drives the TxD output line to the transmit low, which forces the transmitter to generate a break character. Bit D2 = 1 enables the output RxE, which in turn enables the RxR output signal of the receive control. Bit D1 forces the $\overline{DTR}$ output signal on the modem control low. Bit D0 = 1 enables the TxEN, which in turn removes the transmitter buffer TxD line from the marking mode.

The following instructions program the USART to operate and issue $\overline{RTS}$, even parity, 1 stop bit, word length of 8 bits, and clock ratio of 16.

```
MVI    A,40H    D6 = 1
OUT    USART    Reset control
MVI    A,7EH    Stop, parity, etc.
OUT    USART    Load instruction
MVI    A,27H    RTS, T, R, and DTR
OUT    USART    Load command
```

The MVI operator moves the data into the A register. The hexadecimal number 40 = 01000000, which makes bit D6 high and resets the device from the command mode to the instruction mode. The instruction illustrates use of the reset feature.

Table 12.10 8251A Asynchronous Mode Instruction Format

D1 D0	00	01	10	11
TRANSMISSION MODE CLASSIFICATION	SYNC	ASYNC		

D7 D6	00	01	10	11
FRAME CONTROL	UNUSED	1 STOP BIT	1½ STOP BITS	2 STOP BITS

D5 D4	00	10	01	11
PARITY CONTROL	NO PARITY		ODD PARITY	EVEN PARITY

D3 D2	00	01	10	11
CHARACTER LENGTH	5 BITS	6 BITS	7 BITS	8 BITS

D1 D0	00	01	10	11
BAUD RATE SELECTION		x1	x16	x64

The hexadecimal number 7E determines the stop bits, parity, character length, and baud rate.

In the command mode $\overline{\text{RTS}}$ and enable are selected by hexadecimal 27.

The OUT operator directs the data to the proper location.

Synchronous Operation. The USART is also capable of synchronous operation. Under synchronous operation, four control words are loaded to the control register of the control/status register pair. The first three control words must be loaded in sequence after a reset as shown in Figure 7.13. The last control word or command instruction can be loaded or changed at any later time by setting $C/\overline{D} = 1$.

Table 12.11 Command Instruction Format

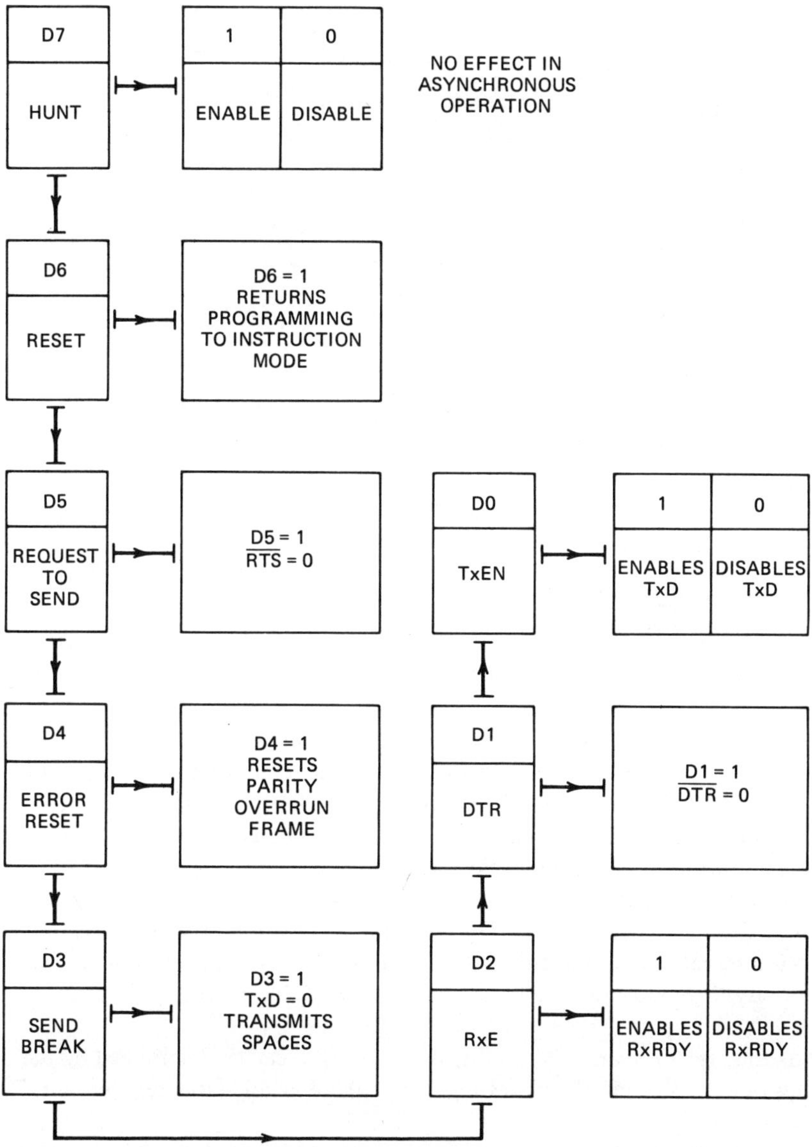

Control word 1 is a mode instruction, control word 2 is synchronizing character 1, and control word 3 is synchronizing character 2. Control word 4 is a command instruction. The command instruction format is as given in Table 12.11. Operation of the synchronous mode instruction format is essentially the same as given in Table 12.9 with the exceptions as given in Table 12.12.

In the synchronous mode instruction format bits D1 D0 = 00 select the synchronous mode. When the synchronous mode is selected, bit D7 is used for character selection. For D7 = 1 synchronizing characters are used, and for D7 = 0 double

Table 12.12 **USART Synchronous Mode: Instruction Format**

D6	→	1	0
SOURCE	→	EXTERNAL SYNC	INTERNAL SYNC

D7	→	1	0
CHARACTER SELECTION	→	SINGLE SYNC CHAR	DOUBLE SYNC CHAR

D1	D0	→	00	01	10	11
TRANSMISSION MODE CLASSIFICATION		→	SYNC	ASYNC		

synchronizing characters are used. Figure 12.11 shows the introduction of synchronizing characters inserted into the transmitter serial output line automatically by the USART. The sync detect (SYNDET) input/output is shown in the receiver control of Figure 12.10. On input an external sync detect is used, and data characters are read on the next $\overline{\text{RxC}}$. For output SYNDET indicates the last sync characters in receive operation. SYNDET is reset by status read. Bit D6 selects the source of synchronization. For D6 = 1 external synchronization is used, and for D6 = 0 internal synchronization is used. Bits D5, D4, D3, and D2 are as described for the asynchronous mode, instruction format of Table 12.10.

Status Register Format. The status of the USART can be determined by the CPU at any time by reading the status register of the control/status register pair. Table

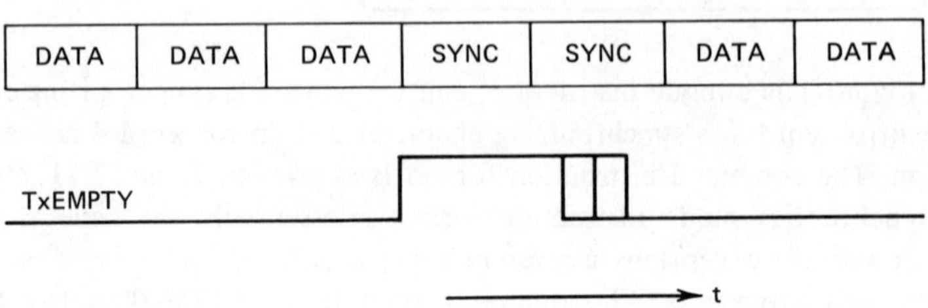

Figure 12.11 Automatic insertion of synchronizing characters.

■ **Table 12.13 Status Register Format**

D7	D6	D5	D4	D3	D2	D1	D0
DSR	SYNET	FE	OE	PE	TxE	RxRDY	TxRDY

12.13 shows this format. This register is accessed by using an $\overline{RD}$ instruction when $C/\bar{D} = 1$. Bit D7 is used to read the status of the DSR. Bit D6 detects the presence of a synchronizing character (SYNDET). It automatically resets the synchronizing character. Bit D5 indicates a framing error. Bit D4 indicates an overrun error. Bit D3 indicates a parity error. Bit D2 gives the status of the transmit control output signal T x EMPTY. Bit D1 gives the status of the receiver output signal R x RDY, and bit D0 gives the status of the transmitter output T x RDY. ■

12.3 PROBLEMS

P12.1 The seven-segment display is given in problem P1.6. Write a subroutine to output the numbers 0 to 4 repeatedly in seven-segment form. Use a commercial interface adapter. Make sure there is a delay of approximately 5 s between each display.

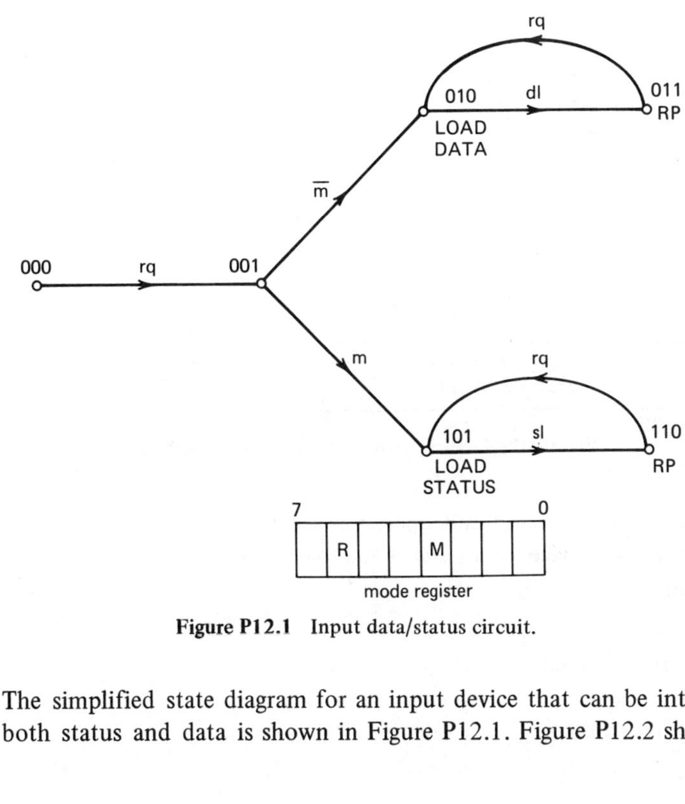

Figure P12.1 Input data/status circuit.

P12.2 The simplified state diagram for an input device that can be integrated for both status and data is shown in Figure P12.1. Figure P12.2 shows a flow

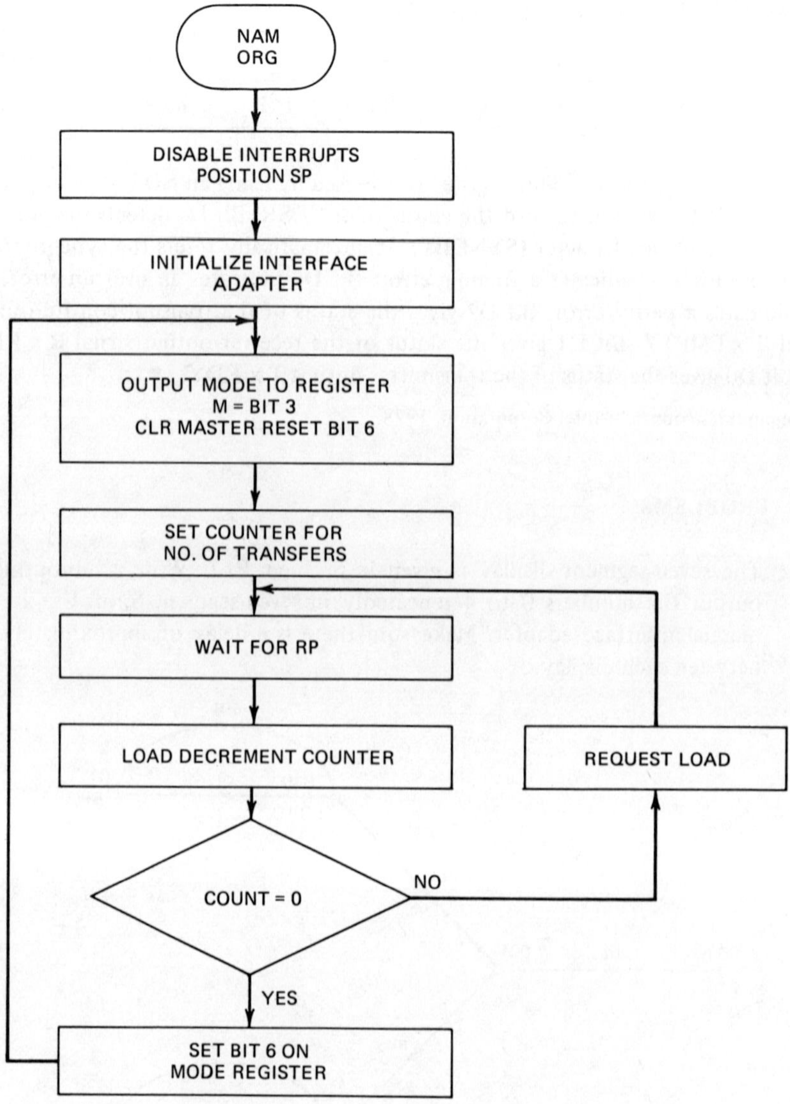

Figure P12.2 Flow chart for data/status interface.

chart for initialization and operation of this circuit. Write a subroutine to initialize the device, check status and load data.

INPUT SYMBOLS

rq request

sl status loaded

dl data loaded

OUTPUT SYMBOLS

RP response

M(m) is bit 3 (mode bit)

R is bit 6 master clear which resets to idle state

P12.3 The state diagram given in Figure P12.3 is a reworked version of Figure 6.16 in that only adapter input/output signals are shown. Write the following subroutine:

(a) Initialize for two input ports A, B, two output control signals GO and CYCOM, and two input control signals CYCLE and IDLE.

(b) Monitor Port A and when bits 4, 3, and 2 are all ones (ATTENTION) issue a GO signal.

(c) Read Port B when the input CYCLE is detected and issue a cycom signal.

(d) When an IDLE is received return to step (b).

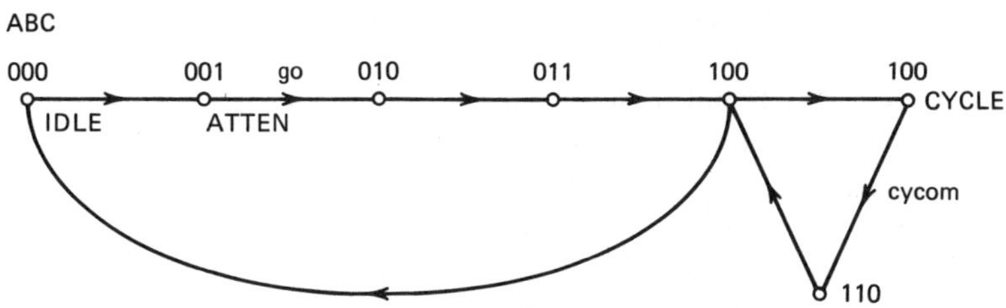

Figure P12.3 Generalized interface for microcomputer.

P12.4 Program a communication interface to transmit the message

"HAVE A GOOD DAY"

Initialize for odd parity, maximum number of stop bits, and slowest speed.

P12.5 Write a program that checks a communicate interface for framing errors and overrun.

CHAPTER 13

Operation and Configuration of Independent Systems

The basic executable instruction set for the microprocessor is described in Chapter 11. The chapter continues with a description of program development by using RAM to simulate ROM, and the methods of executing a program from RAM. This creation of a virtual machine by simulating ROM in RAM has its purposes. Such a system is a good educational device to acquaint the new programmer with the instruction set. The technique also eases the programmers's chores. Program testing and modification is easier using RAM because development and debugging changes can be done from a console conveniently.

However, the development configuration described in Chapter 11 is ill-suited in the processing environment for the following reasons:

- RAM is a poor microcomputer program media because it is expensive, bulky, and insecure.
- The system requires extensive input/output equipment such as CRTs which may not be required in the processing operation. The process may be better suited to the use of LEDs or other inexpensive I/O devices for the communication of intelligence to man.
- Versatility is not required because the tasks of a development system and a process system are different.
- The development system requires many programs (e.g., assemblers, editors, etc.) not needed in a processing system. This may result in an unnecessary component such as a diskette.

The purpose of this chapter is to describe the initial steps in the construction of a process-oriented microprocessor. These steps are:

- Map the system.
- Configure the system.
- Initialize the program.

Details are described in the following sections [31, 32].

13.1 SYSTEM MAP

Mapping a system is the simple procedure of assigning unique addresses to every byte of RAM, every byte of ROM, and every I/O port.

System mapping becomes important because different microcomputer systems have different ways of optimizing access to RAM and to I/O. The system map is the vehicle for this optimization. The following sections give recommended system mapping.

□ 13.1.1 Motorola System Map

A typical Motorola map is shown in Figure 13.1. The Motorola mapping system gives preference to the problems of addressing memory. The direct addressing mode requires that the first 0 through 255 locations in the map be assigned to RAM. As explained in Chapter 11 this results in a saving of execution time for memory reference instructions.

Motorola techniques require that the highest locations in memory is reserved for initializing the CPU and servicing its interrupts. For this reason ROM is assigned the highest addresses in the system map.

In the Motorola system I/O is treated as low-speed memory. I/O is mapped to address locations in the middle, somewhere between RAM and ROM. RAM and ROM are connected as shown in Figure 10.14. That's really all there is to it.

There are some ways of constructing the system map that eliminate subsequent

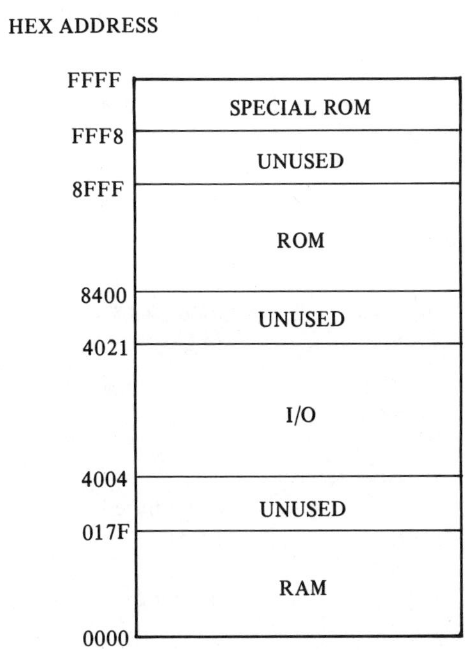

Figure 13.1 Typical Motorola map.

work. Particular attention should be paid to the following

- I/O port address selection.
- Redundancies in mapping.

It is always a problem to know exactly where to locate I/O ports. Considerable effort is saved if exactly the same addresses are allocated to I/O ports system after system. This assumes that the first map is selected with a great deal of care.

The map shown in Figure 13.1 is very sparse. Out of the 65,536 possible address locations, 1024 are used for ROM, 16 are used for I/O, and 384 are used for RAM, for a total of 1424. When dealing with sparse systems such as this, it is not necessary to use all address bus lines. To illustrate this concept, consider the first nibble in an address (that is, A15, A14, A13, and A12). Since the map is sparse, A13 and A12 are not used. The ROM, PIAs, and RAMs are connected as follows:

ROM:	CS	to	A15
PIA:	$\overline{CS2}$	to	A15
	CS1	to	A14
RAM:	$\overline{CS1}$	to	A15
	$\overline{CS2}$	to	A14

To illustrate the effect of not using A13 and A12 consider a four-character hexadecimal address. Assume the last three characters are equal to XXX (some nonchanging value). The following numbers show the wrap-around addressing permitted in sparse maps:

HEX ADDRESS	MAP LOCATION
0XXX	
1XXX	Same byte in
2XXX	RAM
3XXX	
4XXX	
5XXX	Same register
6XXX	in I/O
7XXX	
8XXX	
9XXX	
AXXX	
BXXX	Same byte in
CXXX	ROM
DXXX	
EXXX	
FXXX	

In this way considerable wiring of address bus lines can be avoided. □

■ 13.1.2 Intel System Map

A typical Intel map is shown in Figure 13.2. The Intel system gives preference to the problems of program initialization and I/O. For initialization the first 64 (3F) memory locations are saved for restart functions. However there are 256 (FF) duplicate I/O locations available. I/O of this type is termed isolated. The same address (for example 7F) occurs in both memory and I/O. For improved I/O transfer rates Intel treats memory dedicated instructions and I/O dedicated instructions differently. The difference is that the contents of an isolated I/O can be transferred only to an accumulator, whereas the contents of memory can be transferred using all data transfer instructions.

The I/O can be placed in the memory map. This is referred to as memory mapped I/O. Treating I/O as memory has the advantage of providing a stronger instructional set to handle these ports. It has the disadvantage that memory referenced instructions require more machine cycles than I/O referenced instructions.

Figure 13.3 shows a convenient way to select the RAM, ROM, and memory mapped I/O shown in Figure 13.2, using a DEMUX and two select lines from the address buses A15 and A14. For this sparse map address lines A13, A12, A11, and A10 are simply ignored. As more and more space in the map becomes occupied the address lines, ignored above, must be used as input to the DEMUX select lines.

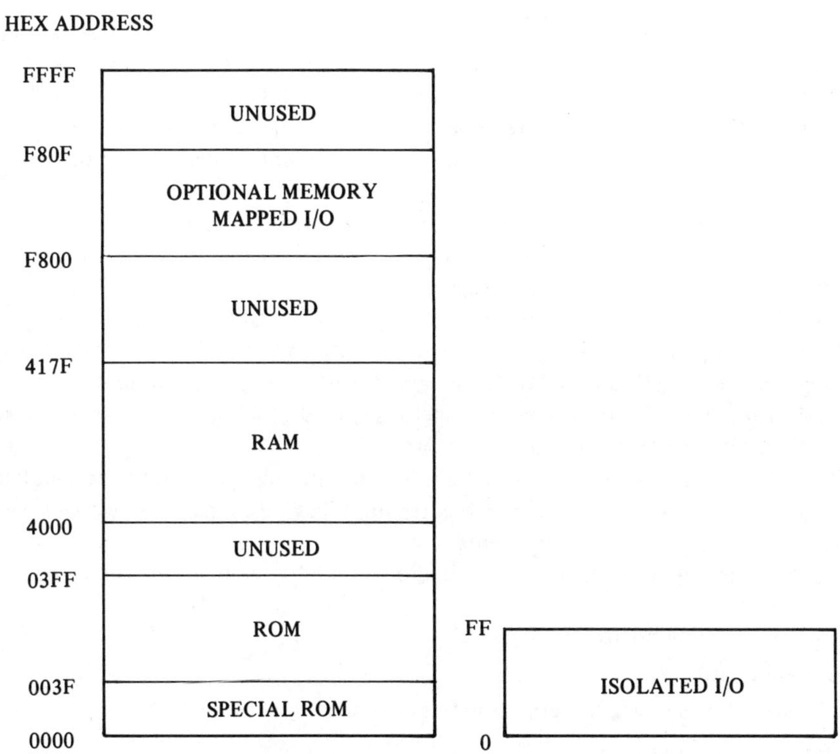

Figure 13.2 Typical Intel map.

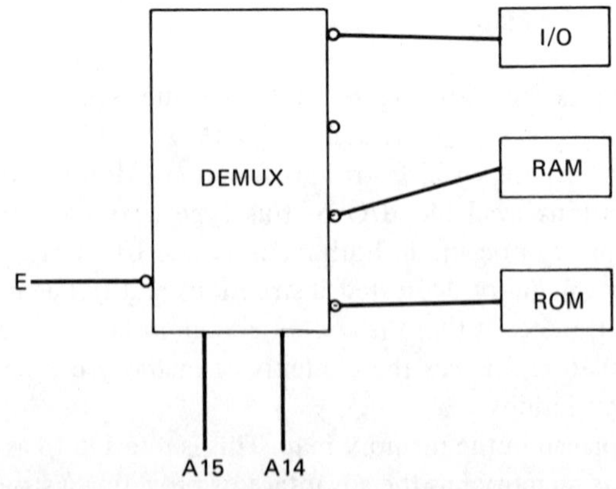

Figure 13.3 Chip select connections for memory maps.

Mapping is an excellent application of the DEMUX circuits given in Section 2.6 of this book. ■

13.2 CONFIGURING MICROCOMPUTER SYSTEMS

Mapping consists of assigning address bits to each chip of the system. Configuring the system consists of laying out boards and connecting the data bus and control bus between the CPU, memory, and I/O. With an address bus, a data bus, a control bus, and a power supply there are approximately 40 lines that must be connected between chips. This means that the bus between the circuit boards in a microcomputer system must have 40 lines.

For systems with a large number of chips it is advisable to use a manufacturer's standard computer board. Boards for systems with fewer chips are conveniently purchased from special vendors or designed by the user.

A small microcomputer system, in at least 99% of the cases, requires only one CPU. ROM chips have high byte densities so ROM requirements are small. Thus, placing the system CPU and ROM on the same board seems almost natural.

RAM requirements can be moderate because small systems seem to suffice on relatively small quantities of temporary storage.

The I/O presents the main problem because conventional I/O modules are limited to two or three ports, and these I/O ports must have external cabling. This can represent a formidable layout problem.

Typical boards required in a compact microcomputer system are the following:

- A CPU and ROM board
- A RAM board
- I/O board for parallel and serial communication
- Power supply board

- Display board
- Operators input board

RAM and I/O boards can be duplicated as the system expands. However, when designing these boards the loading problem should be remembered. The first such board can easily be driven by the CPU. The CPU may not be able to drive two. Thus it is recommended that RAM and I/O boards be buffered to facilitate easy expansion.

Board connections are straightforward and explained in the manufacturers' manuals given in the References.

Very often the system to be built is small. A small system could be described as 1k to 2k of ROM and less than 256 bytes of RAM. It is common for the number of parameters and variables in dedicated processor design to be less than 256. Such systems are unlikely to grow. Chips that contain a CPU and 256 bytes of RAM are available, and they give considerable saving in construction costs and effort. Whenever a well-defined system is to be built, the use of chips with special configurations should be investigated.

13.3 PROGRAM INITIALIZATION

A microcomputer is started when the power switch is turned on. This action forces the PC to be loaded with a specific value. The content of this memory location should be either a jump to start of program or the start of program address if the microcomputer has a built-in automatic jump for start-up. A manually operated reset button should also force load the PC. This feature allows restart without power loss.

This section describes the PC action and gives specific initialization values for the PC.

□ 13.3.1 Motorola Initialization

Figure 13.1 gives the Motorola system map and indicates a special ROM area in memory locations FFF8 to FFFF. This special ROM is used for system start-up. It is also used for interrupt control.

Table 11.17 describes the action of the $\overline{\text{RES}}$, $\overline{\text{NMI}}$, and $\overline{\text{IRQ}}$ pins shown in Figure 10.13. Reset ($\overline{\text{RES}}$, active low) causes PCH ← FFFE, PCL ← FFFF, and a jump instruction to be executed. Place the starting address of your program in FFFE and FFFF and wire $\overline{\text{RES}}$ to the start and restart switches. □

■ 13.3.2 Intel Initialization

Figure 13.2 gives the Intel system map and indicates a special ROM area in memory locations 00 to 3F. This special ROM is used for system start-up. It is also used for interrupt control.

The reset ($\overline{\text{RESET IN}}$, active low) causes the PC to be reset to 00 Hex. Program a jump to the starting address of your program. This programmed jump is placed in address 00.

An interrupt request (INTR) allows a hardware restart (RST) instruction to force load into the CPU when the CPU responds with an $\overline{\text{INTA}}$. A RST instruction has the following code: 11NNN111. Bits 5, 4, and 3 of this instruction are transferred to bits 5, 4, and 3 of the PC. All other PC bits go to 0. Thus excluding 00 (the starting address) there are seven possible restart locations: 08, 10, 18, 20, 28, 30, and 38. Program jumps are required at all used restart locations. A combinational circuit can be designed that has inputs RST (bits 5, 4, and 3) and $\overline{\text{INTA}}$ and for which the output is connected to the data bus. An easier way is to purchase a programmable interrupt controller (PIC), which provides all necessary buffers and so forth. A PIC is programmed using the technique given in Chapter 12. ■

APPENDIX A

Definition of a Boolean Algebra

A Boolean algebra B is a set of elements x, y, z, . . . with the following properties:

- B has two binary operations + and $\cdot$, which satisfy the idempotent, commutative, associative, and distributive laws.
- B has a binary relation denoted by $<$ that is reflexive, antisymmetric, and transitive and satisfies the consistency principle.
- B contains two elements 0 and 1 that are universal bounds and satisfy the intersection and union laws.
- B has a unary operation of complementation $^{-}$, which obeys the complementary, dualization, and involution laws.

It is convenient to read the Boolean operators and relations as follows:

$+$ or
$\cdot$ and
$<$ implies
$-$ not

A symbolic representation of the properties of a Boolean algebra are as follows:

Idempotent	$x \cdot x = x$ and $x + x = x$
Commutative	$x \cdot y = y \cdot x$ and $x + y = y + x$
Associative	$x \cdot (y \cdot z) = (x \cdot y) \cdot z$ and $x + (y + z) = (x + y) + z$
Distributive	$x \cdot (y + z) = (x \cdot y) + (x \cdot z)$ and $x + (y \cdot z) = (x + y) \cdot (x + z)$
Reflexive	$x < x$, for all x
Antisymmetric	$x < y$ and $y < x$, then $x = y$
Transitive	$x < y$ and $y < z$, then $x < z$
Consistency	$x < y$, $x \cdot y = x$, $x + y = y$ are mutually equivalent
Universal bounds	$0 < x < 1$ for all x
Intersection	$0 \cdot x = 0$ and $1 \cdot x = x$

Union	$0 + x = x$ and $1 + x = 1$
Complementary	$x \cdot \bar{x} = 0$ and $x + \bar{x} = 1$
Dualization	$\overline{(x \cdot y)} = \bar{x} + \bar{y}$ and $\overline{(x + y)} = \bar{x} \cdot \bar{y}$
Involution	$\bar{\bar{x}} = x$

Also note that $\cdot$, +, and $<$ are related by a fundamental principle of consistency

$$(y < z) \equiv (y \cdot z = y) \equiv (y + z = z)$$

Signifying the empty set by 0 and the set B with 1, the laws of operation for bounds are

$$0 < x < 1 \qquad \text{for all } x$$
$$0 \cdot x = 0 \qquad 1 \cdot x = x$$
$$0 + x = x \qquad 1 + x = 1$$

0 and 1 are identity elements for these operations. The operations can be compared with addition and multiplication. Subtraction is only meaningful if

$$y < z \quad \text{or} \quad z < y$$

Subtraction is avoided since

$$(y + z) - (y \cdot z)$$

should be written

$$(y \cdot \bar{z}) + (\bar{y} \cdot z)$$

APPENDIX B

Number Systems

Numbering is a fundamental process. It is one of the first concepts taught to a child. It is learned at such an early stage in life that a great number of people assume it to be a physical law. Numbering is not a physical law; it is a concept.

The decimal system, which is on the base 10, has digits from 0 to 9. It is the most popular system. However, there are many other popular number systems.

Base 12 numbers are divisible by 3 or 4. This is why the concept of 12 items to a dozen, 12 troy ounces to a pound, and 12 inches to a foot has been popular throughout history. Despite the fact that there is difficulty dividing 10 by 3 or 4, man seems to prefer the base 10 concept.

Computers are happier with binary numbers. However the binary system is very cumbersome. Thus octal and hexadecimal numbers have become popular with computer-oriented people.

In this appendix an attempt is made to explain the fundamentals behind the numbering systems and to introduce some of the basic numbering systems. Some of these systems are interesting or illustrative, and others are useful in digital circuit design.

B.1 THE DECIMAL SYSTEM

The decimal or base 10 (radix 10) represents a number system in powers of 10. Each power is weighted by an integer between zero and nine, inclusive; for example:

$$329 = 3(10)^2 + 2(10)^1 + 9(10)^0$$

More generally, any integer I is given by

$$\begin{aligned} I = a_p a_{p-1} \cdots a_2 a_1 a_0 = a_p(10)^p + a_{p-1}(10)^{p-1} + \cdots \\ + a_2(10)^2 + a_1(10)^1 \\ + a_0(10)^0 \end{aligned}$$

where $0 \leqslant a_k \leqslant 9$.

The same principle may be extended to fractions by using negative powers of 10. Thus

$$0.485 = 4(10)^{-1} + 8(10)^{-2} + 5(10)^{-3}$$

and any decimal fraction F is given by

$$F = a_{-1}a_{-2} \cdots a_{-q+1}a_{-q} = a_{-1}(10)^{-1} + a_{-2}(10)^{-2} + \cdots + a_{-q+1}(10)^{-q+1} + a_{-q}(10)^{-q}$$

The general decimal number N is thus represented by

$$N = I + F = \sum_{k=-q}^{p} a_k(10)^k; \quad 0 \leqslant a_k \leqslant 9$$

B.2 OTHER RADIX SYSTEMS

For any positive integral radix r, numbers are given by

$$\sum_{k=-q}^{p} a_k r^k; \quad 0 \leqslant a_k \leqslant r - 1$$

The decimal number 329, for example, is given in the system with radix 6 by 1305, where

$$1305 = 1(6)^3 + 3(6)^2 + 0(6)^1 + 5(6)^0$$

The process of conversion is very simple. A decimal integer I is written into radix r by first dividing I by r to obtain a quotient and a remainder. The remainder gives the least significant digit in a. The quotient is again divided by r, and the next most significant digit in a is obtained as the remainder. The process is continued until the final remainder or most significant digit in a appears. The calculation for the example already given is as follows:

```
6)329
  6)54   5
   6)9   0
     1   3
```

In the case where a decimal fraction F is to be converted to a new radix, a process of multiplication is used. The decimal fraction F is multiplied by the radix r. The integer obtained gives the most significant digit in a. The fractional part that results from the multiplication must again be multiplied by the radix r. The integer that results is the next most significant digit in a. The process is repeated until the fractional part becomes zero or until a specific number of digits is reached. For example the decimal fraction 0.46875 becomes 0.132 in the radix 4 system. The calculations are as follows:

```
     46875
         4
     -----
1    87500
         4
     -----
3    500
       4
     ---
2    0
```

Here

$$0.132 = 1(4)^{-1} + 3(4)^{-2} + 2(4)^{-3}$$

B.3 BASIC ARITHMETIC

When carrying out basic arithmetic processes in number systems other than the base 10 system, the digits that are available are 0 to r − 1 inclusive. The digits available in the base 10 system are 0 to 9, and similarly the digits available in the radix 6 system are 0 to 5, so in the radix 3 system there is no digit greater than 2. The simple additional sum

$$\begin{array}{r} 42 \\ \underline{25} \\ 67 \end{array}$$

appears in the radix 3 system as

$$\begin{array}{r} 1120 \\ \underline{221} \\ 2111 \end{array}$$

the multiplication

$$\begin{array}{r} 236 \\ \underline{32} \\ 708 \\ \underline{472} \\ 7552 \end{array}$$

carried out using radix 5 becomes

$$\begin{array}{r} 1421 \\ \underline{112} \\ 1421 \\ 1421 \\ \underline{3342} \\ 220202 \end{array}$$

In arithmetic operations there are r^2 sums and r^2 products of all possible pairs of integers where r is the radix of the number system used; because sums and products are related concepts there are really only $r(r + 1)$ relationships for a system that uses both sums and products. It follows that the smaller the radix system, the easier it is to build a circuit to compute sums and products. Also it is much easier to represent systems with a small radix in an electronic circuit, since each digit in the system requires a different voltage level. Electronically speaking the simplest system of all is the binary system.

B.4 THE BINARY SYSTEM

The binary (modulo 2) system uses only the two digits 0 and 1. The addition and multiplication tables are as follows:

$$0 + 0 = 0$$
$$0 + 1 = 1 + 0 = 1$$
$$1 + 1 = 10$$

and

$$0 \cdot 0 = 0$$
$$0 \cdot 1 = 1 \cdot 0 = 0$$
$$1 \cdot 1 = 1$$

Here is a short table of binary numbers.

1	1
2	10
3	11
4	100
5	101
6	110
7	111
8	1000

This is usually referred to as an 8-4-2-1 code.

B.5 BINARY CODED DECIMAL

Decimal numbers can also be represented by a binary coded decimal (BCD), which avoids calculation in the processes of conversion. In this system four binary digits are used to represent the decimals from 0 to 9 inclusive. Some examples are

23	0010,0011
416	100,0001,0110
3795	11,0111,1001,0101

B.6 NINE'S COMPLEMENT CODE

In some of the early computing systems an attempt was made to use codes such as BCD because of their easy conversion from machine codes to the integer system. However, the use of the BCD code in the process of subtraction is difficult. It is more convenient to use a code where 9 is the complement of 0, 8 is the complement of 1, 7 is the complement of 2, and so on. Negative numbers can be formed simply by complementing and incrementing. Subtraction then reduces to the simple addition process. The following 2 codes are 9's complements:

	5211	2421
0	0000	0000
1	0001	0001
2	0100	0010
3	0110	0011
4	0111	0100
5	1000	1011
6	1001	1100
7	1011	1101
8	1110	1110
9	1111	1111

Note that in the 2421 code the last digit indicates whether the number is odd or even; this is not true in the 5211 code.

An easy code to form from the BCD code is the plus 0011 (excess -3) code:

0	0011
1	0100
2	0101
3	0110
4	0111
5	1000
6	1001
7	1010
8	1011
9	1100

This has the same advantage as other nine's complementation codes. It is easier to compute, and even or odd numbers can be recognized.

B.7 COMPUTER NUMBER SYSTEMS

Since a computer is basically a binary device, it must work in this domain. A byte (8 bits) has the possibility of $2^7 + 2^6 + 2^5 + 2^4 + 2^3 + 2^2 + 2^1 + 2^0$, or 0 to 255 in base 10. A word goes to 2^{15}, or 0 to 65535.

To be able to represent positive and negative numbers, the most significant bit

(MSB) in a byte or word is reserved for a sign bit. In MSB 1 denotes a negative word. The following examples are restricted to bytes, and it is left to the reader to extend the procedure to words.

The number 127 would be represented as 0111 1111.

The number 1 would be represented as 0000 0001.

The sum of + 1 and − 1 is zero. This is achieved in a computer using 2's complement notation. A − 1 is 1111 1111. This means − 127 is represented as 1000 0001.

The reader can move from positive to negative and negative to positive by remembering the following three words:

complement and increment

If unfamiliar, try it on the above examples.

Following are examples of binary addition and subtraction using nibbles (4 bits) and 2's complement notation:

2	0010	2	0010
+5	0101	−5	1011
7	0111	−3	1101

B.8 GRAY CODE

A Gray code is a variation on the binary code. In this code only 1 bit at a time is allowed to change. This code is very convenient in circuit design since it leads to a minimum number of changes. The two codes, binary and Gray, are compared in Table B.1

Table B.1 Binary and Gray Codes

Binary	Gray
000	000
001	001
010	011
011	010
100	110
101	111
110	101
111	100

The reader will note the single bit change in the progression of the Gray code. The distance between code words is given by the number of bits that are different.

B.9 NOTATIONAL CONVENIENCE

Binary notation is very unwieldy in verbal communication, let alone written communication. It is, therefore, convenient to represent numbers in two other notations, that is, octal and hexadecimal. The octal code is repetitive 0 to 7 and the hexadecimal code from 0 to 15. The following table counts from 0 to 20 in decimal, octal, and hexadecimal notation.

Table B.2 Number Conventions

Decimal	Octal	Hexadecimal
0	0	0
1	1	1
2	2	2
3	3	3
4	4	4
5	5	5
6	6	6
7	7	7
8	10	8
9	11	9
10	12	A
11	13	B
12	14	C
13	15	D
14	16	E
15	17	F
16	20	10
17	21	11
18	22	12
19	23	13
20	24	14

The decimal number 255 is 377 in octal and FF in hexadecimal. The reader is asked to work out the hexadecimal for binary 0011 1111 and the octal for binary 1 000 010 000 000.

B.10 INSTRUCTION CODES AND FORMATS

A computer works in binary, and therefore all instructions to it must be coded in binary form. For example, to tell an HP 2100 to add, the instruction code is 010 000 in the six most significant bits of its word; an or function is 011 100, and so forth. The computer takes this code, runs it through a decoder, and energizes the lines telling an ALU to perform the appropriate function. These functions are performed by using the working registers and performing the functions in conjunction with a memory location. The last 10 bits are then available to designate the appropriate memory location. When a machine is set up this way, it is most convenient to use octal numbers, as is done with HP 2100.

Some machines are byte oriented and therefore will use a complete byte, that is 8 bits, for an instruction and 16 bits for the memory location. In such machines it is most common to use hexadecimal notation.

On this basis, the answers to the two problems of Section B.9 are

3F	WAIT	MOTOROLA
102xx	HALT	HP

B.11 ALPHANUMERIC CHARACTERS

Sometimes it is necessary to have the computer represent the characters of the alphabet, the number system, and control characters. A code is used to represent such characters. The most frequently used code to do this is the American Standard Code for Information Interchange (ASCII). The ASCII Code can use 6, 7, or 8 bits. Table B3 gives the characters, numbers, and symbols and their equivalent hexadecimal code.

Table B.3 ASCII Character Code Set (7-Bit Code)

	HEX Character	MS Characters							
		0	1	2	3	4	5	6	7
	0	NUL	DLE	SP	0	@	P	,	p
	1	SOH	DC1	!	1	A	Q	a	q
	2	STX	DC2	"	2	B	R	b	r
	3	ETX	DC3	#	3	C	S	c	s
	4	EOT	DC4	$	4	D	T	d	t
	5	ENQ	NAK	%	5	E	U	e	u
	6	ACK	SYN	&	6	F	V	f	v
LS	7	BEL	ETB	'	7	G	W	g	w
Character	8	BS	CAN	(	8	H	X	h	x
	9	HT	EM	)	9	I	Y	i	y
	A	LF	SUB	*	:	J	Z	j	z
	B	VT	ESC	+	;	K	[	k	{
	C	FF	FS	,	<	L	\	l	:
	D	CR	GS	–	=	M	]	m	}
	E	SO	RS	.	>	N	↑	n	~
	F	SI	VS	/	?	O	↓	o	DEL

B.12 PROBLEMS

PB.1

(a) Convert the following to base 10 integers:

$$(1234)_8, \quad (\text{AD }2)_{16}, \quad (12012)_3$$

(b) Convert to base 8

$$(243.167)_{10}$$

(c) Convert to hexadecimal

$$(1264)_{10}$$

(d) Give the following octal numbers in 2's complement notation:

$$(77), \quad (-26), \quad (35), \quad (-0.125)$$

PB.2 Write the numbers from 0 to 18 in binary and BCD.

PB.3 Convert the following base-10 numbers to base 2:

625, 43.32, 0.51

PB.4 Find the base 8 equivalent of the following base-2 numbers:

10 111 100 101, 1101.101, 1.0111

PB.5 Perform the indicated arithmetic operations after converting the operands to base 2.

$$\begin{array}{r} 695_{10} \\ +272_{10} \\ \hline \end{array} \qquad \begin{array}{r} 695_{10} \\ -272_{10} \\ \hline \end{array} \qquad \begin{array}{r} 272_{10} \\ \times\ 23_{10} \\ \hline \end{array} \qquad \frac{272_{10}}{23_{10}}$$

PB.6 Repeat PB.5 for the following:

$$\begin{array}{r} 97_{10} \\ +83_{10} \\ \hline \end{array} \qquad \begin{array}{r} 131_{10} \\ -219_{10} \\ \hline \end{array} \qquad \begin{array}{r} 97_{10} \\ \times 43_{10} \\ \hline \end{array} \qquad \frac{586_{10}}{39_{10}}$$

PB.7 Carry out the following computations after first converting each negative operand to 2's complement notation. Check your results.

$$\begin{array}{r} .11101 \\ -.10111 \\ \hline \end{array} \qquad \begin{array}{r} .10001 \\ -.11011 \\ \hline \end{array} \qquad \begin{array}{r} -.10101 \\ -.11010 \\ \hline \end{array}$$

PB.8 Convert each of the given numbers directly to binary digits of the indicated number of bits.

(a) 377_8 (8)
(b) 125_8 (8)
(c) 144007_8 (16)

(d) 79_{16} (8)
(e) $1AF_{16}$ (12)
(f) $CDB6_{16}$ (16)

PB.9 Perform the indicated subtractions in binary. Convert the answer to decimal and check.
(a) 17 - 3
(b) 14.625 - 4.0624

PB.10 Find the distance between code words of each of the following pairs:

001	1010	100110
011	1110	101101

PB.11
(a) Perform the addition suggested below by replacing each digit by its BCD code word, adding in binary, correcting as necessary, and checking results.

$$732 + 895 = 1627$$

(b) Convert 732 to binary. Compare with the BCD encoding 732.
(c) Repeat part (a) using the excess-3 encoding. Your answer should be in the excess-3 code.

PB.12 Find a set of rules for generating each Gray code from the corresponding BCD code (0 to 9).

PB.13 Find the base-8 equivalent of the following base-2 numbers:

10111100101 1101.101 1.0111

PB.14 Perform the indicated arithmetic operations in base 5.

$$\begin{array}{r} 143_5 \\ +221_5 \\ \hline \end{array} \qquad \begin{array}{r} 124.22_5 \\ +\ 21.34_5 \\ \hline \end{array} \qquad \begin{array}{r} 413_5 \\ \times\ 32_5 \\ \hline \end{array} \qquad \begin{array}{r} 404_5 \\ -213_5 \\ \hline \end{array}$$

APPENDIX **C**

Motorola's Executable Instructions

C.1 SYMBOLS

The following nomenclature is used in the Motorola Assembler and for operator descriptions [32]:

1. *Operators*

←	becomes
↔	transfer
·	and
+	or
⊕	exclusive or
-	not
#	prefix means immediate addressing mode
$	prefix means hexadecimal

2. *Registers*

A	Accumulator A
B	Accumulator B
CC	Condition code register
X	Index register
PC	Pointer register
SP	Stack pointer
L	Suffixes XL, PCL, and SPL mean lower byte of registers X, PC, and SP respectively
H	Suffixes XH, PCH, and SPH mean higher byte of registers X, PC, and SP respectively

3. *Memory*

M	Contents of memory location M
M( )	Contents of memory location pointed to by register X
#M	The value M
M + 1	Means increment the contents of M by 1
(M + 1)	Means the contents of M + 1

4. *Condition codes*

H	Half carry b5
I	Interrupt mask b4
N	Negative indicator b3
Z	Zero indicator b2
V	Overflow indicator b1
C	Carry–borrow b0

5. *Status*

T	Load ALU result into this bit. However not all Motorola instructions operate in the same way for each test.

In Table C.1 the increment statement V bit test is for the conditions 01111111 prior to execution; the decrement statement V bit test is for the condition 10000000 prior to execution. The rotate statement and the shift statement test for the condition

$$\text{Bit V} = \text{N} \oplus \text{C}$$

In Table C.2 the 2's complement statement V bit test is for the condition 100000000, and the C bit test is for the condition ≠00000000. The decimal adjust add statement C bit test is for the condition of BCD character greater than nine.

In Table C.4 the compare index register N bit test is for the condition MSB = 1, and the V bit test is for the overflow condition on subtraction. The load and store N bit test is MSB (of the word, b15) = 1.

6. *Indexing*

n, X index register offset by n (positive only)

C.2 INSTRUCTION SET

Tables C.1 to C.8 summarize Motorola's instruction set. Table C.9 is intended as a quick reference guide.

Table C.1 Data Transfer Instructions

Operation	Mnemonic	Description	Addressing Mode					Codes	Cycles	Bytes	Status					
			Immed	Direct	Index	Extend	Inher				5 H	4 I	3 N	2 Z	1 V	0 C
CLEAR	CLR	M(X) ← #00			√			6F	7	2						
		M ← #00				√		7F	6	3		0	0	1	0	0
	CLRA	A ← #00					√	4F	2	1						
	CLRB	B ← #00						5F	2	1						
DECREMENT	DEC	M(X) ← M(X) − 1			√			6A	7	2						
		M ← M − 1				√		7A	6	3			T	T	T	
	DECA	A ← A − 1					√	4A	2	1						
	DECB	B ← B − 1					√	5A	2	1						
INCREMENT	INC	M(X) ← M(X) + 1			√			6A	7	2						
		M ← M − 1				√		7A	6	3						
	INCA	A ← A + 1					√	4A	2	1			T	T	T	
	INCB	B ← B + 1					√	5A	2	1						
LOAD ACCUMULATOR	LDAA	A ← #M	√					86	2	2						
		A ← M		√				96	3	2						
		A ← M(X)			√			A6	5	2						
		A ← M				√		B6	4	3			T	T	0	
	LDAB	B ← #M	√					C6	2	2						
		B ← M		√				D6	3	2						
		B ← M(X)			√			E6	5	2						
		B ← M				√		F6	4	3						
STORE ACCUMULATOR	STAA	M ← A		√				97	4	2						
		M(X) ← A			√			A7	6	2						
		M ← A				√		B7	5	3			T	T	T	T

Table C.1 (*continued*)

Operation	Mnemonic	Description	Addressing Mode: Immed	Direct	Index	Extend	Inher	Codes	Cycles	Bytes	Status: 5 H	4 I	3 N	2 Z	1 V	0 C
	STAB	M ← B		√				D7	4	2						
		M(X) ← B			√			E7	6	2						
		M ← B				√		F7	5	3						
TRANSFER	TAB	B ← A					√	16	2	1			T	T	0	
ACCUMULATOR	TBA	A ← B					√	17	2	1						
ROTATE	ROL	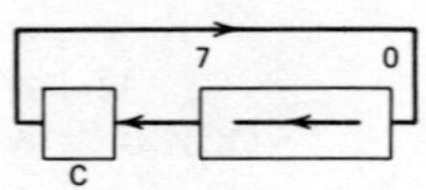			√			69	7	2						
LEFT						√		79	6	3			T	T	T	b7
	ROLA						√	49	2	1						
	ROLB						√	59	2	1						
ROTATE	ROR	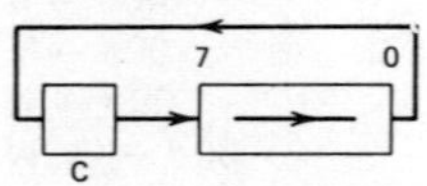			√			66	7	2						
RIGHT						√		76	6	3			T	T	T	b0
	RORA						√	46	2	1						
	RORB						√	56	2	1						
ARITHMETIC	ASL	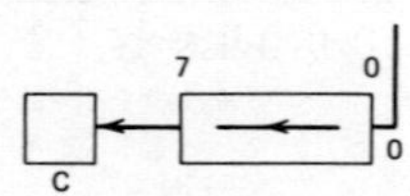			√			68	7	2						
SHIFT						√		78	6	3			T	T	T	b7
LEFT	ASLA						√	48	2	1						
	ASLB						√	58	2	1						
ARITHMETIC	ASR	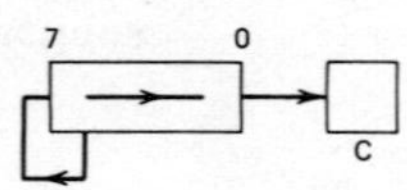			√			67	7	2						
SHIFT						√		77	6	3			T	T	T	b0
RIGHT	ASRA						√	47	2	1						
	ASRB						√	57	2	1						

LOGICAL	LSR	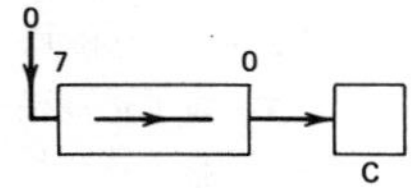	√			64	7	2	
SHIFT				√		74	6	3	0 T T b0
RIGHT	LSRA				√	44	2	1	
	LSRB				√	54	2	1	
PUSH DATA	PSHA	M(SP) ← A							
ONTO STACK		SP ← SP − 1			√	36	4	1	
	PSHB	M(SP) ← B							
		SP ← SP − 1			√	37	4	1	
PULL DATA	PULA	SP ← SP + 1							
		A ← M(SP)			√	32	4	1	
	PULB	SP ← SP + 1							
		B ← M(SP)			√	33	4	1	

Table C.2 Arithmetic Instructions

Operation	Mnemonic	Description	Addressing Mode					Code	Cycles	Bytes	Status					
			Immed	Direct	Index	Extend	Inher				5 H	4 I	3 N	2 Z	1 V	0 C
ADD	ADDA	A ← #M + A	✓					8B	2	2						
		A ← M + A		✓				9B	3	2						
		A ← M(X) + A			✓			AB	5	2						
		A ← M + A				✓		BB	4	3	T		T	T	T	T
	ADDB	B ← #M + B	✓					CB	2	2						
		B ← M + B		✓				DB	3	2						
		B ← M(X) + B			✓			EB	5	2						
		B ← M + B				✓		FB	4	3						
	ABA	A ← A + B					✓	1B	2	1						
ADD WITH CARRY	ADCA	A ← A + #M + C	✓					89	2	2						
		A ← A + M + C		✓				99	3	2						
		A ← A + M(X) + C			✓			A9	5	2						
		A ← A + M + C				✓		B9	4	3	T		T	T	T	T
	ADCB	B ← B + #M + C	✓					C9	2	2						
		B ← B + M + C		✓				D9	3	2						
		B ← B + M(X) + C			✓			E9	5	2						
		B ← B + M + C				✓		F9	4	3						
2's	NEG	M ← 00 − M(X)			✓			60	7	2						

COMPLEMENT		M ← 00 − M				√		70	6	3	T	T	T	T
	NEGA	A ← 00 − A					√	40	2	1				
	NEGB	B ← 00 − B					√	50	2	1				
DECIMAL ADJUST ADD	DAA	[a]					√	19	2	1	T	T	T	T
SUBTRACT	SUBA	A ← A − #M	√					80	2	2				
		A ← A − M		√				90	3	2				
		A ← A − M(X)			√			A0	5	2				
		A ← A − M				√		B0	4	3	T	T	T	T
	SUBB	B ← B − #M	√					C0	2	2				
		B ← B − M		√				D0	3	2				
		B ← B − M(X)			√			E0	5	2				
		B ← B − M				√		F0	4	3				
	SBA	A ← A − B					√	10	2	1				
SUBTRACT WITH CARRY	SBCA	A ← A − #M − C	√					82	2	2				
		A ← A − M − C		√				92	3	2				
		A ← A − M(X) − C			√			A2	5	2	T	T	T	T
		A ← A − M − C				√		B2	4	2				
	SBCB	B ← B − #M − C	√					C2	2	2				
		B ← B − M − C		√				D2	3	2				
		B ← B − M(X) − C			√			E2	5	2				
		B ← B − M − C				√		F2	4	3				

[a]Converts to a BCD format after addition. It regards each byte as two 4-bit BCD numbers.

Table C.3 Logic Operator and Test Data Instructions

Operation	Mnemonic	Description	Addressing Mode					Code	Cycles	Bytes	Status					
			Immed	Direct	Index	Extend	Inher				5 H	4 I	3 N	2 Z	1 V	0 C
AND	ANDA	$A \leftarrow \#M \cdot A$	√					84	2	2						
		$A \leftarrow M \cdot A$		√				94	3	2						
		$A \leftarrow M(X) \cdot A$			√			A4	5	2						
		$A \leftarrow M \cdot A$				√		B4	4	3			T	T	0	
	ANDB	$B \leftarrow \#M \cdot B$	√					C4	2	2						
		$B \leftarrow M \cdot B$		√				D4	3	2						
		$B \leftarrow M(X) \cdot B$			√			E4	5	2						
		$B \leftarrow M \cdot B$				√		F4	4	3						
1's	COM	$M(X) \leftarrow \overline{M}(X)$			√			63	7	2						
COMPLEMENT		$M \leftarrow \overline{M}$				√		73	6	3			T	T	0	1
	COMA	$A \leftarrow \overline{A}$					√	43	2	1						
	COMB	$B \leftarrow \overline{B}$					√	53	2	1						
EXCLUSIVE	EORA	$A \leftarrow \#M \oplus A$	√					88	2	2						
OR		$A \leftarrow M \oplus A$		√				98	3	2						
		$A \leftarrow M(X) \oplus A$			√			A8	5	2						
		$A \leftarrow M \oplus A$				√		B8	4	3						
	EORB	$B \leftarrow \#M \oplus B$	√					C8	2	2			T	T	0	
		$B \leftarrow M \oplus B$		√				D8	3	2						
		$B \leftarrow M(X) \oplus B$			√			E8	5	2						
		$B \leftarrow M \oplus B$				√		F8	4	3						
INCLUSIVE	ORAA	$A \leftarrow \#M + A$	√					8A	2	2						
OR		$A \leftarrow M + A$		√				9A	3	2						
		$A \leftarrow M(X) + A$			√			AA	5	2						
		$A \leftarrow M + A$				√		BA	4	3						

	ORAB	B ← #M + B	√					CA	2	2	T	T	0	
		B ← M + B		√				DA	3	2				
		B ← M(X) + B			√			EA	5	2				
		B ← M + B				√		FA	4	3				
BIT TEST	BITA	#M · A	√					85	2	2				
		M · A		√				95	3	2				
		M(X) · A			√			A5	5	2				
		M · A				√		B5	4	3				
	BITB	#M · B	√					C5	2	2	T	T	0	
		M · B		√				D5	3	2				
		M(X) · B			√			E5	5	2				
		M · B				√		F5	4	3				
COMPARE	CMPA	A – #M	√					81	2	2				
		A – M		√				91	3	2				
		A – M(X)			√			A1	5	2				
		A – M				√		B1	4	3	T	T	T	T
	CMPB	B – #M	√					C1	2	2				
		B – M		√				D1	3	2				
		B – M(X)			√			E1	5	2				
		B – M				√		F1	4	3				
COMPARE ACCUMULATORS	CBA	A – B					√	11	2	1	T	T	T	T
TEST ZERO OR MINUS	TST	M(X) – 00			√			6D	7	2				
		M – 00				√		7D	6	3				
	TSTA	A – 00					√	4D	2	1	T	T	0	0
	TSTB	B – 00					√	5D	2	1				

Table C.4 Index Register and Stack Instructions

			Addressing Mode								Status					
Operation	Mnemonic	Description	Immed	Direct	Index	Extend	Inher	Code	Cycles	Bytes	5 H	4 I	3 N	2 Z	1 V	0 C
COMPARE	CPX	XH − #M	√					8C	3	3						
INDEX		XL − #M + 1														
REGISTER		XH − M		√				9C	4	2						
		XL − (M + 1)											T	T	T	
		XH − M(X)			√			AC	6	2						
		XL − M(X + 1)														
		XH − M				√		BC	5	3						
		XL − (M + 1)														
DECREMENT INDEX REGISTER	DEX	X ← X − 1					√	09	4	1				T		
DECREMENT STACK POINTER	DES	SP ← SP − 1					√	34	4	1						
INCREMENT INDEX REGISTER	INX	X ← X + 1					√	08	4	1				T		
INCREMENT STACK POINTER	INS	SP ← SP + 1					√	31	4	1						
LOAD	LDX	XH ← #M	√					CE	3	3						
INDEX		XL ← #M + 1														
REGISTER		XH ← M		√				DE	4	2						
		XL ← (M + 1)											T	T	T	

		XH ← M(X) XL ← M(X + 1)			√			EE	6	2			
		XH ← M XL ← (M + 1)				√		FE	5	3			
LOAD STACK POINTER	LDS	SPH ← #M SPL ← #M + 1	√					8E	3	3			
		SPH ← M SPL ← (M + 1)		√				9E	4	2			
		SPH ← M(X) SPL ← M(X + 1)			√			AE	6	2	T	T	T
		SPH ← M SPL ← (M + 1)				√		BE	5	3			
STORE INDEX REGISTER	STX	M ← XH (M + 1) ← XL		√				DF	5	2			
		M(X) ← XH M(X + 1) ← XL			√			EF	7	2	T	T	T
		M ← XH (M + 1) ← XL				√		FF	6	3			
STORE	STS	M ← SPH (M + 1) ← SPL		√				9F	5	2			
		M(X) ← SPH M(X + 1) ← SPL			√			AF	7	2	T	T	T
		M ← SPH (M + 1) ← SPL				√		BF	6	3			
TRANSFER FROM INDEX TO STACK	TXS	SP ← X – 1					√	35	4	1			
TRANSFER FROM STACK TO INDEX	TSX	X ← SP + 1					√	30	4	1			

Table C.5 Jump and Branch Instructions

Operation	Mnemonic	Test	Description	Addressing Mode			Codes	Cycles	Bytes
				Rel	Index	Extend			
BRANCH ALWAYS	BRA	1	$PC \leftarrow PC + 2 + REL$	✓			20	4	2
BRANCH IF CARRY CLEAR	BCC	$C = 0$	$PC \leftarrow PC + 2 + REL$	✓			24	4	2
BRANCH IF CARRY SET	BCS	$C = 1$	$PC \leftarrow PC + 2 + REL$	✓			25	4	2
BRANCH IF ZERO SET	BEQ	$Z = 1$	$PC \leftarrow PC + 2 + REL$	✓			27	4	2
BRANCH IF ⩾ ZERO	BGE	$N \oplus V = 0$	$PC \leftarrow PC + 2 + REL$	✓			2C	4	2
BRANCH IF POSITIVE	BGT	$Z + (N \oplus V) = 0$	$PC \leftarrow PC + 2 + REL$	✓			2E	4	2
BRANCH IF HIGHER	BHI	$C \cdot Z = 0$	$PC \leftarrow PC + 2 + REL$	✓			22	4	2
BRANCH IF ⩽ ZERO	BLE	$Z + (N \oplus V) = 1$	$PC \leftarrow PC + 2 + REL$	✓			27	4	2
BRANCH IF LOWER OR SAME	BLS	$C + Z = 1$	$PC \leftarrow PC + 2 + REL$	✓			23	4	2
BRANCH IF < ZERO	BLT	$N \oplus V = 1$	$PC \leftarrow PC + 2 + REL$	✓			20	4	2

BRANCH IF MINUS	BMI	N = 1	PC ← PC + 2 + REL	√			28	4	2
BRANCH IF NOT EQUAL ZERO	BNE	Z = 0	PC ← PC + 2 + REL	√			26	4	2
BRANCH IF OVERFLOW CLEAR	BVC	V = 0	PC ← PC + 2 + REL	√			28	4	2
BRANCH IF OVERFLOW SET	BVS	V = 1	PC ← PC + 2 + REL	√			29	4	2
BRANCH IF PLUS	BPL	N = 0	PC ← PC + 2 + REL	√			2A	4	2
BRANCH TO SUBROUTINE	BSR	1	M(SP) ← PCL + 2 M(SP − 1) ← PCH + 2 SP ← SP − 2 PC ← PC + 2 + REL		√		80	8	2
JUMP	JMP	1	PC ← X		√		6E	4	2
			PCH ← M(PC + 1) PCL ← M(PC + 2)			√	7E	3	3
JUMP TO SUBROUTINE	JSR	1	M(SP) ← PCL + 2 M(SP − 1) ← PCH + 2 PC ← X		√		AD	8	2
			M(SP) ← PCL + 3 M(SP − 1) ← PCH + 3 PCH ← M(PC + 1) PCL ← M(PC + 2)			√	BD	9	3

Table C.6 Hardware Interrupts

Operation	Mnemonic	Description	5 H	4 I	3 N	2 Z	1 V	0 C
			Status					
RESET	$\overline{RES}$	PCH ← FFFE						
(ACTIVE LOW)		PCL ← FFFF		1				
NON-MASKABLE	$\overline{NMI}$	M(SP) ← PCL						
INTERRUPT		M(SP – 1) ← PCH						
(ACTIVE LOW)		M(SP – 2) ← XL						
		M(SP – 3) ← XH						
		M(SP – 4) ← A		1				
		M(SP – 5) ← B						
		M(SP – 6) ← CC						
		SP ← SP – 7						
		PCH ← FFFC						
		PCL ← FFFD						
INTERRUPT	$\overline{IRQ}$	M(SP) ← PCL						
REQUEST		M(SP – 1) ← PCH						
(ACTIVE LOW)		M(SP – 2) ← XL						
		M(SP – 3) ← XH		*a*				
		M(SP – 4) ← A						
		M(SP – 5) ← B						
		M(SP – 6) ← CC						
		SP ← SP – 7						
		PCH ← FFF8						
		PCL ← FFF9						

[a] If set, IRQ is ignored; if not set, IRQ is executed and status bit 4 is set.

Table C.7 Interrupt Instructions

Operation	Mnemonic	Description	Inherent Add Mode	Code	Cycles	Bytes	Status 5 H	4 I	3 N	2 Z	1 V	0 C
NO OPERATION	NOP	PC ← PC + 1	✓	01	2	1						
RETURN FROM INTERRUPT	RTI	CC ← M(SP + 1) B ← M(SP + 2) A ← M(SP + 3) XH ← M(SP + 4) XL ← M(SP + 5) PCH ← M(SP + 6) PCL ← M(SP + 7) SP ← SP + 7	✓	3B	10	1						
RETURN FROM SUBROUTINE	RTS	PCH ← M(SP + 1) PCL ← M(SP + 2)	✓	39	5	1						
SOFTWARE INTERRUPT	SWI	PC ← PC + 1 M(SP) ← PCL M(SP – 1) ← PCH M(SP – 2) ← XL M(SP – 3) ← XH M(SP – 4) ← A M(SP – 5) ← B M(SP – 6) ← CC SP ← SP – 7 PCH ← FFFA PCL ← FFFB	✓	3F	12	1						
WAIT FOR INTERRUPT	WAI	PC ← PC + 1 M(SP) ← PCL M(SP – 1) ← PCH M(SP – 2) ← XL M(SP – 3) ← XH M(SP – 4) ← A M(SP – 5) ← B M(SP – 6) ← CC WHEN INTERRUPT ARRIVES STATUS BIT 4 IS SET PCH ← FFF8 PCL ← FFF9		3E	9	1		T				

Table C.8 Control Instructions

Operation	Mnemonic	Description	Inherent Add Mode	Code	Cycles	Bytes	Status					
							5 H	4 I	3 N	2 Z	1 V	0 C
CLEAR CARRY	CLC	$C \leftarrow 0$	√	0C	2	1						0
CLEAR INTERRUPT MASK	CLI	$I \leftarrow 0$	√	0E	2	1		0				
CLEAR OVERFLOW	CLV	$V \leftarrow 0$	√	0A	2	1					0	
SET CARRY	SEC	$C \leftarrow 1$	√	0D	2	1						1
SET INTERRUPT MASK	SEI	$I \leftarrow 1$	√	0F	2	1		1				
SET OVERFLOW	SEV	$V \leftarrow 1$	√	0B	2	1					1	
LOAD CC WITH ACCA	TAP	$CC \leftarrow A$	√	06	2	1	T		T	T	T	T
LOAD ACCA WITH CC	TPA	$A \leftarrow CC$	√	07	2	1						

Table C.9 Quick Reference Guide Data Transfer Instructions

Operation	Mnemonic	Description	Addressing Mode: Immed	Direct	Index	Extend	Inher	Codes	Cycles	Bytes	Status: 5 H	4 I	3 N	2 Z	1 V	0 C
CLEAR	CLR	M(X) ← #00			✓			6F	7	2						
		M ← #00				✓		7F	6	3		0	0	1	0	0
	CLRA	A ← #00					✓	4F	2	1						
	CLRB	B ← #00						5F	2	1						
DECREMENT	DEC	M(X) ← M(X) − 1			✓			6A	7	2						
		M ← M − 1				✓		7A	6	3			T	T	T	
	DECA	A ← A − 1					✓	4A	2	1						
	DECB	B ← B − 1					✓	5A	2	1						
INCREMENT	INC	M(X) ← M(X) + 1			✓			6A	7	2						
		M ← M − 1				✓		7A	6	3						
	INCA	A ← A + 1					✓	4A	2	1			T	T	T	
	INCB	B ← B + 1					✓	5A	2	1						
LOAD	LDAA	A ← #M	✓					86	2	2						
ACCUMULATOR		A ← M		✓				96	3	2						
		A ← M(X)			✓			A6	5	2						
		A ← M				✓		B6	4	3			T	T	0	
	LDAB	B ← #M	✓					C6	2	2						
		B ← M		✓				D6	3	2						
		B ← M(X)			✓			E6	5	2						
		B ← M				✓		F6	4	3						
STORE	STAA	M ← A		✓				97	4	2						
ACCUMULATOR		M(X) ← A			✓			A7	6	2						
		M ← A				✓		B7	5	3			T	T	T	T
	STAB	M ← B		✓				D7	4	2						
		M(X) ← B			✓			E7	6	2						
		M ← B				✓		F7	5	3						
TRANSFER	TAB	B ← A					✓	16	2	1			T	T	0	
ACCUMULATOR	TBA	A ← B					✓	17	2	1						
ROTATE	ROL	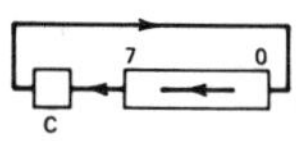			✓			69	7	2						
LEFT						✓		79	6	3			T	T	T	b7
	ROLA						✓	49	2	1						
	ROLB						✓	59	2	1						
ROTATE	ROR	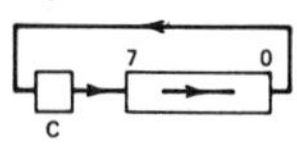			✓			66	7	2						
RIGHT						✓		76	6	3			T	T	T	b0
	RORA						✓	46	2	1						
	RORB						✓	56	2	1						
ARITHMETIC	ASL	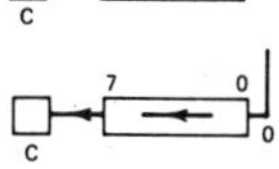			✓			68	7	2						
SHIFT						✓		78	6	3			T	T	T	b7
LEFT	ASLA						✓	48	2	1						
	ASLB						✓	58	2	1						
ARITHMETIC	ASR	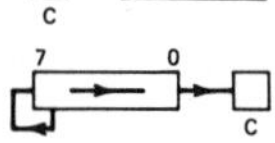			✓			67	7	2						
SHIFT						✓		77	6	3			T	T	T	b0
RIGHT	ASRA						✓	47	2	1						
	ASRB						✓	57	2	1						
LOGICAL	LSR	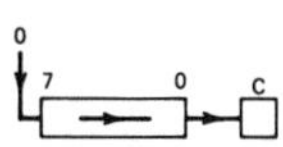			✓			64	7	2						
SHIFT						✓		74	6	3			0	T	T	b0
RIGHT	LSRA						✓	44	2	1						
	LSRB						✓	54	2	1						
PUSH DATA	PSHA	M(SP) ← A														
ONTO STACK		SP ← SP − 1					✓	36	4	1						
	PSHB	M(SP) ← B														
		SP ← SP − 1					✓	37	4	1						
PULL DATA	PULA	SP ← SP + 1														
		A ← M(SP)					✓	32	4	1						
	PULB	SP ← SP + 1														
		B ← M(SP)					✓	33	4	1						

Logic Operator Instructions

Operation	Mnemonic	Description	Addressing Mode: Immed	Direct	Index	Extend	Inher	Code	Cycles	Bytes	Status: 5 H	4 I	3 N	2 Z	1 V	0 C
AND	ANDA	A ← #M · A	✓					84	2	2						
		A ← M · A		✓				94	3	2						
		A ← M(X) · A			✓			A4	5	2						
		A ← M · A				✓		B4	4	3						
	ANDB	B ← #M · B	✓					C4	2	2			T	T	0	
		B ← M · B		✓				D4	3	2						
		B ← M(X) · B			✓			E4	5	2						
		B ← M · B				✓		F4	4	3						
1's	COM	$M(X) \leftarrow \overline{M}(X)$			✓			63	7	2						
COMPLEMENT		$M \leftarrow \overline{M}$				✓		73	6	3						
	COMA	$A \leftarrow \overline{A}$					✓	43	2	1			T	T	0	1
	COMB	$B \leftarrow \overline{B}$					✓	53	2	1						
EXCLUSIVE	EORA	A ← #M ⊕ A	✓					88	2	2						
OR		A ← M ⊕ A		✓				98	3	2						
		A ← M(X) ⊕ A			✓			A8	5	2						
		A ← M ⊕ A				✓		B8	4	3			T	T	0	
	EORB	B ← #M ⊕ B	✓					C8	2	2						
		B ← M ⊕ B		✓				D8	3	2						
		B ← M(X) ⊕ B			✓			E8	5	2						
		B ← M ⊕ B				✓		F8	4	3						
INCLUSIVE	ORAA	A ← #M + A	✓					8A	2	2						
OR		A ← M + A		✓				9A	3	2						
		A ← M(X) + A			✓			AA	5	2						
		A ← M + A				✓		BA	4	3						
	ORAB	B ← #M + B	✓					CA	2	2			T	T	0	
		B ← M + B		✓				DA	3	2						
		B ← M(X) + B			✓			EA	5	2						
		B ← M + B				✓		FA	4	3						

Arithmetic Instructions

Operation	Mnemonic	Description	Addressing Mode: Immed	Direct	Index	Extend	Inher	Code	Cycle	Bytes	Status: 5 H	4 I	3 N	2 Z	1 V	0 C
ADD	ADDA	A ← #M + A	✓					8B	2	2						
		A ← M + A		✓				9B	3	2						
		A ← M(X) + A			✓			AB	5	2						
		A ← M + A				✓		BB	4	3	T		T	T	T	T
	ADDB	B ← #M + B	✓					CB	2	2						
		B ← M + B		✓				DB	3	2						
		B ← M(X) + B			✓			EB	5	2						
		B ← M + B				✓		FB	4	3						
	ABA	A ← A + B					✓	1B	2	1						
ADD WITH CARRY	ADCA	A ← A + #M + C	✓					89	2	2						
		A ← A + M + C		✓				99	3	2						
		A ← A + M(X) + C			✓			A9	5	2						
		A ← A + M + C				✓		B9	4	3	T		T	T	T	T
	ADCB	B ← B + #M + C	✓					C9	2	2						
		B ← B + M + C		✓				D9	3	2						
		B ← B + M(X) + C			✓			E9	5	2						
		B ← B + M + C				✓		F9	4	3						
2's COMPLEMENT	NEG	M ← 00 - M(X)			✓			60	7	2						
		M ← 00 - M				✓		70	6	3			T	T	T	T
	NEGA	A ← 00 - A					✓	40	2	1						
	NEGB	B ← 00 - B					✓	50	2	1						
DECIMAL ADJUST ADD	DAA	[a]					✓	19	2	1			T	T	T	T
SUBTRACT	SUBA	A ← A - #M	✓					80	2	2						
		A ← A - M		✓				90	3	2						
		A ← A - M(X)			✓			A0	5	2						
		A ← A - M				✓		B0	4	3						
	SUBB	B ← B - #M	✓					C0	2	2			T	T	T	T
		B ← B - M		✓				D0	3	2						
		B ← B - M(X)			✓			E0	5	2						
		B ← B - M				✓		F0	4	3						
	SBA	A ← A - B					✓	10	2	1						
SUBTRACT WITH CARRY	SBCA	A ← A - #M - C	✓					82	2	2						
		A ← A - M - C		✓				92	3	2						
		A ← A - M(X) - C			✓			A2	5	2						
		A ← A - M - C				✓		B2	4	2			T	T	T	T
	SBCB	B ← B - #M - C	✓					C2	2	2						
		B ← B - M - C		✓				D2	3	2						
		B ← B - M(X) - C			✓			E2	5	2						
		B ← B - M - C				✓		F2	4	3						

[a] Converts to a BCD format after addition. It regards each byte as two 4-bit BCD numbers.

Jump and Branch Instructions

Operation	Mnemonic	Test	Description	Addressing Mode: Rel	Index	Extend	Codes	Cycles	Bytes
BRANCH ALWAYS	BRA	1	PC ← PC + 2 + REL	✓			20	4	2
BRANCH IF CARRY CLEAR	BCC	C = 0	PC ← PC + 2 + REL	✓			24	4	2
BRANCH IF CARRY SET	BCS	C = 1	PC ← PC + 2 + REL	✓			25	4	2
BRANCH IF ZERO SET	BEQ	Z = 1	PC ← PC + 2 + REL	✓			27	4	2
BRANCH IF ≥ ZERO	BGE	N ⊕ V = 0	PC ← PC + 2 + REL	✓			2C	4	2
BRANCH IF POSITIVE	BGT	Z + (N ⊕ V) = 0	PC ← PC + 2 + REL	✓			2E	4	2
BRANCH IF HIGHER	BHI	C · Z = 0	PC ← PC + 2 + REL	✓			22	4	2
BRANCH IF ≤ ZERO	BLE	Z + (N ⊕ V) = 1	PC ← PC + 2 + REL	✓			27	4	2
BRANCH IF LOWER OR SAME	BLS	C + Z = 1	PC ← PC + 2 + REL	✓			23	4	2
BRANCH IF < ZERO	BLT	N ⊕ V = 1	PC ← PC + 2 + REL	✓			20	4	2
BRANCH IF MINUS	BMI	N = 1	PC ← PC + 2 + REL	✓			28	4	2
BRANCH IF NOT EQUAL ZERO	BNE	Z = 0	PC ← PC + 2 + REL	✓			26	4	2
BRANCH IF OVERFLOW CLEAR	BVC	V = 0	PC ← PC + 2 + REL	✓			28	4	2
BRANCH IF OVERFLOW SET	BVS	V = 1	PC ← PC + 2 + REL	✓			29	4	2
BRANCH IF PLUS	BPL	N = 0	PC ← PC + 2 + REL	✓			2A	4	2
BRANCH TO SUBROUTINE	BSR	1	M(SP) ← PCL + 2 M(SP - 1) ← PCH + 2 SP ← SP - 2 PC ← PC + 2 + REL		✓		80	8	2
JUMP	JMP	1	PC ← X		✓		6E	4	2
			PCH ← M(PC + 1) PCL ← M(PC + 2)			✓	7E	3	3
JUMP TO SUBROUTINE	JSR	1	M(SP) ← PCL + 2 M(SP - 1) ← PCH + 2 PC ← X		✓		AD	8	2
			M(SP) ← PCL + 3 M(SP - 1) ← PCH + 3 PCH ← M(PC + 1) PCL ← M(PC + 2)			✓	BD	9	3

Index Register and Stack Instructions

Operation	Mnemonic	Description	Immed	Direct	Index	Extend	Inher	Code	Cycles	Bytes	5 H	4 I	3 N	2 Z	1 V	0 C
COMPARE INDEX REGISTER	CPX	XH − #M XL − #M + 1	√					8C	3	3			T	T	T	
		XH − M XL − (M + 1)		√				9C	4	2						
		XH − M(X) XL − M(X + 1)			√			AC	6	2						
		XH − M XL − (M + 1)				√		BC	5	3						
DECREMENT INDEX REGISTER	DEX	X ← X − 1					√	09	4	1				T		
DECREMENT STACK POINTER	DES	SP ← SP − 1					√	34	4	1						
INCREMENT INDEX REGISTER	INX	X ← X + 1					√	08	4	1				T		
INCREMENT STACK POINTER	INS	SP ← SP + 1					√	31	4	1						
LOAD INDEX REGISTER	LDX	XH ← #M XL ← #M + 1	√					CE	3	3			T	T	T	
		XH ← M XL ← (M + 1)		√				DE	4	2						
		XH ← M(X) XL ← M(X + 1)			√			EE	6	2						
		XH ← M XL ← (M + 1)				√		FE	5	3						
LOAD STACK POINTER	LDS	SPH ← #M SPL ← #M + 1	√					8E	3	3			T	T	T	
		SPH ← M SPL ← (M + 1)		√				9E	4	2						
		SPH ← M(X) SPL ← M(X + 1)			√			AE	6	2						
		SPH ← M SPL ← (M + 1)				√		BE	5	3						
STORE INDEX REGISTER	STX	M ← XH (M + 1) ← XL		√				DF	5	2			T	T	T	
		M(X) ← XH M(X + 1) ← XL			√			EF	7	2						
		M ← XH (M + 1) ← XL				√		FF	6	3						
STORE	STS	M ← SPH (M + 1) ← SPL		√				9F	5	2			T	T	T	
		M(X) ← SPH M(X + 1) ← SPL			√			AF	7	2						
		M ← SPH (M + 1) ← SPL				√		BF	6	3						
TRANSFER FROM INDEX TO STACK	TXS	SP ← X − 1					√	35	4	1						
TRANSFER FROM STACK TO INDEX	TSX	X ← SP + 1					√	30	4	1						

Status Control Instructions

Operation	Mnemonic	Description	Inherent Add Mode	Code	Cycles	Bytes	5 H	4 I	3 N	2 Z	1 V	0 C
CLEAR CARRY	CLC	C ← 0	√	0C	2	1						0
CLEAR INTERRUPT MASK	CLI	I ← 0	√	0E	2	1		0				
CLEAR OVERFLOW	CLV	V ← 0	√	0A	2	1					0	
SET CARRY	SEC	C ← 1	√	0D	2	1						1
SET INTERRUPT MASK	SEI	I ← 1	√	0F	2	1		1				
SET OVERFLOW	SEV	V ← 1	√	0B	2	1					1	
LOAD CC WITH ACCA	TAP	CC ← A	√	06	2	1	T		T	T	T	T
LOAD ACCA WITH CC	TPA	A ← CC	√	07	2	1						

Interrupt Instructions

Operation	Mnemonic	Description	Inherent Add Mode	Code	Cycles	Bytes	Status 5 H	4 I	3 N	2 Z	1 V	0 C
NO OPERATION	NOP	PC ← PC + 1	√	01	2	1						
RETURN FROM INTERRUPT	RTI	CC ← M(SP + 1) B ← M(SP + 2) A ← M(SP + 3) XH ← M(SP + 4) XL ← M(SP + 5) PCH ← M(SP + 6) PCL ← M(SP + 7) SP ← SP + 7	√	3B	10	1						
RETURN FROM SUBROUTINE	RTS	PCH ← M(SP + 1) PCL ← M(SP + 2)	√	39	5	1						
SOFTWARE INTERRUPT	SWI	PC ← PC + 1 M(SP) ← PCL M(SP - 1) ← PCH M(SP - 2) ← XL M(SP - 3) ← XH M(SP - 4) ← A M(SP - 5) ← B M(SP - 6) ← CC SP ← SP - 7 PCH ← FFFA PCL ← FFFB	√	3F	12	1						
WAIT FOR INTERRUPT	WAI	PC ← PC + 1 M(SP ← PCL M(SP - 1) ← PCH M(SP - 2) ← XL M(SP - 3) ← XH M(SP - 4) ← A M(SP - 5) ← B M(SP - 6) ← CC WHEN INTERRUPT ARRIVES STATUS BIT 4 IS SET PCH ← FFF8 PCL ← FFF9		3E	9	1		T				

Hardware Interrupts

Operation	Mnemonic	Description	Status 5 H	4 I	3 N	2 Z	1 V	0 C
RESET (ACTIVE LOW)	$\overline{RES}$	PCH ← FFFE PCL ← FFFF		1				
NON-MASKABLE INTERRUPT (ACTIVE LOW)	$\overline{NMI}$	M(SP) ← PCL M(SP - 1) ← PCH M(SP - 2) ← XL M(SP - 3) ← XH M(SP - 4) ← A M(SP - 5) ← B M(SP - 6) ← CC SP ← SP - 7 PCH ← FFFC PCL ← FFFD		1				
INTERRUPT REQUEST (ACTIVE LOW)	$\overline{IRQ}$	M(SP) ← PCL M(SP - 1) ← PCH M(SP - 2) ← XL M(SP - 3) ← XH M(SP - 4) ← A M(SP - 5) ← B M(SP - 6) ← CC SP ← SP - 7 PCH ← FFF8 PCL ← FFF9		*a*				

[a]If set, IRQ is ignored; if not set, IRQ is executed and status bit 4 is set.

APPENDIX **D**

Intel's Executable Instructions

D.1 SYMBOLS

The following nomenclature is used in the Intel assembler and for operator descriptions [33]:

1. *Operators*

$\leftarrow$	becomes
$\leftrightarrow$	transfer
$\cdot$	and
+	or
$\oplus$	exclusive or
-	not
H	suffix in operand means hexadecimal

2. *Registers*

A, B, C, D, E, H, L	Accumulators
PC	Pointer register
SP	Stack pointer
L	Suffix PCL and SPL means lower byte of PC and SP respectively
H	Suffix PCH and SPH means higher byte of PC and SP respectively

3. *Memory*

M	Contents of memory location M
M(SP)	Contents of memory location pointed to by bracketed pointer
M + 1	Means increment the contents of M by 1
(M + 1)	Means the contents of M + 1

4. *Processor word status*

S	MSB = 1	bit 7
Z	Zero	bit 6
AC	half carry	bit 4
P	parity	bit 2
CY	carry	bit 0

5. *Status*

T Load ALU result into this bit

6. *Register source and destination designation*

SSS source
DDD destination

SSS or DDD	Register
111	A
000	B
001	C
010	D
011	E
100	H
101	L

D.2 INSTRUCTION SET

Tables D.1 to D.8 summarizes Intel's instruction set. Table D.9 is intended as a quick reference guide.

Table D.1 Intel Data Transfer Group

Operation	Mnemonic and Operand	Description	Add Mode: Immediate	Direct	Indirect	Inherent	Cycles	Bytes	Code
MOVE	MOV r1, r2	r1 ← r2				√	1	1	01DDDSSS
	MOV r, M	r ← M(HL)		√			2	1	01DDD110
	MOV M, r	M(HL) ← r		√			2	1	01110SSS
	MVI r, data	r ← byte 2	√				2	2	00DDD110
	MVI m, data	M(HL) ← byte 2	√				3	2	36
LOAD REGISTER PAIR	LXI B	B ← byte 3 C ← byte 2	√				3	3	01
	LXI D	D ← byte 3 E ← byte 2	√				3	3	11
	LXI H	H ← byte 3 L ← byte 2	√				3	3	21
LOAD ACCUMULATOR	LDA M	A ← M		√			4	3	3A
	LDAX B	A ← M(BC)			√		2	1	0A
	LDAX D	A ← M(DE)			√		2	1	1A
STORE ACCUMULATOR	STA M	M ← A		√			4	3	32
	STA B	M(BC) ← A			√		2	1	02
	STA D	M(DE) ← A			√		2	1	12
LOAD H and L	LHLD	L ← M H ← M + 1		√			5	3	2A
STORE H and L	SHLD	M ← L (M + 1) ← H		√			5	3	22
EXCHANGE HL WITH DE	XCHG	H ↔ D L ↔ E				√	1	1	EB

Table D.2 Intel Add and Subtract Instructions

Operation	Mnemonic and Operand	Description	Add Mode: Immediate	Direct	Indirect	Inherent	Cycles	Bytes	Code
ADD	ADD r	$A \leftarrow A + r$				√	1	1	10000SSS
	ADD M	$A \leftarrow A + M(HL)$			√		2	1	86
	ADI data	$A \leftarrow A +$ byte 2	√				2	2	C6
ADD WITH CARRY	ADC r	$A \leftarrow A + r + CY$				√	1	1	10001SSS
	ADC M	$A \leftarrow A + M(HL) + CY$			√		2	1	8E
	ACI data	$A \leftarrow A +$ byte 2 $+ CY$	√				2	2	CE
SUBTRACT	SUB r	$A \leftarrow A - r$				√	1	1	10010SSS
	SUB M	$A \leftarrow A - M(HL)$			√		2	1	96
	SUI data	$A \leftarrow A -$ byte 2	√				2	2	D6
SUBTRACT WITH BORROW	SBB r	$A \leftarrow A - r - CY$				√	1	1	10011SSS
	SBB M	$A \leftarrow A - M(HR) - CY$			√		2	1	9E
	SBI data	$A \leftarrow A -$ byte 2 $- CY$	√				2	2	DE
DECIMAL ADJUST ACCUMULATOR	DDA	Converts the contents of A from HEX to BCD				√	1	1	27
ADD REGISTER PAIR TO HL	DAD B	$HL \leftarrow HL + BC$				√	3	1	09
	DAD D	$HL \leftarrow HL + DE$				√	3	1	19
	DAD H	$HL \leftarrow HL + HL$				√	3	1	29
	DAD SP	$HL \leftarrow HL + SP$				√	3	1	39

Table D.3 Intel Increment and Decrement Instructions

Operation	Mnemonic and Operand	Description	Add Mode: Immediate	Direct	Indirect	Inherent	Cycles	Bytes	Code
INCREMENT	INR r	r ← r + 1				√	1	1	00DDD100
	INR M	M(HL) ← M(HL) + 1			√		3	1	34
DECREMENT	DCR r	r ← r − 1				√	1	1	00DDD101
	DCR M	M(HL) ← M(HL) − 1			√		3	1	35
INCREMENT	INX B	BC ← BC + 1				√	1	1	03
	INX D	DE ← DE + 1				√	1	1	13
	INX H	HL ← HL + 1				√	1	1	23
DECREMENT	DCX B	BC ← BC − 1				√	1	1	0B
	DCX D	DE ← DE − 1				√	1	1	1B
	DCX H	HL ← HL − 1				√	1	1	2B

Table D.4 Intel Logic Operator and Compare Instructions

Operation	Mnemonic and Operand	Description	Add Mode: Immediate	Direct	Indirect	Inherent	Cycles	Bytes	Code
AND	ANA e	A ← A · r				√	1	1	10100SSS
	ANA M	A ← A · M(HL)			√		2	1	A6
	AN1 data	A ← A · byte 2	√				2	2	E6
EXCLUSIVE OR	XRA r	A ← A ⊕ r				√	1	1	10101SSS
	XRA M	A ← A ⊕ M(HL)			√		2	1	AE
	XRI data	A ← A ⊕ byte 2	√				2	2	EE
INCLUSIVE OR	ORA r	A ← A + r				√	1	1	10110SSS
	ORA M	A ← A + M(HL)			√		2	1	B6
	ORI data	A ← A + byte 2	√				2	2	F6
COMPARE	CMP r	A – r				√	1	1	10111SSS
	CMP M	A – M(HL)			√		2	1	BE
	CMP data	A – byte 2	√				2	2	FE
ROTATE A LEFT	RLC	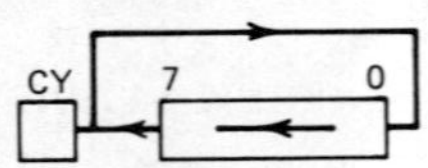				√	1	1	07
ROTATE A RIGHT	RRC	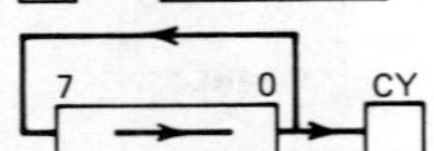				√	1	1	0F
ROTATE A LEFT THRU CARRY	RAL	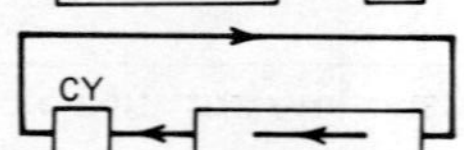				√	1	1	17
ROTATE A RIGHT THRU CARRY	RAR	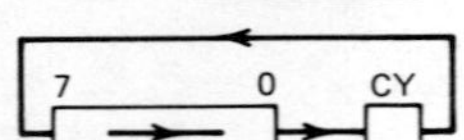				√	1	1	1F
COMPLEMENT CARRY	CMC	CY ← $\overline{CY}$				√	1	1	3F
SET CARRY	STC	CY ← 1				√	1	1	37
COMPLEMENT ACCUMULATOR	CMA	A ← $\overline{A}$				√	1	1	2F

Table D.5 Intel Stack-Related Instructions

Operation	Mnemonic and Operand	Description	Add Mode			Cycles	Bytes	Code
			Direct	Indirect	Inherent			
PUSH ACCUMULATORS ONTO STACK	PUSH B	M(SP – 1) ← B M(SP – 2) ← C SP ← SP – 2		√		3	1	C5
	PUSH D	M(SP – 1) ← D M(SP – 2) ← + SP ← SP – 2		√		3	1	D5
	PUSH H	M(SP – 1) ← H M(SP – 2) ← L SP ← SP – 2		√		3	1	E5
PUSH PSW ONTO STACK	PUSH PSW	M(SP – 1) ← A M(SP – 2) ← PSW SP ← SP – 2		√		3	1	F5
POP FROM STACK TO ACCUMULATORS	POP B	C ← M(SP) B ← M(SP + 1) SP ← SP + 2		√		3	1	C1
	POP D	E ← M(SP) D ← M(SP + 1) SP ← SP + 2		√		3	1	D1
	POP H	L ← M(SP) H ← M(SP + 1) SP ← SP + 2		√		3	1	E1
POP STATUS FROM STACK	POP PSW	PSW ← M(SP)			√	3	1	F1

Table D.6 Intel Branch Instructions

Operation	Mnemonic	Test	Description	Add Mode			Cycles	Bytes	Code
				Direct	Indirect	Inherent			
JUMP	JMP	1	PCH ← byte 3	√			3	3	C3
	JC	CY = 1	PCL ← byte 2	√			3	3	DA
	JNC	CY = 0		√			3	3	D2
	JZ	Z = 1		√			3	3	CA
	JNZ	Z = 0		√			3	3	C2
	JP	S = 1		√			3	3	F2
	JM	S = 0		√			3	3	FA
	JPE	P = 1		√			3	3	EA
	JPO	P = 0		√			3	3	E2
CALL	CALL	1	M(SP – 1) ← PCH	√			5	3	CD
	CC	CY = 1	M(SP – 2) ← PCL	√			5	3	DC
	CNC	CY = 0	SP ← SP – 2	√			5	3	D4
	CZ	Z = 1	PCH ← byte 3	√			5	3	CA
	CNZ	Z = 0	PCL ← byte 2	√			5	3	C4
	CP	S = 1		√			5	3	F4
	CM	S = 0		√			5	3	FC
	CPE	P = 1		√			5	3	EC
	CPO	P = 0		√			5	3	E4
RETURN	RET	1	PCL ← M(SP)		√		3	1	C9
	RC	CY = 1	PCH ← M(SP + 1)		√		3	1	D8
	RNC	CY = 0	SP ← SP + 2		√		3	1	D0
	RZ	Z = 1			√		3	1	C8
	RNZ	Z = 0			√		3	1	C0
	RP	S = 1			√		3	1	F0
	RM	S = 0			√		3	1	F8
	RPE	P = 1			√		3	1	E8
	RDO	P = 0			√		3	1	E0
JUMP INDIRECT MOVE HL TO PC	PCHL	1	PC ← HL			√	1	1	E9

Table D.7 Intel Hardware Interrupts

Operation	Mnemonic and Operand	Description
NONMASKABLE INTERRUPT	TRAP	PC ← data data = 24H
MASKABLE INTERRUPT	RST 5.5	PC ← data data = 2CH
MASKABLE INTERRUPT	RST 6.5	PC ← data data = 34H
MASKABLE INTERRUPT	RST 7.5	PC ← data data = 3CH

Table D.8 Intel I/O and Special Instructions

Operation	Mnemonic and Operand	Description	Add Mode: Direct	Add Mode: Indirect	Add Mode: Inherent	Cycles	Bytes	Code
INPUT	IN port	$A \leftarrow port$	✓			3	2	DB
OUTPUT	OUT port	$port \leftarrow A$	✓			3	2	D3
ENABLE INTERRUPTS	EI	$I \leftarrow 0$			✓	1	1	FB
DISABLE INTERRUPTS	DI	$I \leftarrow 1$			✓	1	1	F3
READ INTERRUPT MASK	RIM	$A \leftarrow flags$[a]			✓	1	1	20
SET INTERRUPT MASK	SIM	$flags \leftarrow A$[a]			✓	1	1	30
HALT	HLT	$PC \leftarrow PC$			✓	1	1	76
NO OPERATION	NOP	$PC \leftarrow PC + 1$			✓	1	1	00

[a]See Table 11.22 page 241.

Table D.9 Quick Reference Guide
Data Transfer Group

Operation	Mnemonic and Operand	Description	Add Mode: Immediate	Direct	Indirect	Inherent	Cycles	Bytes	Code
MOVE	MOV r1, r2	r1 ← r2				√	1	1	01DDDSSS
	MOV r, M	r ← M(HL)		√			2	1	01DDD110
	MOV M, r	M(HL) ← r		√			2	1	01110SSS
	MVI r, data	r ← byte 2	√				2	2	00DDD110
	MVI m, data	M(HL) ← byte 2	√				3	2	36
LOAD REGISTER PAIR	LXI B	B ← byte 3 C ← byte 2	√				3	3	01
	LXI D	D ← byte 3 E ← byte 2	√				3	3	11
	LXI H	H ← byte 3 L ← byte 2	√				3	3	21
LOAD ACCUMULATOR	LDA M	A ← M		√			4	3	3A
	LDAX B	A ← M(BC)			√		2	1	0A
	LDAX D	A ← M(DE)			√		2	1	1A
STORE ACCUMULATOR	STA M	M ← A		√			4	3	32
	STA B	M(BC) ← A			√		2	1	02
	STA D	M(DE) ← A			√		2	1	12
LOAD H and L	LHLD	L ← M H ← M + 1		√			5	3	2A
STORE H and L	SHLD	M ← L (M + 1) ← H		√			5	3	22
EXCHANGE HL WITH DE	XCHG	H ↔ D L ↔ E				√	1	1	EB

Add and Subtract Instructions

Operation	Mnemonic and Operand	Description	Add Mode: Immediate	Direct	Indirect	Inherent	Cycles	Bytes	Code
ADD	ADD r	A ← A + r				√	1	1	10000SSS
	ADD M	A ← A + M(HL)			√		2	1	86
	ADI data	A ← A + byte 2	√				2	2	C6
ADD WITH CARRY	ADC r	A ← A + r + CY				√	1	1	10001SSS
	ADC M	A ← A + M(HL) + CY			√		2	1	8E
	ACI data	A ← A + byte 2 + CY	√				2	2	CE
SUBTRACT	SUB r	A ← A – r				√	1	1	10010SSS
	SUB M	A ← A – M(HL)			√		2	1	96
	SUI data	A ← A – byte 2	√				2	2	D6
SUBTRACT WITH BORROW	SBB r	A ← A – r – CY				√	1	1	10011SSS
	SBB M	A ← A – M(HR) – CY			√		2	1	9E
	SBI data	A ← A – byte 2 – CY	√				2	2	DE
DECIMAL ADJUST ACCUMULATOR	DDA	Converts the contents of A from HEX to BCD				√	1	1	27
ADD REGISTER PAIR TO HL	DAD B	HL ← HL + BC				√	3	1	09
	DAD D	HL ← HL + DE				√	3	1	19
	DAD H	HL ← HL + HL				√	3	1	29
	DAD SP	HL ← HL + SP				√	3	1	39

Stack-Related Instructions

Operation	Mnemonic and Operand	Description	Add Mode: Direct	Add Mode: Indirect	Add Mode: Inherent	Cycles	Bytes	Code
PUSH ACCUMULATORS ONTO STACK	PUSH B	M(SP − 1) ← B M(SP − 2) ← C SP ← SP − 2		✓		3	1	C5
	PUSH D	M(SP − 1) ← D M(SP − 2) ← E SP ← SP − 2		✓		3	1	D5
	PUSH H	M(SP − 1) ← H M(SP − 2) ← L SP ← SP − 2		✓		3	1	E5
PUSH PSW ONTO STACK	PUSH PSW	M(SP − 1) ← A M(SP − 2) ← PSW SP ← SP − 2		✓		3	1	F5
POP FROM STACK TO ACCUMULATORS	POP B	C ← M(SP) B ← M(SP + 1) SP ← SP + 2		✓		3	1	C1
	POP D	E ← M(SP) D ← M(SP + 1) SP ← SP + 2		✓		3	1	D1
	POP H	L ← M(SP) H ← M(SP + 1) SP ← SP + 2		✓		3	1	E1
POP STATUS FROM STACK	POP PSW	PSW ← M(SP) A ← SP + 1 SP ← SP + 2			✓	3	1	F1

Increment and Decrement Instructions

Operation	Mnemonic and Operand	Description	Add Mode: Immediate	Add Mode: Direct	Add Mode: Indirect	Add Mode: Inherent	Cycles	Bytes	Code
INCREMENT	INR r	r ← r + 1				✓	1	1	00DDD100
	INR M	M(HL) ← M(HL) + 1			✓		3	1	34
DECREMENT	DCR r	r ← r − 1				✓	1	1	00DDD101
	DCR M	M(HL) ← M(HL) − 1			✓		3	1	35
INCREMENT	INX B	BC ← BC + 1				✓	1	1	03
	INX D	DE ← DE + 1				✓	1	1	13
	INX H	HL ← HL + 1				✓	1	1	23
DECREMENT	DCX B	BC ← BC − 1				✓	1	1	0B
	DCX D	DE ← DE − 1				✓	1	1	1B
	DCX H	HL ← HL − 1				✓	1	1	2B

Logic Operator and Compare Instructions

Operation	Mnemonic and Operand	Description	Add Mode: Immediate	Direct	Indirect	Inherent	Cycles	Bytes	Code
AND	ANA e	A ← A · r				√	1	1	10100SSS
	ANA M	A ← A · M(HL)			√		2	1	A6
	AN1 data	A ← A · byte 2	√				2	2	E6
EXCLUSIVE OR	XRA r	A ← A $\oplus$ r				√	1	1	10101SSS
	XRA M	A ← A $\oplus$ M(HL)			√		2	1	AE
	XRI data	A ← A $\oplus$ byte 2	√				2	2	EE
INCLUSIVE OR	ORA r	A ← A + r				√	1	1	10110SSS
	ORA M	A ← A + M(HL)			√		2	1	B6
	ORI data	A ← A + byte 2	√				2	2	F6
COMPARE	CMP r	A − r				√	1	1	10111SSS
	CMP M	A − M(HL)			√		2	1	BE
	CMP data	A − byte 2	√				2	2	FE
ROTATE A LEFT	RLC	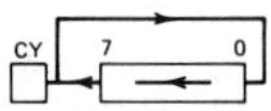				√	1	1	07
ROTATE A RIGHT	RRC	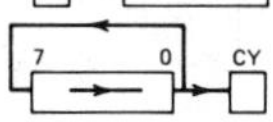				√	1	1	0F
ROTATE A LEFT THRU CARRY	RAL	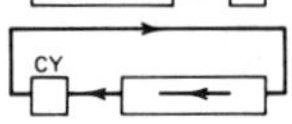				√	1	1	17
ROTATE A RIGHT THRU CARRY	RAR	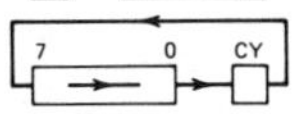				√	1	1	1F
COMPLEMENT CARRY	CMC	CY ← $\overline{CY}$				√	1	1	3F
SET CARRY	STC	CY ← 1				√	1	1	37
COMPLEMENT ACCUMULATOR	CMA	A ← $\overline{A}$				√	1	1	2F

I/O and Special Instructions

Operation	Mnemonic and Operand	Description	Add Mode: Direct	Indirect	Inherent	Cycles	Bytes	Code
INPUT	IN port	A ← port	√			3	2	DB
OUTPUT	OUT port	port ← A	√			3	2	D3
ENABLE INTERRUPTS	EI	I ← 0			√	1	1	FB
DISABLE INTERRUPTS	DI	I ← 1			√	1	1	F3
READ INTERRUPT MASK	RIM	A ← flags			√	1	1	20
SET INTERRUPT MASK	SIM	flags ← A			√	1	1	30
HALT	HLT	PC ← PC			√	1	1	76
NO OPERATION	NOP	PC ← PC + 1			√	1	1	00

Branch Instructions

Operation	Mnemonic	Test	Description	Add Mode Direct	Add Mode Indirect	Add Mode Inherent	Cycles	Bytes	Code
JUMP	JMP	1	PCH ←byte 3	√			3	3	C3
	JC	CY = 1	PCL ←byte 2	√			3	3	DA
	JNC	CY = 0		√			3	3	D2
	JZ	Z = 1		√			3	3	CA
	JNZ	Z = 0		√			3	3	C2
	JP	S = 1		√			3	3	F2
	JM	S = 0		√			3	3	FA
	JPE	P = 1		√			3	3	EA
	JPO	P = 0		√			3	3	E2
CALL	CALL	1	M(SP - 1) PCH	√			5	3	CD
	CC	CY = 1	M(SP - 2) PCL	√			5	3	DC
	CNC	CY = 0	SP ←SP - 2	√			5	3	D4
	CZ	Z = 1	PCH ←byte 3	√			5	3	CA
	CNZ	Z = 0	PCL ←byte 2	√			5	3	C4
	CP	S = 1		√			5	3	F4
	CM	S = 0		√			5	3	FC
	CPE	P = 1		√			5	3	EC
	CPO	P = 0		√			5	3	E4
RETURN	RET	1	PCL ←M(SP)		√		3	1	C9
	RC	CY = 1	PCH ←M(SP + 1)		√		3	1	D8
	RNC	CY = 0	SP ←SP + 2		√		3	1	DO
	RZ	z = 1			√		3	1	C8
	RNZ	Z = 0			√		3	1	CO
	RP	s = 1			√		3	1	FO
	RM	S = 0			√		3	1	F8
	RPE	P = 1			√		3	1	E8
	RDO	P = 0			√		3	1	EO
JUMP INDIRECT MOVE HL TO PC	PCHL	1	PC ←HL			√	1	1	E9

Hardware Interrupts

Operation	Mnemonic and Operand	Description
NONMASKABLE INTERRUPT	TRAP	PC ←data data = 24H
MASKABLE INTERRUPT	RST 5.5	PC ←data data = 2CH
MASKABLE INTERRUPT	RST 6.5	PC ←data data = 34H
MASKABLE INTERRUPT	RST 7.5	PC ←data data = 3CH

APPENDIX **E**

Motorola Directives

In Table E.1 of this appendix a list of fundamental directives is given [32]. In the description of these directives the symbol LC means the location counter. The location counter points to the current memory address during program assembling. The symbol *** means label not allowed.

Table E.1 Motorola Directives

Label	Mnemonic	Operand	Description	Purpose
***	END		End of program	Last statement of a source program
PIA	EQU	$400A	PIA ← 400A	Assigns value to symbol.
	FCB	$A2	M(LC) ← A2 LC ← LC + 1	Puts data bytes in memory. Many have more than one operand separated by commas.
NUM2	FCB	$0F, +6F, $C1	NUM2 ← LC M(LC) ← 0F M(LC+1) ← 6F M(LC+2) ← C1 LC ← LC + 3	
NUM4	FDB	A2	NUM4 ← LC M(LC) ← 00A2 LC ← LC + 1	Puts data word in memory. Many have more than one operand separated by commas.
	FDB	$0F, $6FC1	M(LC) ← 000F M(LC+1) ← 6FC1 LC ← LC + 2	
MSG3	FCC	1PEACE1	MSG3 ← LC M(LC) ← 50 M(LC+1) ← 45 M(LC+2) ← 41 M(LC+3) ← 43 M(LC+4) ← 45 LC ← LC + 6	Translates ASCII CHARACTERS to CODE text between identical delimiters of some character, delimiter number text cannot begin with comma. Text after character count and comma. Space valid character.
MSG4	FCC	6, PEACE	MSG4 ← LC M(LC) ← 50 M(LC+1) ← 45 M(LC+2) ← 41 M(LC+3) ← 43 M(LC+4) ← 45 M(LC+5) ← 20 LC ← LC + 6	
	RMB	6	LC ← LC + 6	

Table E.1 Motorola Directives (*Continued*)

Label	Mnemonic	Operand	Description	Purpose
VOLTS	RMB	3	VOLTS ← LC LC ← LC + 3	
***	ORG	$600	LC ← 600	
***	NAM	HEAD		Top of page heading for listing.
***	PAGE			Advance paper to top of next page.
***	SPC	N		line ← line + N
***	OPT	See list below.		

These options are as follows with the necessary part of symbol in uppercase and optional part in lowercase.

Generate	Give code generated by FCC directive (selected by default)
List	Lists the assembly in long format (selected by default)
Otape = (FILENAME)	Object program is saved in an object file tape named (FILENAME). Overwrites any existing file of the same type and the same name.
Symbol	Symbol table is given in listing

APPENDIX F
Intel Directives

In Table F.1 of this appendix a list of fundamental directives is given [33]. In the description of these directives the symbol LC means the location counter. The location counter points to the current memory address during program assembling. A comment field must start with a semicolon (;). The description is not intended as comment field entries.

Table F.1 Intel Directives

Label	Mnemonic	Operand	Description	Purpose
	END	N	PC ← N	Last statement of source program.
PPIU	EQU	1CH	PPIU ← 1C	Assigns value to symbol.
AMPS	SET	5CH	AMPS ← 5C	Assigns value to symbol; can be changed any time.
	DB	2CH	M(LC) ← 2C LC ← LC + 1	Put data bytes in memory. Many have more than one operand separated by commas. Also translates characters in operand between delimiters to ACSII code.
COUN:	DB	0FH, 17H	COUN ← LC M(LC) ← 0F M(LC+1) ← 17 LC ← LC + 2	
MSG6:	DB	'PEACE'	MSG6 ← LC M(LC) ← 50 M(LC+1) ← 45 M(LC+2) ← 41 M(LC+3) ← 43 M(LC+4) ← 45 LC ← LC + 5	
ICOU:	DW	1BH, 2C6AH	ICOU ← LC M(LC) ← 1B00 M(LC+2) ← 6A2C LC ← LC + 4	Put data words in memory in the reverse order. Many have more than one operand.
	DW	0FH	M(LC) ← 0F00 LC ← LC + 2	
	DS	2	LC ← LC + 2	Move location counter in data memory and name memory.
NEW1:	DS	6	NEW1 ← LC LC ← LC + 6	
PAGE2:	ORG	0IFFH	PAGE2 ← LC LC ← 0IFF	Move location counter in program memory and name memory.
	NAME	MODULE2		Names a program.

APPENDIX G

Motorola Programming Aid Commands

The modular configuration of a system suited to programming aids is given in Figure G.1. The operation of this system is given in the following sections [35].

G.1 SYSTEM START-UP

The Motorola executive module is entered using the following steps:

1. Power up all system components, diskettes first.
2. Insert system diskette into drive 0 and close door.
3. Insert your diskette into drive 1 (or any other) and close doors.
4. System response on console display (EXBUG 1 : 1).
5. Enter MAID on console.
6. System responds with (CR) and the prompt *. The character * signifies the

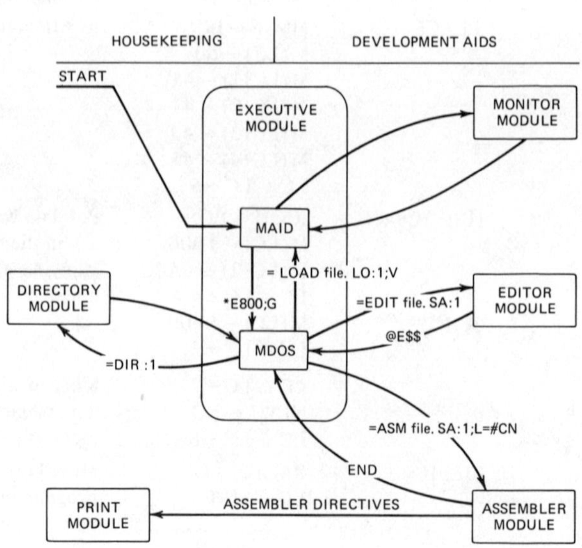

Figure G.1 Motorola Programming aid modules.

MAID module segment. This segment is used to run programs. It can run your program or the MDOS system program.

7. Enter E800;G on console. This loads the MDOS bootstrap.
8. System responds with (CR) and the prompt =. The character = signifies the MDOS module segment. This segment is used to execute commands.

G.2 MAID MODULE COMMANDS

The maid module is identified by the * prompt starting each line. Two important functions of the MAID module are

- Run MDOS (E800;G)
- Run your program (n:G where n is program starting address)

The MAID module can perform other tasks. The following is a list of commands and their descriptions:

Command	Description
n/	Open byte addressed by n
(LF)	Open next sequential location
(CR)	Close open location
(UA)	Open previous sequential location
X	Return to EXbug scan loop
n;V	Set breakpoint at location n
;U	Remove all breakpoints
n;U	Remove breakpoint at location n
n:W	Search for n bit pattern
;G	Execute target program from restart vector
n;G	Execute target program from location n
;P	Continue executing from encountered breakpoint
n;P	Continue executing until breakpoint found n times
n;O	Calculate offset from current location to n
;N	Trace one instruction
n;N	Trace n instructions
$V	Display breakpoints
$M	Display search mask and limits
$R	Display/change target program registers
$T	Set trace mode and set trace to address
$S	Set stop on address compare
;T	Reset trace mode
;S	Reset stop on address compare
#n=	Convert decimal to hexadecimal
#$n=	Convert hexadecimal to decimal
#@n=	Convert octal to hexadecimal

(LF)= line feed
(UA)= up arrow Λ
(CR)= carriage return

G.3 EDITOR MODULE COMMANDS

This editor module is entered from MDOS using the command

=EDIT file.SA:1(CR)

where

EDIT is the command
file is your file name
SA designates source file
:1 designates drive No. 1.

The editor module is identified by the @ prompt at the start of each line.

The editor module responds to character commands. Commands can be grouped together. Commands are executed when double escape $$ characters are entered. Character strings are terminated using one escape $ character. Grouped editor commands may be separated by one escape character, but it is not necessary. For example,

@2K$-3L$10T$$ or 2K-3L10T$$

are equally valid.

@S line $0LT$$ requires the $ after line to delimit the character string.

The following is a list of commands and their description

Command	Description
B	Pointer to the beginning of text.
nL	Moves pointer n lines from its present position (positive or negative). If n equals 0, the workspace character pointer is positioned to the beginning of the line in which the workspace character pointer presently resides.
nT	Displays from 1 to 254 lines positive or negative.
nK	Kills from 1 to 254 lines (up to and including CR) positive or negative.
Z	Positions pointer to the end of the contents.
nM	Positions pointer n characters from its present position positive or negative.
nP	Displays from 1 to 254 lines from the beginning. Output is to the punch device. Lines output are deleted from the workspace.
nD	Deletes from 1 to 254 characters positive or negative.
Sstring	Searches for string in the workspace buffer. Positions character pointer after found string.
Istring	Inserts string at the present position of the pointer.
Cstring1$string2	Searches for string1 and, if found, replaces (changes) string1 with string2. String2 need not be the same length as string1.
E	Ends the edit operation and transfers the contents of the workspace buffer to the file.

Command	Description
F	Outputs 6 inches of blank leader/trailer code to the punch device.
A	Appends an input string from the reader device to the workspace. Ignores nulls and rubouts. Terminates on E0F character (1A Hex), workspace full, or after 50 lines.

The A command is a sleeper. It is usually the first command required after the = EDIT command.

G.4 ASSEMBLER MODULE COMMANDS

The assembler module is entered from MDOS. Its purpose is to assemble a source file stored on diskettes. An MDOS command to initiate this operation is

=ASM file.SA: 1;L=#CN(CR)

where

ASM is the command

file is your file name

SA designates source file

:1 designates drive No. 1

L=#CN causes listing on console.

The assembler creates a paper tape listing. Use the command

=EXBIN file.LX:1(CR)

where LX designates a file in paper tape format to create object code.

G.5 DIRECTORY MODULE COMMAND

There are a number of types of files that can be listed. However a simple command that gives a directory listing is

=DIR :1(CR)

where

DIR is the command

:1 is the drive No. 1

You may wish to list drive 0. Use =DIR. Watch out. Don't ask for 0 (blank) when you want 1. There is an automatic return to MDOS.

APPENDIX H
Intel Programming Aid Commands

The modular configuration of a system suited to programming aids is given in Figure H.1. The operation of this system is given in the following sections [34].

H.1 SYSTEM START-UP

The Intel executive module is entered using the following steps:

1. Power up all system components.
2. Insert system diskette into Drive 0 and close door.
3. Insert your diskette into Drive 1 (or any other drive) and close door.
4. Set bootstrap switch on.
5. Press reset.
6. Select console using space key.
7. Set bootstrap switch off.
8. The executive module ISIS-II responds with a (CR) and a prompt character -. ISIS-II uses the prompt - at the start of each line to identify itself.

The executive module uses commands to control the editor, assembler, monitor, directory, and display formats. Some systems (MDS 230) require steps 1, 2, 3, 5, and 7 only.

In the following commands the symbol (CR) is used for carriage return.

H.2 EDITOR MODULE COMMANDS

The editor module is entered from ISIS-II using the command

-EDIT file.ASM :F1 (CR)

where

- \- is the prompt
- EDIT is the command

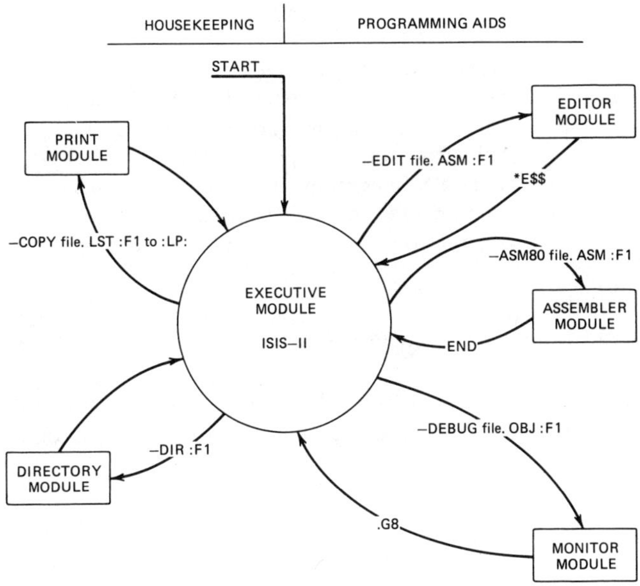

Figure H.1 Intel programming aid modules.

file is your file name
ASM designates source file
:F1 designates drive No. 1.

The editor module is identified by the * prompt at the start of each line.

The editor module responds to single character commands. Commands can be grouped together. Commands are executed when double escape ($$) characters are entered. Grouped editor commands may be separated by $ characters but it is not necessary. Character strings must be terminated using a single $ character. For example:

*Fstring$0L10T$$

The $ character after string is required to indicate termination of the character string.

The following is a list of commands and their description.

Command	Description
A	Append text from disc to workspace
B	Set editor buffer pointer to beginning of workspace
C	Move buffer pointer ±N characters
D	Delete ±N characters
E	End-terminate editor; update file on disc
F	Find a character string
I	Insert text string
K	Kill ±N lines

Command	Description
L	Move buffer pointer ±N lines
M	Print available free space in workspace memory
Q	Quit without changing files on disc
S	Substitute one string for another
T	Type ±N lines from current pointer position
W	Write N lines to destination file
Z	Move buffer pointer to end of text buffer

H.3 ASSEMBLER MODULE COMMANDS

The assembler module is entered from ISIS-II. Its purpose is to assemble a source file stored on diskette. The ISIS-II command to initiate this operation is

-ASM80 file.ASM :F1 (CR)

where

- is the prompt
ASM85 is the command
file is your file name
ASM designates source file
:F1 designates drive No. 1

H.4 MONITOR MODULE COMMANDS

The monitor module is entered from ISIS-II using the command

-DEBUG file.OBJ :F1 (CR)

where

- is the prompt
DEBUG is the command
file is your file name
OBJ designates object file
:F1 designates drive 1

or by pressing the interrupt pushbutton on the console. The purpose of the monitor module is to run programs. This module is identified by the • prompt at the start of each line. Some of the commands that help in program debugging are the following:

- ASSIGN I/O DEVICE

- DISPLAY CONTENTS OF MEMORY
- FILL RAM WITH CONSTANT
- EXECUTE PROGRAM (GO)
- SUBSTITUTE MEMORY
- I/O STATUS QUERY
- EXAMINE/MODIFY REGISTERS
- MOVE DATA BLOCK

H.5 DIRECTORY MODULE COMMAND

A simple command that will give a directory listing is

-DIR :F1(CR)

where

- is the prompt
DIR is the command
:F1 designates drive 1.

If you wish to list drive 0, use

-DIR

A common mistake is to forget :F1 when wishing to get a directory for drive 1.

H.6 PRINT MODULE COMMAND

A simple command that will list a program is

-COPY file.LST :F1 to :LP:(CR)

where

- is the prompt
COPY is the command
file is your file name
LST is the assembly listing
:F1 designates drive 1
:LP: is the line printer.

APPENDIX I

Motorola Interface Adapters

Interface adapters are initialized to perform a specific function. The function is selected according to the code stored to the adapter control register. In this appendix a brief description of the formation of the control bytes for the Motorola PIA and ACIA is given [30].

I.1 PIA REGISTERS

The data direction register (DDR) contents establish the direction of the data lines. Zeros designate input. Ones designate output.

The control register (CR) contents shown in Figure I.1 establish the operation of control lines CA1 and CA2. Tables I.1 to I.4 give CR coding.

I.2 ACIA REGISTERS

The code stored in the control register of the control/status register pair establishes the transmission rates, parity, and so on. The operational status format is shown in Table I.5. The control register format is shown in Table I.6.

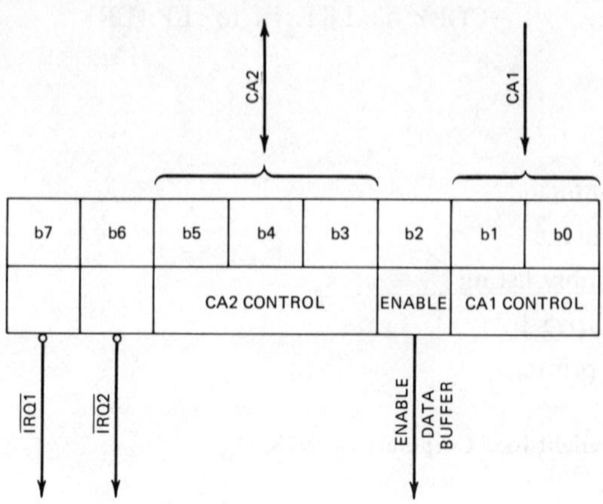

Figure I.1 Control register.

Table I.1 Input Programming for CA1

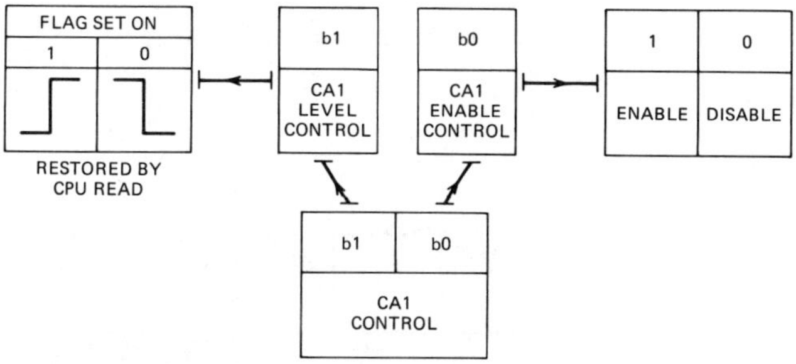

Table I.2 Input Programming for CA2

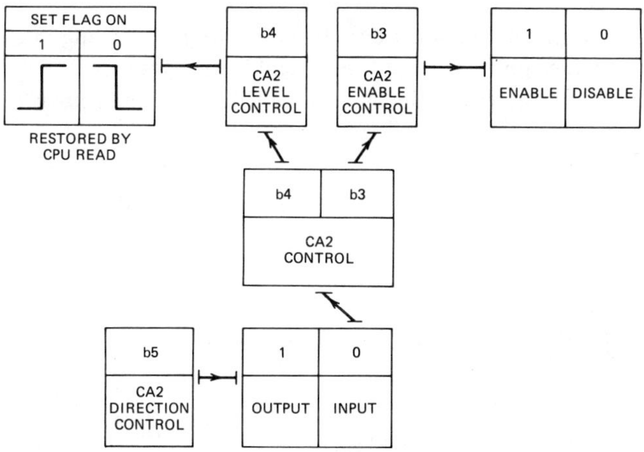

Table I.3 Output Programming for CA2

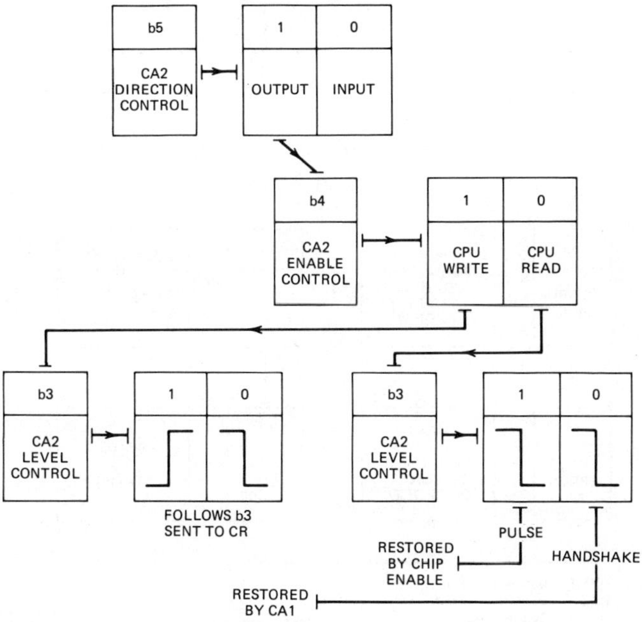

Table I.4 Output Programming for Bit b4 of CB2 Control

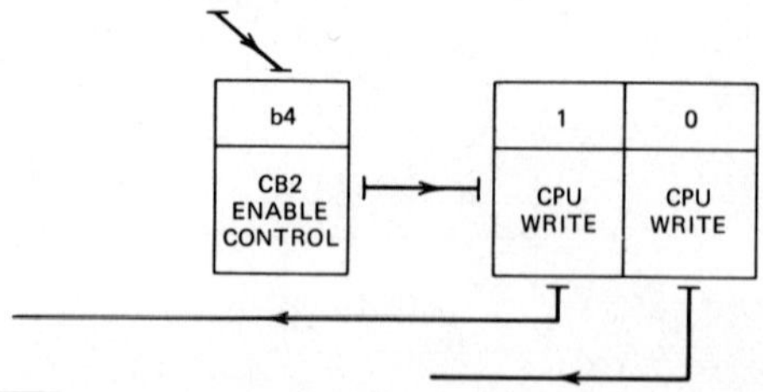

In addition to eight bidirectional data lines and four addressing lines, the ACIA has the following lines for control and peripheral interfacing:

There are three control lines connected to the CPU bus. These are as follows:

$\overline{IRQ}$, interrupt request.

R/$\overline{W}$, read/write, which selects one of the registers in a register pair.

E, enable.

Table I.5 Status Register Byte Format

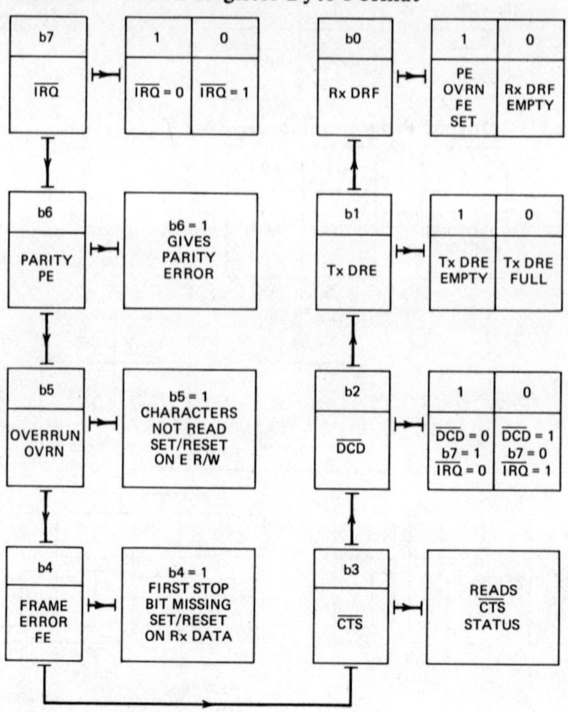

DRF/E	Data register FULL/EMPTY
DCD	Data carrier detect
CTS	Clear to send

Table I.6 Control Register Instruction Format

b7	1	0
RECEIVE INTERR ENABLE CONTROL	ENABLE $\overline{IRQ}$	DISABLE $\overline{IRQ}$

RIE – receive interrupt enable
TIE – transmit interrupt enable

b6 b5	00	01	10	11
TRANSMIT CONTROL	$\overline{RTS}$ = 0 TIE = 0	$\overline{RTS}$ = 0 TIE = 1	$\overline{RTS}$ = 1 TIE = 0	$\overline{RTS}$ = 0 TIE = 0 SPACES

b4 b3 b2: WORD LENGTH, PARITY, STOP BITS

VALUE	PARITY	STOP BITS	WORD LENGTH
000	EVEN	2	7
001	ODD		
010	EVEN	1	
011	ODD		
100	NONE	2	8
101	NONE	1	
110	EVEN		
111	ODD		

b1 b0	00	01	10	11
MASTER RESET AND COUNT RATIO	÷ 1	÷ 16	÷ 64	MASTER RESET

On the peripheral side of the ACIA there are two serial data lines as follows:

Tx Data, output that is a serial transmit.
Rx Data, input that is the serial receive line.

There are two clock inputs as follows:

Clk Rx, receive side clock.
Clk Tx, transmit side clock.

There are three control signals used for handshake as follows:

$\overline{\text{CTS}}$, clear to send.
$\overline{\text{DCD}}$, data carry detect.
$\overline{\text{RTS}}$, request to send.

APPENDIX J

Intel Peripheral Interfaces

Interfaces are initialized to perform a specific function. The function is selected according to the code written into the control register. In this appendix a brief description of the formation of the control bytes for the PPIU and USART is given [31].

J.1 PPIU REGISTERS

There are four ways to use the PPIU port lines

- Bit set/reset. Port C lines controlled individually as shown in Table J.1.
- Mode 0. All 24 port lines used for data. They are programmable for data direction in groups of two 8-bit ports (A and B) and two 4-bit ports (upper and lower C) shown in Table J.2.
- Mode 1. In this mode Port C lines are used to provide automatic handshake for data Ports A and B. Control word format is shown in Table J.3.
- Mode 2. In this mode bidirectional control is provided for Port A and unidirectional control for Port B. Port C lines are used for handshake. Control word format is shown in Table J.3.

J.2 USART REGISTER

A USART is programmed for asynchronous operation by sending two consecutive words to the control register. The first word sent to the control register determines the mode of operation. Table J.4 shows the mode format. The second word sent to the control register determines the command. Table J.5 shows the command format.

Table J.5 can also be used to program USART synchronous operation with exceptions given in Table J.6.

The status of the USART can be *read* from the control register. Table J.7 gives the status format.

Table J.1 Bit Set/Reset Format

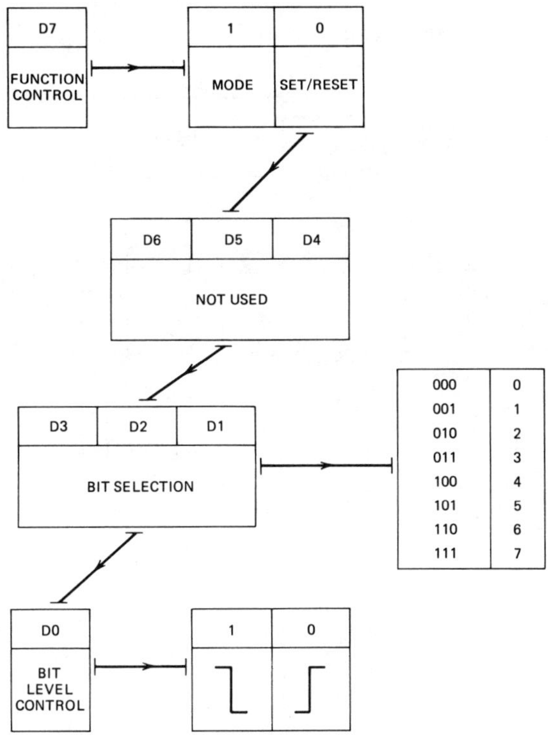

Table J.2 Mode 0 Control Word Format

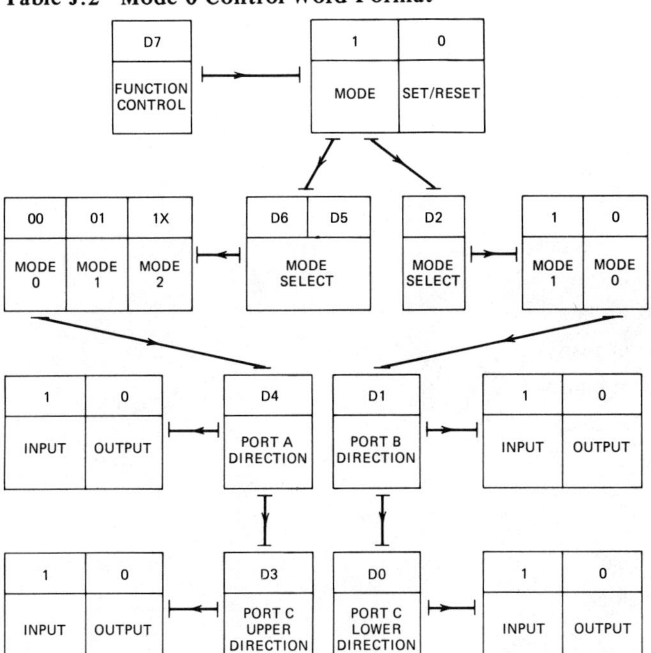

Table J.3 Modes 1 and 2 Control Word Format

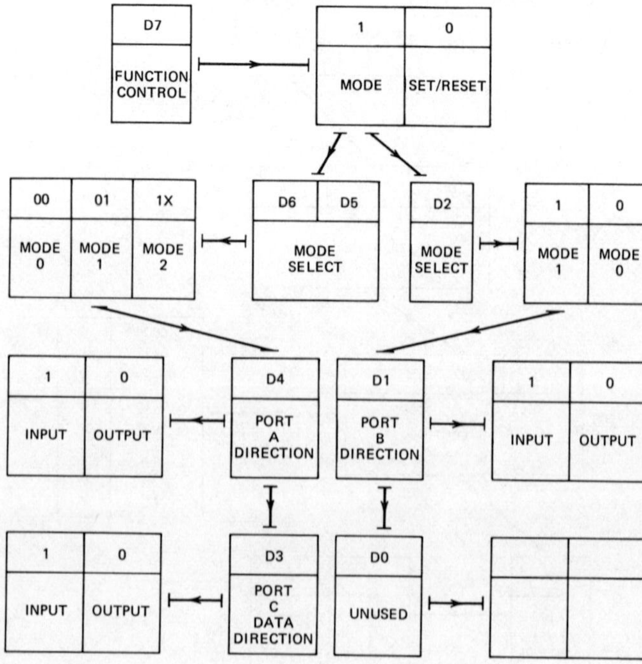

MODEM
$\overline{\text{DSR}}$ data set ready
$\overline{\text{DTR}}$ data terminal ready
$\overline{\text{RTS}}$ request to send
$\overline{\text{CTS}}$ clear to send

Transmitter
TxD transmitter data
TxR transmitter ready
TxE transmitter empty
$\overline{\text{TxC}}$ transmitter clock

Receiver
RxD receiver data
RxR receiver ready
$\overline{\text{RxC}}$ receiver clock

Table J.4 8257A Asynchronous Mode Instruction Format

D1 D0	TRANSMISSION MODE CLASSIFICATION	00: SYNC	01: ASYNC	10: ASYNC	11: ASYNC
D7 D6	FRAME CONTROL	00: UNUSED	01: 1 STOP BIT	10: 1½ STOP BITS	11: 2 STOP BITS
D5 D4	PARITY CONTROL	00: NO PARITY	10: NO PARITY	01: ODD PARITY	11: EVEN PARITY
D3 D2	CHARACTER LENGTH	00: 5 BITS	01: 6 BITS	10: 7 BITS	11: 8 BITS
D1 D0	BAUD RATE SELECTION	00:	01: x1	10: x16	11: x64

Table J.5 Command Instruction Format

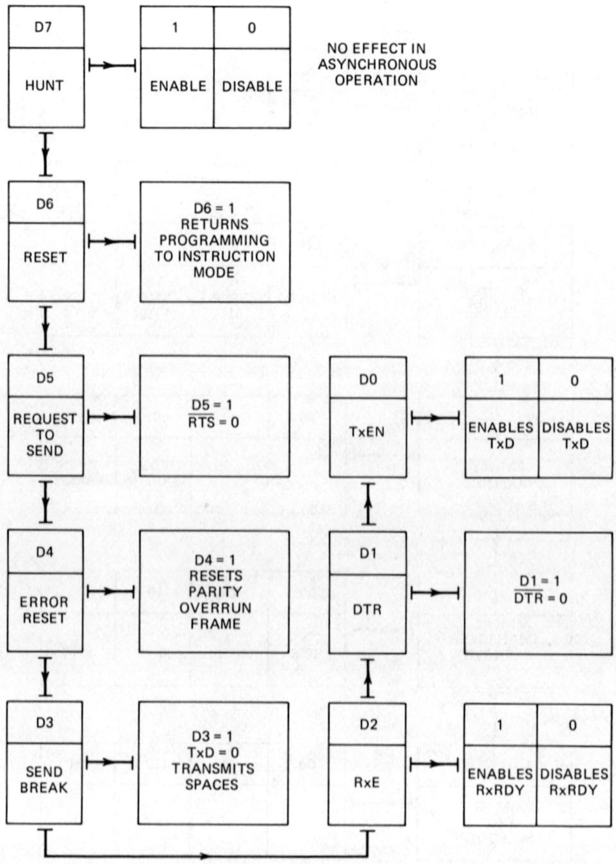

Table J.6 USART Synchronous Mode Instruction Format

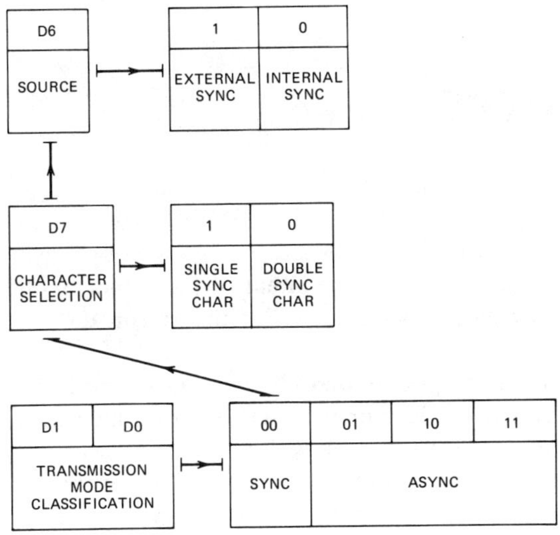

Table J.7 Status Register Format

D7	D6	D5	D4	D3	D2	D1	D0
DSR	SYNET	FE	OE	PE	TxE	RxRDY	TxRDY

References

In this book the approach is more toward step by step or procedural techniques in logic design than is usual. At least the intent of the author has been to fill a void and stress technique more than theory. For this reason the use of references is limited.

It is pointed out in the preface that there are many excellent textbooks in the logic and microcomputer field. For the student who is interested in all aspects of the subject a list of references is given with a very brief description. No attempt is made to be complete or to indicate the intended level of the work. Of the many books the author considers excellent, personal fancy has been taken to those by Blakeslee, Mowle, and Zissos.

1. G. Boole, *An Investigation of the Laws of Thought*, Dover, New York, 1958.
   Classic.
2. F. Hill and G. Peterson, *Introduction to Switching Theory and Logical Design*, John Wiley & Sons, Inc., New York, 1968.
   Numbers, maps, codes, sequential circuits.
3. D. Dietmeyer, *Logic Design of Digital Systems*, Allyn & Bacon Inc., Boston, 1971.
   Numbers, combinational, flip-flops, spaces, addressing, microprogramming.
4. R. M. Kline, *Digital Computer Design*, Prentice-Hall Inc., Englewood Cliffs, New Jersey, 1977.
   Combinational, electronics, sequential, architecture.
5. E. McCluskey, *Introduction to the Theory of Switching Circuits*, McGraw-Hill Book Company, New York, 1965.
   Classic.
6. H. T. Nagle, B. D. Carroll, J. D. Irwin, *An Introduction to Computer Logic*, Prentice-Hall Inc., Englewood Cliffs, New Jersey, 1975.
   Numbers, sequential, computer-aided design.
7. Y. Chu, *Digital Computer Design Fundamentals,* McGraw-Hill Book Company, New York, 1962.
   Combinational, electronics, matrices, elements.
8. D. Lewin, *Logical Design of Switching Circuits*, Nelson, London, 1968.
   Combinational, counters, reduction.
9. F. J. Mowle, *A Systematic Approach to Digital Logic Design*, Addison-Wesley Publishing Company, Reading, Massachusetts, 1976.
   Combinational, MSI, registers, FFs, and special circuits.
10. T. R. Blakeslee, *Digital Design with Standard MSI and LSI*, John Wiley & Sons, Inc., New York, 1975.
    Combinational and clocked MSI, microcomputer programming, and interrupt.

11. M. Marcus, *Switching Circuits for Engineers*, Prentice-Hall Inc., Englewood Cliffs, New Jersey, 1962.
Fundamentals of sequential circuits.

12. D. Zissos, *Problems and Solutions in Logic Design*, 2nd Edition, Oxford University Press, London, 1979.
Reduction, charts, flip-flops, sequential circuits.

13. L. F. Lind and J. C. C. Nelsen, *Analysis and Design of Sequential Digital Systems*, The Macmillan Press Ltd., London, 1977.
Combinational, sequential, systems.

14. R. Oberman, *Disciplines in Combinational and Sequential Circuit Design*, McGraw-Hill Book Company, New York, 1970.
Shift registers and codes.

15. J. Warfield, *Principles of Logic Design*, Ginn and Company, Boston, 1963.
Charts, distances, and sequential circuits.

16. W. Wicks, *Logic Design with Integrated Circuits*, John Wiley & Sons Inc., New York, 1968.
Combinational and counters.

17. J. Peatman, *The Design of Digital Systems*, McGraw-Hill Book Company, New York, 1972.
Combinational circuits and clocks. IC devices and special circuits.

18. Z. Kohavi, *Switching and Finite Automata Theory*, McGraw-Hill Book Company, New York, 1978.
Numbers, machines, information.

19. F. Hill and G. Peterson, *Digital Systems: Hardware Organization and Design*, John Wiley & Sons, Inc., New York, 1978.
Organization, components, microprocessors, interrupts, large systems.

20. H. Katzan, Jr., *Microprogramming Primer*, McGraw-Hill Book Company, New York, 1977.
Structures, programming.

21. A. Osborne, *An Introduction to Microcomputers*, Adam Osborne and Associates Inc., Berkeley, California, 1977.
Describes many micros specifically.

22. A. Osborne, *8080 Programming for Logic Design*, Adam Osborne and Associates Inc., Berkeley, California, 1976.
8080 registers, instructions, construction.

23. A. Osborne, *6800 Programming for Logic Design*, Adam Osborne and Associates Inc., Berkeley, California, 1977.

24. R. S. Sandige, *Digital Concepts using Standard Integrated Circuits*, McGraw-Hill Book Company, New York, 1978.
Codes, combinational, FFs, and counters.

25. A Tanenbaum, *Structural Computer Organization*, Prentice-Hall Inc., Englewood Cliffs, New Jersey, 1976.
Computer systems, assemblers, and virtuals.

26. B. Ward, *Microprocessor/Microprogramming Handbook*, Tab Books, Blue Ridge Summitt, Pennsylvania, 1975.
Architecture, 6800, 4004, programming.

27. D. Zissos, *System Design with Microprocessors*, Academic Press, London, 1978.
Interrupt, interfaces, and 8080A.

28. E. E. Klingman, *Microprocessor Systems Design*, Prentice-Hall Inc., Englewood Cliffs, New Jersey, 1977.
Devices and systems.

Manuals

29. *The TTL Data Book for Design Engineers*, Texas Instruments Incorporated, Dallas, Texas, 1973, No. CC-411.
30. *M6800 Microprocessor Applications Manual*, Motorola Semiconductor Products Inc., Scotsdale, Arizona, 1975. No. 9703-5.
31. *MCS-85 User's Manual*, Intel Corporation, Santa Clara, California, 1978, No. 9800366D.
32. *M6800 Programming Reference Manual*, Motorola Semiconductor Products Inc., Scotsdale, Arizona, 1976, No. M68PRM(D).
33. *8080/8085 Assembly Language Programming Manual*, Intel Corporation, Santa Clara, California, 1977, No. 8900301C.
34. *ISIS-II User's Guide*, Intel Corporation, Santa Clara, California, 1976, No. 9800306D.
35. *Exordisk II/3 Operating System*, Motorola Semiconductor Products Inc., Scotsdale, Arizona, 1978, No. M68MDOS3.

Index

Accumulator, 206, 208
Action:
  concept, 70
  JKFF, 107
  map, 68, 107
  RSFF, 69
Adapter:
  ACIA, 275, 352
  general purpose, 261
  PIA, 263, 352
  PPIU, 269, 356
  USART, 279, 356
Addition, binary, 25
Address:
  bus, 181
  lines, 178
  mode, 204
American Standard Code for Information Interchange (ASCII), 201, 244, 248, 274, 304, 341, 343
Analog-to-digital converter, 101
And:
  definition, 1, 295
  map, 8
  symbol, 8
Annunciator, 67, 89
Area:
  maximum, 2
  minimum, 1
Arithmetic instructions, 215
Arithmetic logic unit, 174, 178, 303
Arrow:
  input, 148
  transition, 49, 67, 142
Assembler:
  module, 251
  program, 197
Asynchronous communication interface adapter (ACIA), 275, 352
Automatic start up, 72, 91

Binary:
  addition, 25
  coded decimal (BCD), 31, 300
  counting registers, 170
BiQuinary code, 47
Boolean algebra, 1, 295
Bounce, contacts, 65
Branch instructions, 233
Break, 136, 283
Byte, 181

Car race, 100
Cells:
  adjacent, 15
  description, 14
Central processing unit (CPU), 176, 292
Chip select, 275, 290, 292
Circle:
  logic symbols, 70
  mode diagram, 49
  state diagram, 68, 143
Circuit:
  combinational, 1
  nested, 121
  programmable, 119, 196
  sequential, 49
  types, 108
Clock:
  JKFF, 107
  microprocessor, 182
Clocked circuits:
  JKFF, 108
  MSI, 129
Code:
  ASCII, 201, 244, 248, 274, 304, 341, 343
  binary, 301
  BiQuinary, 47
  Gray, 302
  Hamming, 46, 162, 174

Haxadecimal, 289, 303
machine, 197
nine's complement, 300
octal, 303
register, 161
UPC, 156
Coffee machine, 83
Command:
instruction, 283
system, 252, 255, 358, 363
Comment field, 200
Configuration, 280
Contact bounce, 65
Control:
register, 262, 269
set/reset, 270
word (CW), 119, 272
Counter:
BCD, 111
design, 121
operation of HSI, 128
program, 180, 206
ripple, 111
state diagram, 109
synchronous, 104, 111, 128
Cycle:
instruction, 183
machine, 183

Data:
acquisition system (DAS), 101
bus, 181
direction register (DDR), 264
directive, 239, 341, 343
transfer instructions, 210
Decimal, 297
Decoder:
instruction, 180
seven segment, 31
Delay:
electronic, 23
equalization, 86
microprocessor, 257
DeMorgan's theorem, 13
Demultiplexer:
description, 44
design, 96, 133, 291
symbol, 45
Design steps in, combinational, 25
JKFF, 80
multiplexer, 43, 87
RSFF, 60

Dice, electronic, 98
Diesel, 146
Direct addressing mode, 204
Direct memory access, 141, 152, 157
Directives, 239, 341, 343
Directory module, 251, 347, 351
Distributed rule, 12
Don't care, 18
Duality, 143

Editor module, 251, 346, 348
Elevator control, 82
Enable, 45, 189, 265
End directive, 239
EPROM, 177
Equate, 243, 247, 341, 343
Event driven:
state diagram, 144
synthesis of, 67, 71
Excess 3 code, 47
Excitation:
J, 116
K, 116
R, 75
S, 75
Y, 51, 56, 68
Exclusive or:
definition of, 21
map, 22
multiplexer realization, 36
symbol, 22
Extended addressing mode, 204

Feedback, 67, 84
see also Stability
Field, 200
File, 197
Flag, 261
Flow table, 51
Format:
assembler, 199
bit set/reset, 270
control word, 272, 277
directive, 242
Frequency divider, 82
Fundamental mode:
description, 49
design, 51
Future state, 68

Gates, 13
Generator:
diesel, 146

sine, 27
Glitches, 97

Hamming code, 46, 162, 174
Handshake, 139, 273

Immediate addressing mode, 204
Indexed addressing mode, 205
Indirect addressing mode, 205
Inherent mode, 205
Input:
  programming, 266
  register, 265
Input/output:
  instructions, 236
  isolated, 291
  multiplexer, 35
  ports, 176, 268
  system, 188
Instruction:
  command, 283
  description, 183
  index, 223
  indirect, 223
  set, 196, 208, 308, 328
Interface:
  adaptive, 188, 261
  computer-to-computer, 103, 114
  generalized, 115, 134
  microprocessor, 123
  minicomputer, 99, 141
  modes, 124
  parallel, 189
  programming, 280
  serial, 139, 189
Interrupt, 181, 191, 261, 276

JK flip-flop (JKFF):
  construction, 80
  design, 80
  excitation, 116
Jump instructions, 233

Karnaugh map:
  definition, 3
  multiple output, 80

Label, 200
Location directive, 242, 328, 341
Logic instructions, 217

Map, 7, 128
Mark, 135, 283
Matrix:
  input, 36, 131
  loop, 132
  selector, 36
Maxterm, 5
Medium scale integration (MSI), 34, 128
Merging, 54
Microcomputer:
  interface, 99, 114
  systems, 190
Microprocessor, 176
Minimization, 13
Minterm, 4
Mnemonic, 200
Mode:
  asynchronous, 282, 359
  interface, 124, 271, 356
  multiple channel, 102
  test, 102
  see also Fundamental mode
Modem, 280
Module configuration, 252, 255, 344
Monitor module, 251, 345, 349
Multiplexer:
  analog, 101
  application guidelines, 97
  definition, 34
  design notation, 35
  design steps, 43, 87
  feedback, 84
  forward, 130
  loop, 130
  state diagram, 86
  symbol, 35
  time, 138
Multiplication, 36, 84, 89
Multiplier, 30
Mux, *see* Multiplexer

Nand, 9
Nor, 9
Not, 2, 295
Numbers, 297

Object file, 197
Operand, 200
Or, 2, 8, 295
Origin (ORG), 246, 250, 342, 343
Output:
  gate, 18, 57
  multiple, 80
  multiplexer, 43

programming, 267
register, 265
serial, 136

Parity, 47, 277, 282, 355, 359
Periodic sequence, 166
Peripheral device, 188
Peripheral interface adapter, 263
Pointer, 180, 206
Polynomial, 14
Printed circuit, 137
Product, sum of, 6
Program initialization, 288
Programmable circuits, 119
Programmable interface, 121
Programmable peripheral interface unit (PPIU), 269
read only memory (PROM), 177
Propositional calculus, 1, 13
Pseudoequivalence, 54

Random access memory (RAM), 176, 192, 288
Read only memory (ROM), 176, 192, 288
Reduction, 23
Redundancy, 53, 290
Register:
binary counting, 170
code generation, 161
complementing, 172
control, 262, 269, 276, 279, 352, 355
data, 158, 262, 276, 352
data direction, 264
index, 206
instruction, 183, 307, 327
multifunction, 168
shift, 158
status, 179, 207, 278, 308, 328
working, 180
Relative addressing mode, 205
Reset:
excitation, 71
master, 277
signal, 100, 191, 193, 264
Ringing, 97
RSFF, *see* Set reset flip flop

Sample hold (S/H), 101
Secondaries:
description, 50, 67
excitation, 51, 56, 68
Select lines, 34
Sequence:
circuit, 49
maximum length, 175
periodic, 166
Set:
control, 63, 81
excitation, 71
Set/reset control, 270
Set reset flip flop (RSFF):
circuit, 62
design, 60
map, 69
misbehaviour, 105
state representation, 69
symbol, 70
Seven segment decoder, 31
Shannon, 1
Ship-torpedo, 100
Sine generator, 27
Source, 197
Stability, 50
State:
assignment of y's, 50, 55
concept, 67
diagram, 67, 86, 142
duality, 143
Status, register, 179, 207, 278, 308, 328
instruction, 236
Subtractor, 30
Sum, 3
Sums, product of, 6
Symbols:
assembler directive, 202, 239, 307, 327
logic, 6
table, 198
Symmetry, 22
Synthesis, 58, 71, 87, 108

Traffic control, 81, 97
Transition, 49, 55, 116
Tristate, 181, 191
Truth table, 1
Turbine control, 72, 91
2's complement, 301
Type of circuit, 108

Universal produce code (UPC), 156

Venn diagram, 1